D0200221

MAGIC
SQUARES AND CUBES

BY

W. S. ANDREWS

WITH CHAPTERS BY OTHER WRITERS

SECOND EDITION. REVISED AND ENLARGED

DOVER PUBLICATIONS, INC.
NEW YORK

Published in Canada by General Publishing Company, Ltd., 30 Lesmill Road, Don Mills, Toronto, Ontario.

Published in the United Kingdom by Constable and Company, Ltd., 10 Orange Street, London WC 2.

This Dover edition, first published in 1960, is an unabridged and unaltered republication of the second edition as published by the Open Court Publishing Company in 1917.

International Standard Book Number: 0-486-20658-0

Manufactured in the United States of America

Dover Publications, Inc.

180 Varick Street

New York, N. Y. 10014

TABLE OF CONTENTS.

PAGE

Publishers' Preface v

Introduction. By Paul Carus vii

Chapter I. Magic Squares. By W. S. Andrews 1
The Essential Characteristics of Magic Squares 1
Associated or Regular Magic Squares of Odd Numbers 2
Associated or Regular Magic Squares of Even Numbers 18
Construction of Even Magic Squares by De La Hire's Method 34
Composite Magic Squares 44
Concentric Magic Squares 47
General Notes on the Construction of Magic Squares 54

Chapter II. Magic Cubes. By W. S. Andrews 64
The Essential Characteristics of Magic Cubes 64
Associated or Regular Magic Cubes of Odd Numbers 65
Associated or Regular Magic Cubes of Even Numbers 76
General Notes on Magic Cubes 84

Chapter III. The Franklin Squares. By W. S. Andrews................ 89
An Analysis of the Franklin Squares, By Paul Carus 96

Chapter IV. Reflections on Magic Squares. By Paul Carus 113
The Order of Figures 113
Magic Squares in Symbols 120
The Magic Square in China 122
The Jaina Square 125

Chapter V. A Mathematical Study of Magic Squares. By L. S. Frierson. 129
A New Analysis 129
Notes on Number Series Used in the Construction of Magic Squares 137

Chapter VI. Magics and Pythagorean Numbers. By C. A. Browne 146
Mr. Browne's Square and *lusus numerorum*. By Paul Carus 158

Chapter VII. Some Curious Magic Squares and Combinations. By W. S. Andrews 163

Chapter VIII. Notes on Various Constructive Plans by which Magic Squares May be Classified. By W. S. Andrews 178
The Mathematical Value of Magic Squares. By W. S. Andrews... 187

PAGE

Chapter IX. Magic Cubes of the Sixth Order 189
A "Franklin" Cube of Six. By H. M. Kingery 189
A Magic Cube of Six. By Harry A. Sayles 196
Magic Cube of Six. By John Worthington 201

Chapter X. Various Kinds of Magic Squares 207
Overlapping Magic Squares. By D. F. Savage 207
Oddly-Even Magic Squares. By D. F. Savage 217
Notes on Oddly-Even Magic Squares. By W. S. Andrews 225
Notes on Pandiagonal and Associated Magic Squares. By L. S. Frierson 229
Serrated Magic Squares. By Harry A. Sayles 241
Lozenge Magic Squares. By Harry A. Sayles 244

Chapter XI. Sundry Constructive Methods 248
A New Method for Making Magic Squares of Odd Orders. By L. S. Frierson 248
The Construction of Magic Squares and Rectangles by the Method of Complementary Differences. By W. S. Andrews 257
Notes on the Construction of Magic Squares of Orders in which n is of the General Form $4p+2$. By W. S. Andrews and L. S. Frierson 267
Notes on the Construction of Magic Squares of Orders in which n is of the General Form $8p+2$. By Harry A. Sayles 277
Geometric Magic Squares and Cubes. By Harry A. Sayles 283

Chapter XII. The Theory of Reversions. By Dr. C. Planck 295

Chapter XIII. Magic Circles, Spheres and Stars 321
Magic Circles. By Harry A Sayles 321
Magic Spheres. By Harry A. Sayles 331
Magic Stars. By W. S. Andrews 339

Chapter XIV. Magic Octahedroids 351
Magic in the Fourth Dimension. By H. M. Kingery 351
Fourfold Magics. By Dr. C. Planck 363

Chapter XV. Ornate Magic Squares 376
General Rule for Constructing Ornate Magic Squares of Orders $\equiv 0 \pmod 4$. By Dr. C. Planck 376
Ornate Magic Squares of Composite Odd Orders. By Dr. C. Planck 383
The Construction of Ornate Magic Squares of Orders 8, 12 and 16 by Tables. By Frederic A. Woodruff 390
The Construction of Ornate Magic Squares of Order 16 by Magic Rectangles. By W. S. Andrews 404
Pandiagonal-Concentric Magic Squares of Orders $4m$. By Harry A. Sayles 410

Index 415

Diagrams of Completed Magics 419

PUBLISHERS' PREFACE.

THE essays which comprise this volume appeared first in *The Monist* at different times during the years 1905 to 1916, and under different circumstances. Some of the diagrams were photographed from the authors' drawings, others were set in type, and different authors have presented the results of their labors in different styles. In compiling all these in book form the original presentation has been largely preserved, and in this way uniformity has been sacrificed to some extent. Clarity of presentation was deemed the main thing, and so it happens that elegance of typographical appearance has been considered of secondary importance. Since mathematical readers will care mainly for the thoughts presented, we hope they will overlook the typographical shortcomings. The first edition contained only the first eight chapters, and these have now been carefully revised. The book has been doubled in volume through the interest aroused by the first edition in mathematical minds who have contributed their labors to the solution of problems along the same line.

In conclusion we wish to call attention to the title vignette which is an ancient Tibetan magic square borne on the back of the cosmic tortoise.

INTRODUCTION.

THE peculiar interest of magic squares and all *lusus numerorum* in general lies in the fact that they possess the charm of mystery. They appear to betray some hidden intelligence which by a preconceived plan produces the impression of intentional design, a phenomenon which finds its close analogue in nature.

Although magic squares have no immediate practical use, they have always exercised a great influence upon thinking people. It seems to me that they contain a lesson of great value in being a palpable instance of the symmetry of mathematics, throwing thereby a clear light upon the order that pervades the universe wherever we turn, in the infinitesimally small interrelations of atoms as well as in the immeasurable domain of the starry heavens, an order which, although of a different kind and still more intricate, is also traceable in the development of organized life, and even in the complex domain of human action.

Pythagoras says that number is the origin of all things, and certainly the law of number is the key that unlocks the secrets of the universe. But the law of number possesses an immanent order, which is at first sight mystifying, but on a more intimate acquaintance we easily understand it to be intrinsically necessary; and this law of number explains the wondrous consistency of the laws of nature. Magic squares are conspicuous instances of the intrinsic harmony of number, and so they will serve as an interpreter of the cosmic order that dominates all existence. Though they are a mere intellectual play they not only illustrate the nature of mathematics, but also, incidentally, the nature of existence dominated by mathematical regularity.

In arithmetic we create a universe of figures by the process of counting; in geometry we create another universe by drawing lines in the abstract field of imagination, laying down definite directions; in algebra we produce magnitudes of a still more abstract nature, expressed by letters. In all these cases the first step producing the general conditions in which we move, lays down the rule to which all further steps are subject, and so every one of these universes is dominated by a consistency, producing a wonderful symmetry.

There is no science that teaches the harmonies of nature more clearly than mathematics, and the magic squares are like a mirror which reflects the symmetry of the divine norm immanent in all things, in the immeasurable immensity of the cosmos and in the construction of the atom not less than in the mysterious depths of the human mind.

PAUL CARUS.

MAGIC
SQUARES AND CUBES

CHAPTER I.

MAGIC SQUARES.

THE study of magic squares probably dates back to prehistoric times. Examples have been found in Chinese literature written about A. D. 1125* which were evidently copied from still older documents. It is recorded that as early as the ninth century magic squares were used by Arabian astrologers in their calculations of horoscopes etc. Hence the probable origin of the term "magic" which has survived to the present day.

THE ESSENTIAL CHARACTERISTICS OF MAGIC SQUARES.

A magic square consists of a series of numbers so arranged in a square, that the sum of each row and column and of both the corner diagonals shall be the same amount which may be termed the *summation* (S). Any square arrangement of numbers that fulfils these conditions may properly be called a magic square. Various features may be added to such a square which may enhance its value as a mathematical curio, but these must be considered non-essentials.

There are thus many different kinds of magic squares, but this chapter will be devoted principally to the description of *associated* or *regular* magic squares, in which the sum of any two numbers that are located in cells diametrically equidistant from the center of the square equals the sum of the first and last terms of the series, or $n^2 + 1$.

Magic squares with an odd number of cells are usually con-

---

* See page 19 of *Chinese Philosophy* by Paul Carus.

structed by methods which differ from those governing the construction of squares having an even number of cells, so these two classes will be considered under separate headings.

ASSOCIATED OR REGULAR MAGIC SQUARES OF ODD NUMBERS.

The square of 3×3 shown in Fig. 1 covers the smallest aggregation of numbers that is capable of magic square arrangement, and it is also the only possible arrangement of nine different numbers, relatively to each other, which fulfils the required conditions. It will be seen that the sum of each of the three vertical, the three horizontal, and the two corner diagonal columns in this square is 15, making in all eight columns having that total: also that the sum of any two opposite numbers is 10, which is twice the center number, or $n^2 + 1$.

The next largest odd magic square is that of 5×5, and there are a great many different arrangements of twenty-five numbers,

8	1	6
3	5	7
4	9	2

S = 15.

Fig. 1.

17	24	1	8	15
23	5	7	14	16
4	6	13	20	22
10	12	19	21	3
11	18	25	2	9

S = 65.

Fig. 2.

which will show magic results, each arrangement being the production of a different constructive method. Fig. 2 illustrates one of the oldest and best known arrangements of this square.

The sum of each of the five horizontal, the five vertical, and the two corner diagonal columns is 65, and the sum of any two numbers which are diametrically equidistant from the center number is 26, or twice the center number.

In order intelligently to follow the rule used in the construction of this square it may be conceived that its upper and lower edges are bent around backwards and united to form a horizontal cylinder with the numbers on the outside, the lower line of figures thus coming next in order to the upper line. It may also be conceived that the square is bent around backwards in a direction at right angles to that which was last considered, so that it forms a vertical cylinder with the extreme right- and left-hand columns adjacent to each other.

An understanding of this simple conception will assist the student to follow other methods of building odd magic squares that are to be described. which are based on a right- or left-hand diagonal formation.

Referring to Fig. 2, it will be seen that the square is started by writing unity in the center cell of the upper row, the consecutive numbers proceeding diagonally therefrom in a right-hand direction. Using the conception of a horizontal cylinder, 2 will be located in the lower row, followed by 3 in the next upper cell to the right. Here the formation of the vertical cylinder being conceived, the next upper cell will be where 4 is written, then 5; further progress being here blocked by 1 which already occupies the next upper cell in diagonal order.

When a block thus occurs in the regular spacing (which will be at every fifth number in a 5×5 square) the next number must in this case be written in the cell vertically below the one last filled, so that 6 is written in the cell below 5, and the right-hand diagonal order is then continued in cells occupied by 7 and 8. Here the horizontal cylinder is imagined, showing the location of 9, then the conception of the vertical cylinder will indicate the location of 10; further regular progression being here once more blocked by 6, so 11 is written under 10 and the diagonal order continued to 15. A mental picture of the combination of vertical and horizontal cylinders will here show that further diagonal progress is blocked by 11, so 16 is written under 15. The vertical cylinder will then indicate the cell in which 17 must be located, and the horizontal cylinder will show the next cell diagonally upwards to the right to be occupied by 18, and so on until the final number 25 is reached and the square completed.

Fig. 3 illustrates the development of a 7×7 square constructed according to the preceding method, and the student is advised to follow the sequence of the numbers to impress the rule on his memory. A variation of the last method is shown in Fig. 4, illustrating another 7×7 square. In this example 1 is placed in the next cell horizonally to the right of the center cell, and the consecutive numbers proceed diagonally upward therefrom, as before, in a right-hand direction until a block occurs. The next number is then written in the second cell horizontally to the right of the last cell filled (instead of the cell below as in previous examples) and the upward diagonal order is resumed until the next block occurs.

30	39	48	1	10	19	28
38	47	7	9	18	27	29
46	6	8	17	26	35	37
5	14	16	25	34	36	45
13	15	24	33	42	44	4
21	23	32	41	43	3	12
22	31	40	49	2	11	20

Fig. 3.

S = 175

4	29	12	37	20	45	28
35	11	36	19	44	27	3
10	42	18	43	26	2	34
41	17	49	25	1	33	9
16	48	24	7	32	8	40
47	23	6	31	14	39	15
22	5	30	13	38	21	46

Fig. 4.

10	18	1	14	22
11	24	7	20	3
17	5	13	21	9
23	6	19	2	15
4	12	25	8	16

S = 65.

Fig. 5.

Then two cells to the right again, and regular diagonal order continued, and so on until all the cells are filled.

The preceding examples may be again varied by writing the numbers in left-hand instead of right-hand diagonal sequence, making use of the same spacing of numbers as before when blocks occur in the regular sequence of construction.

We now come to a series of very interesting methods for building odd magic squares which involve the use of the knight's move in chess, and it is worthy of note that the squares formed by these methods possess curious characteristics in addition to those previously referred to. To chess-players the knight's move will require no comment, but for those who are not familiar with this game it may be explained as a move of two cells straight forward in any direction and one cell to either right or left.

The magic square of 5 × 5 illustrated in Fig. 5 is started by placing 1 in the center cell of the upper row, and the knight's move employed in its construction will be two cells upward and one cell to the right.

Using the idea of the horizontal cylinder 2 must be written in the second line from the bottom, as shown, and then 3 in the second line from the top. Now conceiving a combination of the horizontal and vertical cylinders, the next move will locate 4 in the extreme lower left-hand corner, and then 5 in the middle row. We now find that the next move is blocked by 1, so 6 is written below 5, and the knight's moves are then continued, and so until the last number, 25, is written in the middle cell of the lower line, and the square is thus completed.

In common with the odd magic squares which were previously described, it will be found that in this square the sum of each of the five horizontal, the five perpendicular, and the two corner diagonal columns is 65, also that the sum of any two numbers that are diagonally equidistant from the center is 26, or twice the number in the center cell, thus filling all the qualifications of an associated magic square.

In addition, however, to these characteristics it will be noted that each spiral row of figures around the horizontal and vertical cylinders traced either right-handed or left-handed also amounts to 65. In the vertical cylinder, there are five right-hand, and five left-hand spirals, two of which form the corner diagonal columns across the square, leaving eight new combinations. The same number of combinations will also be found in the horizontal cylinder. Counting therefore five horizontal columns, five vertical columns, two corner diagonal columns, and eight right- and left-hand spiral columns, there are in all twenty columns each of which will sum up to 65, whereas in the 5 × 5 square shown in Fig. 2 there will be found only sixteen columns that will amount to that number.

This method of construction is subject to a number of variations. For example, the knight's move may be upwards and to the left hand instead of to the right, or it may be made downward and either to the right or left hand, and also in other directions. There are in fact eight different ways in which the knight's move may be started from the center cell in the upper line. Six of these moves are indicated by figure 2's in different cells of Fig. 6, and each of these moves if continued in its own direction, varied by the breaks as before described, will produce a different but associated square. The remaining two possible knight's moves, indicated by cyphers, will not produce magic squares under the above rules.

		1		
0				0
	2		2	
	2		2	
2				2

Fig. 6.

		19	2	15	23	
	12	25	8		4	
10	18	1	14	22	10	
11	24	7	20	3		
17	5	13	21	9	17	
23	6	19	2	15		
4	12	25	8	16		

Fig. 7.

It may here be desirable to explain another method for locating numbers in their proper cells which some may prefer to that which involves the conception of the double cylinder. This method consists in constructing parts of auxiliary squares around two or more sides of the main square, and temporarily writing the numbers in the cells of these auxiliary squares when their regular placing carries them outside the limits of the main square. The temporary location of these numbers in the cells of the auxiliary squares will then indicate into which cells of the main square they must be permanently transferred.

Fig. 7 shows a 5×5 main square with parts of three auxiliary squares, and the main square will be built up in the same way as Fig. 5.

Starting with 1 in the center of the top line, the first knight's move of two cells upward and one to the right takes 2 across the top margin of the main square into the second cell of the second line from the bottom in one of the auxiliary squares, so 2 must be transferred to the same relative position in the main square. Starting again from 2 in the main square, the next move places 3 within the main square, but 4 goes out of it into the lower left-hand corner of an auxiliary square, from which it must be transferred to the same location in the main square, and so on throughout.

The method last described and also the conception of the double cylinders may be considered simply as aids to the beginner. With a little practice the student will be able to select the proper cells in the square as fast as figures can be written therein.

Having thus explained these specific lines of construction, the general principles governing the development of odd magic squares by these methods may now be formulated.

1. The center cell in the square must always contain the middle number of the series of numbers used, i. e., a number which is equal to one-half the sum of the first and last numbers of the series, or $n^2 + 1$.
2. No associated magic square can therefore be started from its center cell, but it may be started from any cell other than the center one.
3. With certain specific exceptions which will be referred to later on, odd magic squares may be constructed by either right- or left-hand diagonal sequence, or by a number of so-called knight's moves, varied in all cases by periodical and well defined departures from normal spacing.
4. The directions and dimensions of these departures from normal spacing, or "break-moves," as they may be termed, are governed by the relative spacing of cells occupied by the first and last numbers of the series, and may be determined as follows:

RULE: Place the first number of the series in any desired cell (excepting the center one) and the last number of the series in the cell which is diametrically opposite to the cell containing the first number. The relative spacing between the cell that contains the last number of the series and the cell that contains the first number of the series must then be repeated whenever a block occurs in the regular progression.

EXAMPLES.

Using a blank square of 5 × 5, 1 may be written in the middle cell of the upper line. The diametrically opposite cell to this being the middle cell in the lower line, 25 must be written therein. 1 will therefore be located four cells above in the middle vertical column, or what is the same thing, and easier to follow, one cell below 25.

		1		15
	5			16
	6		20	
10			21	
11		25		

Fig. 8.

	16			15
20	21			
	25	1		
		5		6
11			10	

Fig. 9.

When, therefore, a square of 5 × 5 is commenced with the first number in the middle cell of the upper line, the break-move will be one cell downward, irrespective of the method of regular advance. Fig. 8 shows the break-moves in a 5 × 5 square as above described using a right-hand upward diagonal advance.

Again using a blank 5 × 5 square, 1 may be written in the cell immediately to the right of the center cell, bringing 25 into the cell to the left of the center cell. The break-moves in this case will therefore be two cells to the right of the last cell occupied, irrespective of the method used for regular advance. Fig. 9 illustrates the break-moves in the above case, when a right-hand upward diagonal advance is used. The positions of these break-moves in the square will naturally vary with the method of advance, but the relative spacing of the moves themselves will remain unchanged.

> NOTE: The foregoing break-moves were previously described in several specific examples (See Figs. 1, 2, 3, 4, and 5) and the reader will now observe how they agree with the general rule.

Once more using a blank square of 5 × 5, 1 may be written in the upper left-hand corner and 25 in the lower right-hand corner. 1 will then occupy a position four cells removed from 25 in a left-hand upward diagonal, or what is the same thing and easier to follow, the next cell in a right-hand downward diagonal. This will therefore be the break-move whenever a block occurs in the regular spacing. Fig. 10 shows the break-moves which occur when a

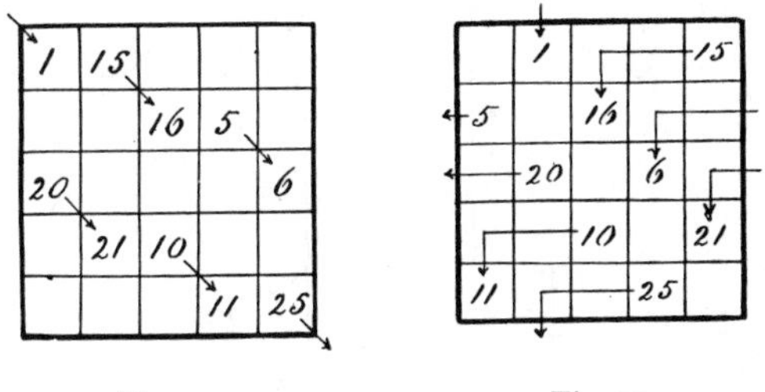

Fig. 10. Fig. 11.

knight's move of two cells to the right and one cell upward is used for the regular advance.

As a final example we will write 1 in the second cell from the left in the upper line of a 5 × 5 square, which calls for the placing of 25 in the second square from the right in the lower line. The place relation between 25 and 1 may then be described by a knight's move of two cells to the left and one cell downward, and this will be the break-move whenever a block occurs in the regular spacing. The break-moves shown in Fig. 11 occur when an upward right-hand diagonal sequence is used for the regular advance.

As before stated odd magic squares may be commenced in any cell excepting the center one, and associated squares may be built up from such commencements by a great variety of moves, such as right-hand diagonal sequence, upward or downward, left-hand diagonal sequence upward or downward, or a number of knight's moves in various directions. There are four possible moves from each cell in diagonal sequence, and eight possible moves from each cell by the knight's move. Some of these moves will produce associated magic squares, but there will be found many exceptions which can be shown most readily by diagrams.

Fig. 12 is a 5×5 square in which the pointed arrow heads indicate the directions of diagonal sequence by which associated squares may be constructed, while the blunt arrow heads show the directions of diagonal sequence which will lead to imperfect results. Fig. 13 illustrates the various *normal* knight's moves which may be

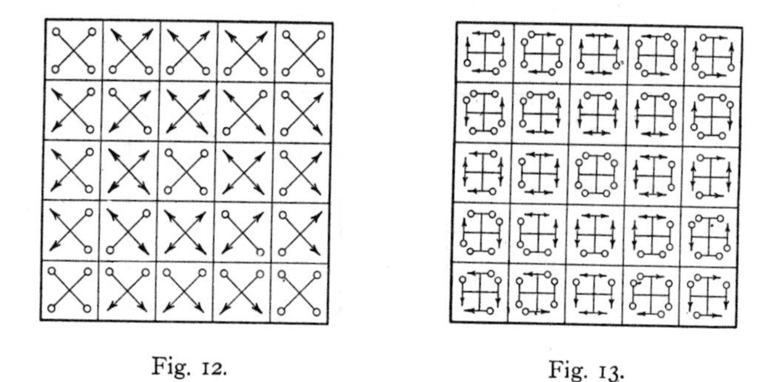

Fig. 12. Fig. 13.

started from each cell and also indicates with pointed and blunt arrow heads the moves which will lead to perfect or imperfect results. For example it will be seen from Fig. 12 that an associated 5×5 square cannot be built by starting from either of the four corner cells in any direction of diagonal sequence, but Fig. 13 shows four different normal knight's moves from each corner cell, any of which will produce associated squares. It also shows four other normal knight's moves which produce imperfect squares.

EXAMPLES OF 5×5 MAGIC SQUARES.

Figs. 14 and 15 show two 5×5 squares, each having 1 in the upper left-hand corner cell and 25 in the lower right-hand corner cell, and being constructed with different knight's moves. Fig. 16 shows a similar square in which an elongated knight's move

1	15	24	8	17
23	7	16	5	14
20	4	13	22	6
12	21	10	19	3
9	18	2	11	25

Fig. 14.

1	24	17	15	8
14	7	5	23	16
22	20	13	6	4
10	3	21	19	12
18	11	9	2	25

Fig. 15.

1	18	10	22	14
20	7	24	11	3
9	21	13	5	17
23	15	2	19	6
12	4	16	8	25

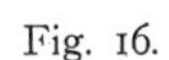

Fig. 16.

is used for regular advance. The break-move is necessarily the same in each example. (See Fig. 10.)

8	1	24	17	15
5	23	16	14	7
22	20	13	6	4
19	12	10	3	21
11	9	2	25	18

Fig. 17.

15	1	17	8	24
23	14	5	16	7
6	22	13	4	20
19	10	21	12	3
2	18	9	25	11

Fig. 18.

Figs. 17, 18, 19 and 20 show four 5×5 squares, each having 1 in the second cell from the left in the upper line and 25 in the

22	1	10	14	18
11	20	24	3	7
5	9	13	17	21
19	23	2	6	15
8	12	16	25	4

Fig. 19.

23	1	9	12	20
15	18	21	4	7
2	10	13	16	24
19	22	5	8	11
6	14	17	25	3

Fig. 20.

second cell from the right in the lower line, and being built up respectively with right- and left-hand upward diagonal sequence and upward right- and downward left-hand knight's moves, and with similar break-moves in each example. See Fig. 11.)

Figs. 21, 22, and 23 illustrate three 5×5 squares, each having 1 in the upper right-hand corner and 25 in the lower left-hand corner, and being built up respectively with upward and downward right-hand normal knight's moves, and a downward right-hand elongated knight's move.

For the sake of simplicity these examples have been shown in 5×5 squares, but the rules will naturally apply to all sizes of odd magic squares by using the appropriate numbers. The explanations have also been given at some length because they cover general and comprehensive methods, a good understanding of which is desirable.

It is clear that no special significance can be attached to the

18	10	22	14	1
11	3	20	7	24
9	21	13	5	17
2	19	6	23	15
25	12	4	16	8

Fig. 21.

9	12	20	23	1
18	21	4	7	15
2	10	13	16	24
11	19	22	5	8
25	3	6	14	17

Fig. 22.

12	23	9	20	1
4	15	21	7	18
16	2	13	24	10
8	19	5	11	22
25	6	17	3	14

Fig. 23.

so-called knight's move, *per se,* as applied to the construction of magic squares, it being only one of many methods of regular spacing, all of which will produce equivalent results. For example, the 3×3 square shown in Fig. 1 may be said to be built up by a succession of abbreviated knight's moves of one cell to the right and one cell upwards. Squares illustrated in Figs. 2, 3, and 4 are also constructed by this abbreviated knight's move, but the square illustrated in Fig. 5 is built up by the normal knight's move.

It is equally easy to construct squares by means of an elongated knight's move, say, four cells to the right and one cell upwards as shown in Fig. 24, or by a move consisting of two cells to the right and two cells downwards, as shown in Fig. 25, the latter being equivalent to a right hand downward diagonal sequence wherein alternate cells are consecutively filled.

There are in fact almost innumerable combinations of moves by which these odd magic squares may be constructed.

The foregoing method for building odd magic squares by a

80	58	45	23	1	69	47	34	12
9	68	46	33	11	79	57	44	22
10	78	56	43	21	8	67	54	32
20	7	66	53	31	18	77	55	42
30	17	76	63	41	19	6	65	52
40	27	5	64	51	29	16	75	62
50	28	15	74	61	39	26	4	72
60	38	25	3	71	49	36	14	73
70	48	35	13	81	59	37	24	2

S = 369.

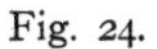

Fig. 24.

continuous process involves the regular spacing of consecutive numbers varied by different well defined break-moves, but other methods of construction have been known for many years.

39	34	20	15	1	77	72	58	53
49	44	30	25	11	6	73	68	63
59	54	40	35	21	16	2	78	64
69	55	50	45	31	26	12	7	74
79	65	60	46	41	36	22	17	3
8	75	70	56	51	37	32	27	13
18	4	80	66	61	47	42	28	23
19	14	9	76	71	57	52	38	33
29	24	10	5	81	67	62	48	43

S = 369.

Fig. 25.

One of the most interesting of these other methods involves the use of two or more primary squares, the sums of numbers in similarly located cells of which constitute the correct numbers for transfer into the corresponding cells of the magic square that is to be constructed therefrom.

This method has been ascribed primarily to De la Hire but has been more recently improved by Prof. Scheffler.

It may be simply illustrated by the construction of a few 5×5 squares as examples. Figs. 26 and 27 show two simple primary squares in which the numbers 1 to 5 are so arranged that like numbers occur once and only once in similarly placed cells in the two squares; also that pairs of unlike numbers are not repeated in the same order in any similarly placed cells. Thus, 5 occupies the extreme right-hand cell in the lower line of each square, but this combination does not occur in any of the other cells. So also in Fig. 27 4 occupies the extreme right-hand cell in the upper line, and in Fig.

1	5	4	3	2
3	2	1	5	4
5	4	3	2	1
2	1	5	4	3
4	3	2	1	5

Fig. 26.

1	3	5	2	4
5	2	4	1	3
4	1	3	5	2
3	5	2	4	1
2	4	1	3	5

Fig. 27.

26 this cell contains 2. No other cell, however, in Fig. 27 that contains 4 corresponds in position with a cell in Fig. 26 that contains 2. Leaving the numbers in Fig. 26 unaltered, the numbers in Fig. 27 must now be changed to their respective root numbers, thus producing the root square shown in Fig. 28. By adding the cell numbers of the primary square Fig. 26 to the corresponding cell numbers

Primary numbers .... 1, 2, 3, 4, 5.
Root numbers .... 0, 5, 10, 15, 20.

of the root square Fig. 28, the magic square shown in Fig. 29 is formed, which is also identical with the one previously given in Fig. 14.

The simple and direct formation of Fig. 14 may be thus compared with the De la Hire method for arriving at the same result.

It is evident that the root square shown in Fig. 28 may be dispensed with by mentally substituting the root numbers for the primary numbers given in Fig. 27 when performing the addition, and by so doing only two primary squares are required to construct the magic square. The arrangement of the numbers 1 to 5 in the two primary squares is obviously open to an immense number of variations, each of which will result in the formation of a different but associated magic square. Any of these squares, however, may be readily constructed by the direct methods previously explained.

0	10	20	5	15
20	5	15	0	10
15	0	10	20	5
10	20	5	15	0
5	15	0	10	5

Fig. 28.

1	15	24	8	17
23	7	16	5	14
20	4	13	22	6
12	21	10	19	3
9	18	2	11	25

Fig. 29.

A few of these variations are given as examples, the root numbers remaining unchanged. The root square Fig. 32 is formed from the primary square Fig. 31, and if the numbers in Fig. 32 are added to those in the primary square Fig. 30, the magic square Fig. 33 will be produced. This square will be found identical with that shown in Fig. 15.

1	4	2	5	3
4	2	5	3	1
2	5	3	1	4
5	3	1	4	2
3	1	4	2	5

Fig. 30.

1	5	4	3	2
3	2	1	5	4
5	4	3	2	1
2	1	5	4	3
4	3	2	1	5

Fig. 31.

As a final example the magic square shown in Fig. 37, previously given in Fig. 17, is made by the addition of numbers in the primary square Fig. 34 to the numbers occupying similar cells in root square Fig. 36, the latter being derived from the primary square Fig. 35. If the root square shown in Fig. 38 is now constructed

0	20	15	10	5
10	5	0	20	15
20	15	10	5	0
5	0	20	15	10
15	10	5	0	20

Fig. 32.

1	24	17	15	8
14	7	5	23	16
22	20	13	6	4
10	3	21	19	12
18	11	9	2	25

Fig. 33.

from the primary square Fig. 34 and the root numbers therein added to the primary numbers in Fig. 35, the magic square shown in Fig. 39 is obtained, showing that two different magic squares may be

3	1	4	2	5
5	3	1	4	2
2	5	3	1	4
4	2	5	3	1
1	4	2	5	3

Fig. 34.

2	1	5	4	3
1	5	4	3	2
5	4	3	2	1
4	3	2	1	5
3	2	1	5	4

Fig. 35.

5	0	20	15	10
0	20	15	10	5
20	15	10	5	0
15	10	5	0	20
10	5	0	20	15

Fig. 36.

8	1	24	17	15
5	23	16	14	7
22	20	13	6	4
19	12	10	3	21
11	9	2	25	18

Fig. 37.

10	0	15	5	20
20	10	0	15	5
5	20	10	0	15
15	5	20	10	0
0	15	5	20	10

Fig. 38.

12	1	20	9	23
21	15	4	18	7
10	24	13	2	16
19	8	22	11	5
3	17	6	25	14

Fig. 39.

made from any two primary squares by forming a root square from each of them in turn. Fig. 39 has not been given before in this book, but it may be directly produced by an elongated knight's move consisting of two cells to the right and two downward, using the normal knight's move of two cells to the left and one cell downward as a break-move at every block in the regular spacing.

It will be observed in all the preceding examples that the number 3 invariably occupies the center cell in all 5 × 5 primary squares, thus bringing 10 in the center of the root squares, and 13 in the center of the magic squares, no other number being admissible in the center cell of an associated 5 × 5 magic square. A careful study of these examples should suffice to make the student familiar with the De la Hire system for building odd magic squares, and

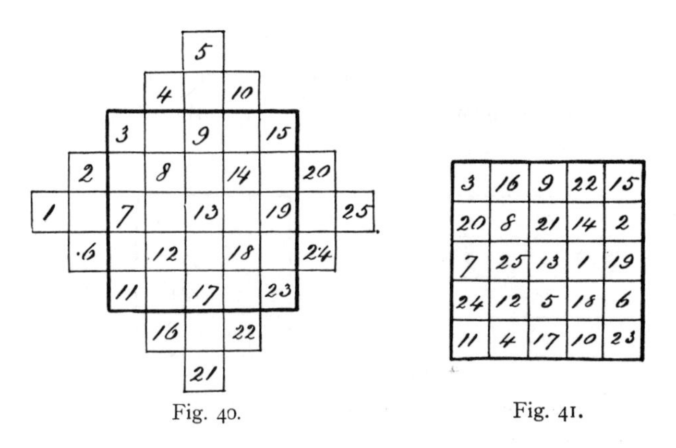

Fig. 40.

3	16	9	22	15
20	8	21	14	2
7	25	13	1	19
24	12	5	18	6
11	4	17	10	23

Fig. 41.

this knowledge is desirable in order that he may properly appreciate the other methods which have been described.

Before concluding this branch of the subject, mention may be made of another method for constructing odd magic squares which is said to have been originated by Bachet de Méziriac. The application of this method to a 5 × 5 square will suffice for an example.

The numbers 1 to 25 are written consecutively in diagonal columns, as shown in Fig. 40, and those numbers which come outside the center square are transferred to the empty cells on the opposite sides of the latter without changing their order. The result will be the magic square of 5 × 5 shown in Fig. 41. It will be seen that the arrangement of numbers in this magic square is similar to that in the 7×7 square shown in Fig. 4, which was built by writing the numbers 1 to 49 consecutively according to rule. The 5×5 square shown in Fig. 41 may also be written out directly by the same rule without any preliminary or additional work.

ASSOCIATED OR REGULAR MAGIC SQUARES OF EVEN NUMBERS.

The numbers in the two corner diagonal columns in these magic squares may be determined by writing the numbers of the series in arithmetical order in horizontal rows, beginning with the first number in the left-hand cell of the upper line and writing line after line as in a book, ending with the last number in the right-hand cell of the lower line. The numbers then found in the two diagonal columns will be in magic square order, but the position of the other numbers must generally be changed.

1	15	14	4
12	6	7	9
8	10	11	5
13	3	2	16

Fig. 42.

1	2	3	4
5	6	7	8
9	10	11	12
13	14	15	16

Fig. 43.

The smallest even magic square that can be built is that of 4×4, and one of its forms is shown in Fig. 42. It will be seen that the sum of each of the four horizontal, the four vertical, and the two corner diagonal columns in this square is 34, making in all ten columns having that total; also that the sum of any two diametrically opposite numbers is 17, which is the sum of the first and last numbers of the series. It is therefore an associated square of 4×4.

The first step in the construction of this square is shown in Fig. 43, in which only the two corner diagonal columns, which are written in heavy figures, have the correct summation. The numbers in these two columns must therefore be left as they are, but the location of all the other numbers, which are written in light figures, must be changed. A simple method for effecting this change consists in substituting for each number the complement between it and 17. Thus, the complement between 2 and 17 is 15, so 15 may be written in the place of 2, and so on throughout. All of the light figure

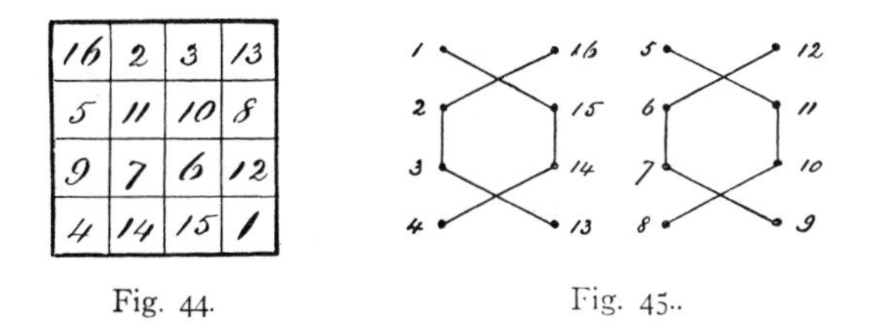

16	2	3	13
5	11	10	8
9	7	6	12
4	14	15	1

Fig. 44. Fig. 45.

numbers being thus changed, the result will be the magic square shown in Fig. 42.

The same relative arrangement of figures may be attained by leaving the light figure numbers in their original positions as shown in Fig. 43, and changing the heavy figure numbers in the two corner diagonal columns to their respective complements with 17. It will be seen that this is only a reversal of the order of the figures

1	35	34	3	32	6
30	8	28	27	11	7
24	23	15	16	14	19
13	17	21	22	20	18
12	26	9	10	29	25
31	2	4	33	5	36

Fig. 46.

1	2	3	4	5	6
7	8	9	10	11	12
13	14	15	16	17	18
19	20	21	22	23	24
25	26	27	28	29	30
31	32	33	34	35	36

Fig. 47.

in the two corner diagonal columns, and the resulting magic square which is shown in Fig. 44 is simply an inversion of Fig. 42.

Fig. 45 is a geometrical diagram of the numbers in Fig. 42, and it indicates a regular law in their arrangement, which also holds good in many larger even squares, as will be seen later on.

There are many other arrangements of sixteen numbers which will fulfil the required conditions but the examples given will suffice to illustrate the principles of this square.

The next even magic square is that of 6 × 6, and one of its many variations is shown in Fig. 46. An analysis of this square

1	35	34	33	32	6
30	8	28	27	11	25
24	23	15	16	20	19
18	17	21	22	14	13
12	26	10	9	29	7
31	5	4	3	2	36

Fig. 48.

with the aid of geometrical diagrams will point the way not only to its own reconstruction but also to an easy method for building other 6 × 6 squares of this class.

Fig. 47 shows a 6 × 6 square in which all the numbers from

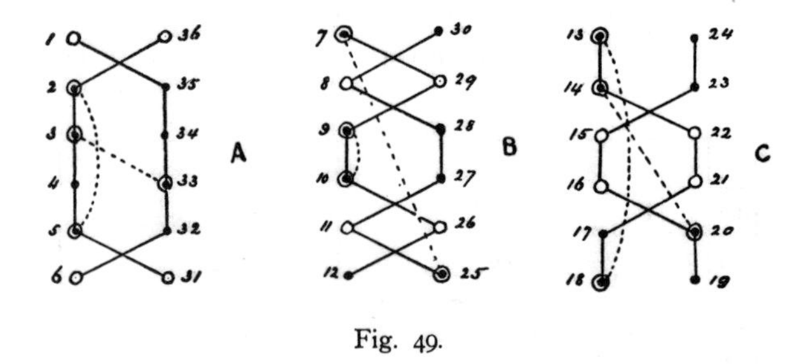

Fig. 49.

1 to 36 are written in arithmetical sequence, and the twelve numbers in the two corner diagonal columns will be found in magic square order, all other numbers requiring rearrangement. Leaving therefore the numbers in the diagonal columns unchanged, the next step will be to write in the places of the other numbers their complements with 37, making the square shown in Fig. 48. In this square twenty-four numbers (written in heavy figures) out of the total of

thirty-six numbers, will be found in magic square order, twelve numbers (written in light figures) being still incorrectly located. Finally, the respective positions of these twelve numbers being reversed in pairs, the magic square given in Fig. 46 will be produced.

Fig. 50 shows the geometrical diagrams of this square, A being a diagram of the first and sixth lines, B of the second and fifth lines, and C of the third and fourth lines. The striking irregularity of these diagrams points to the irregularity of the square which they represent, in which, although the sum of each of the two corner diagonal, the six horizontal, and the six perpendicular columns is 111, yet only in the two diagonal columns does the sum of any two numbers which occupy diametrically opposite cells, amount to 37, or the sum of the first and last numbers of the series. Owing to their pronounced irregularities, these diagrams convey

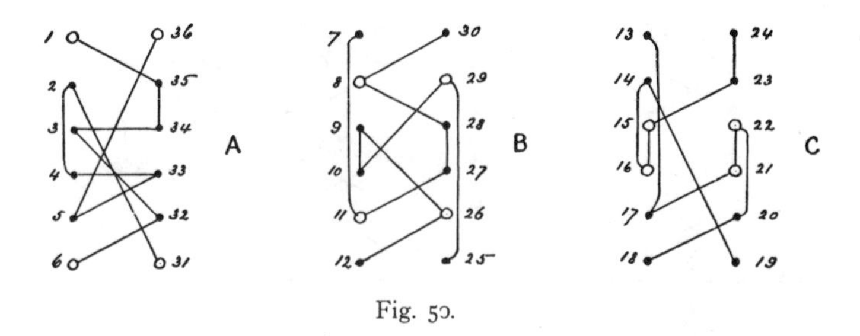

Fig. 50.

but little meaning, and in order to analyze their value for further constructive work it will be necessary to go a step backwards and make diagrams of the intermediate square Fig. 48. These diagrams are shown in Fig. 49, and the twelve numbers therein which must be transposed (as already referred to) are marked by small circles around dots, each pair of numbers to be transposed in position being connected by a dotted line. The numbers in the two corner diagonal columns which were permanently located from the beginning are marked with small circles.

We have here correct geometrical figures with definite and well defined irregularities. The series of geometrical figures shown in A, B, and C remain unchanged in shape for all variations of these 6×6 squares, but by modifying the irregularities we may readily obtain the data for building a large number of variants, all showing, however, the same general characteristics as Fig. 46.

A series of these diagrams, with some modifications of their irregularities, is given in Fig. 51, and in order to build a variety of 6 × 6 magic squares therefrom it is only necessary to select three

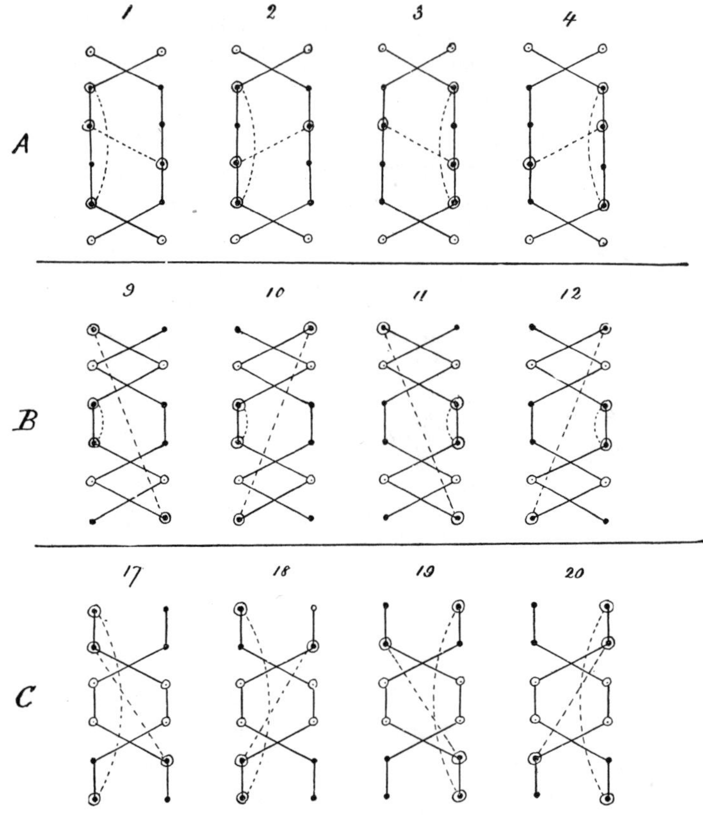

Fig. 51 (First Part).

diagrams in the order A, B, and C, *which have each a different form of irregularity,* and after numbering them in arithmetical sequence from 1 to 36, as shown in Fig. 49, copy the numbers in *diagrammatic order* into the cells of a 6 × 6 square.

It must be remembered that the cells in the corner diagonal

columns of these even magic squares may be correctly filled by writing the numbers in arithmetical order according to the rule previously given, so in beginning any new even square it will be found helpful to first write the numbers in these columns, and they will then serve as guides in the further development of the square.

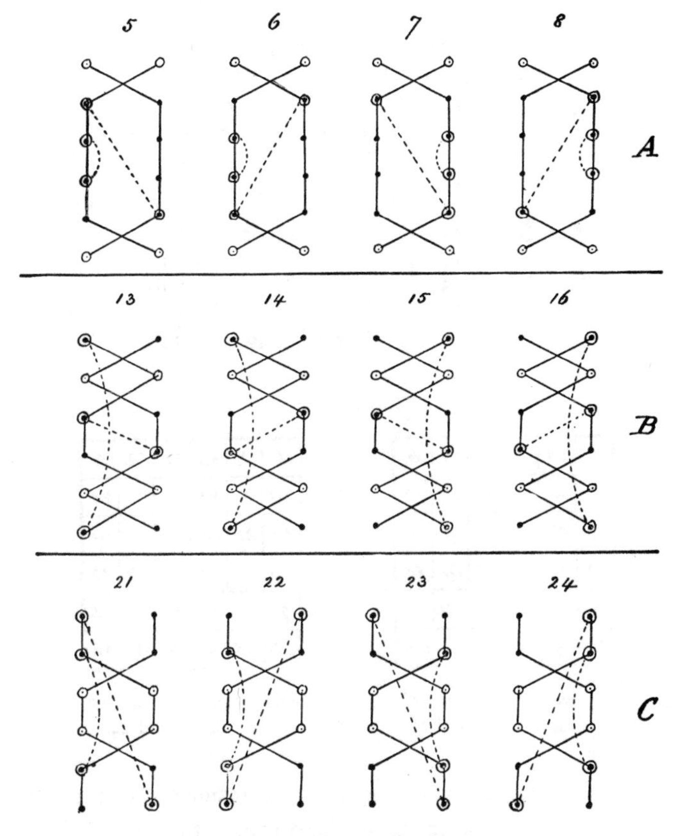

Fig. 51 (Second Part).

Taking for example the 6 × 6 magic square shown in Fig. 46, it will be seen from Fig. 49 that it is constructed from the diagrams marked 1—9 and 17 in Fig. 51. Comparing the first line of Fig. 46 with diagram A, Fig. 49, the sequence of numbers is 1,—35,—34 in unbroken order; then the diagram shows that 33 and 3 must be transposed, so 3 is written next (instead of 33) then 32 and 6 in unbroken order. In the last line of this square (still using diagram A) 31 comes first, then, seeing that 5 and 2 must be transposed, 2 is written instead of 5; then 4; then as 3 and 33 must be transposed, 33 is written instead of 3, 5 instead of 2, and the line is finished with 36. Diagram B gives the development of the second

TABLE SHOWING 128 CHANGES WHICH MAY BE RUNG ON THE TWENTY-FOUR DIAGRAMS IN FIG. 51.

A	B	C	
1, 2, 3 or 4	9	17, 18, 19 or 20	= 16 changes
" " " "	10	" " " "	= 16 "
" " " "	11	" " " "	= 16 "
" " " "	12	" " " "	= 16 "
5, 6, 7 or 8	13	21, 22, 23 or 24	= 16 "
" " " "	14	" " " "	= 16 "
" " " "	15	" " " "	= 16 "
" " " "	16	" " " "	= 16 "
		Total changes	= 128 "

EXAMPLES.

1	35	4	33	32	6
12	8	28	27	11	25
24	17	15	16	20	19
13	23	21	22	14	18
30	26	9	10	29	7
31	2	34	3	5	36

Square derived from diagrams 2, 10, and 18.

1	5	33	34	32	6
30	8	28	9	11	25
18	23	15	16	20	19
24	14	21	22	17	13
7	26	10	27	29	12
31	35	4	3	2	36

Square derived from diagrams 8, 13, and 22.

and fifth lines of the square in the same manner, and diagram C the development of the third and fourth lines, thus completing the square.

The annexed table shows 128 changes which may be rung on the twenty-four diagrams shown in Figure 51, each combination giving a different 6×6 square, and many others might be added to the list.

The next size of even magic square is that of 8×8, and instead of presenting one of these squares ready made and analyzing it, we will now use the information which has been offered by previous examples in the construction of a new square of this size.

Referring to Fig. 45, the regular geometrical diagrams of the 4×4 square naturally suggest that an expansion of the same may be utilized to construct an 8×8 square. This expanded diagram

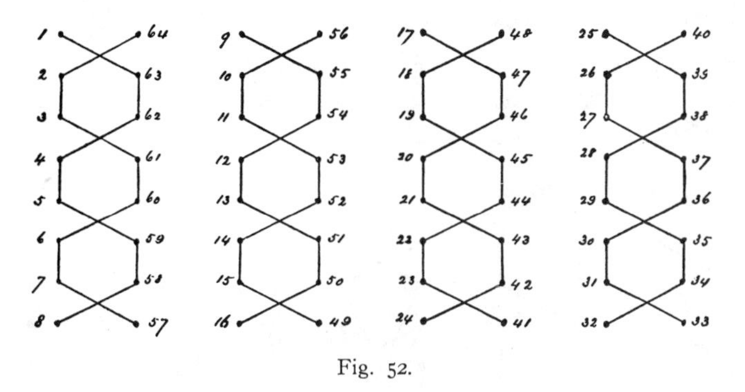

Fig. 52.

is accordingly shown in Fig. 52, and in Fig. 53 we have the magic square that is produced by copying the numbers in diagrammatic order.

1	63	62	4	5	59	58	8
56	10	11	53	52	14	15	49
48	18	19	45	44	22	23	41
25	39	38	28	29	35	34	32
33	31	30	36	37	27	26	40
24	42	43	21	20	46	47	17
16	50	51	13	12	54	55	9
57	7	6	60	61	3	2	64

Totals = 260.

Fig. 53.

As might be anticipated, this square is associated and the ease with which it has been constructed points to the simplicity of the method employed.

The magic square shown in Fig. 53 is, however, only one of a multitude of 8 × 8 squares, all of which have the same general characteristics and may be constructed with equal facility from

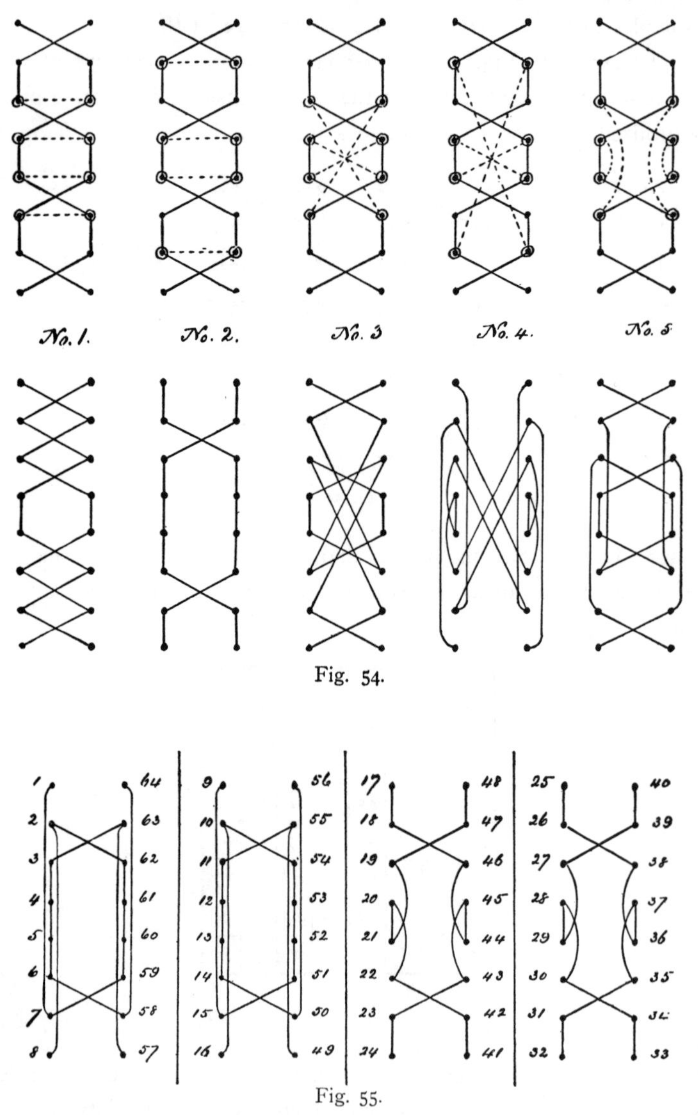

Fig. 54.

Fig. 55.

various regular diagrams that can be readily derived from transpositions of Fig. 52. Five of these variations are illustrated in Fig.

54, which also show the transpositions by which they are formed from the original diagrams. To construct an associated magic square from either of these variations it is only necessary to make four copies of the one selected, annex the numbers 1 to 64 in arithmetical

1	7	59	60	61	62	2	8
16	10	54	53	52	51	15	9
48	47	19	21	20	22	42	41
33	34	30	28	29	27	39	40
25	26	38	36	37	35	31	32
24	23	43	45	44	46	18	17
56	50	14	13	12	11	55	49
57	63	3	4	5	6	58	64

Totals = 260.

Fig. 56.

order as before explained, and then copy the numbers in diagrammatic sequence into the cells of an 8 × 8 square.

It will be noted in the construction of the 4 × 4 and 8 × 8

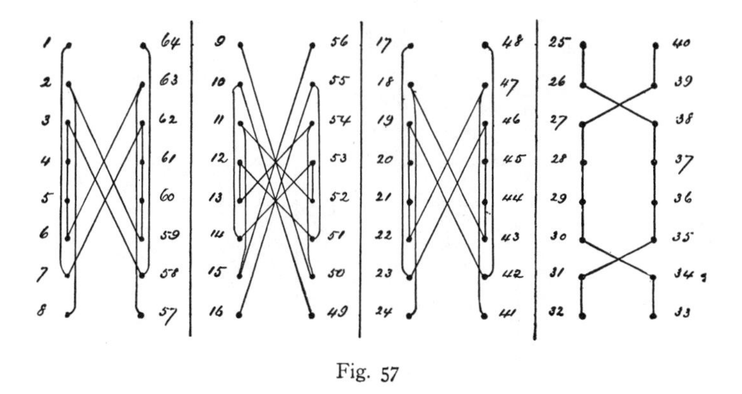

Fig. 57

squares that only one form of diagram has been hitherto used for each square, whereas three different forms were required for the 6 × 6 square. It is possible, however, to use either two, three, or four different diagrams in the construction of an 8 × 8 square, as

1	7	62	61	60	59	2	8
49	10	14	53	52	11	15	56
48	42	19	20	21	22	47	41
40	39	27	28	29	30	34	33
32	31	35	36	37	38	26	25
24	18	43	44	45	46	23	17
9	50	54	13	12	51	55	16
57	63	6	5	4	3	58	64

Totals = 260.

Fig. 58.

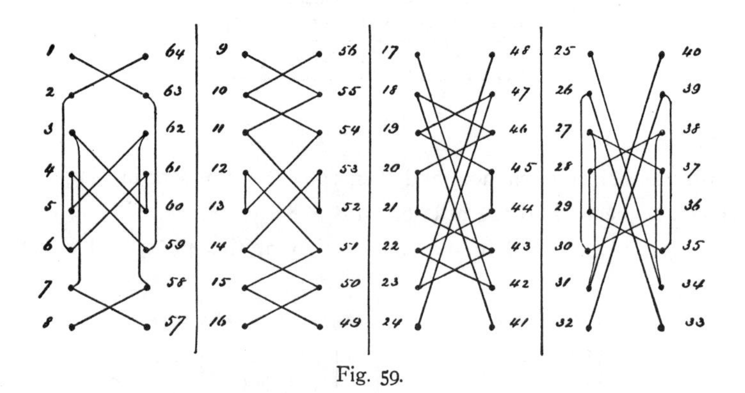

Fig. 59.

1	63	59	4	5	62	58	8
56	10	54	13	12	51	15	49
24	47	19	45	44	22	42	17
25	34	38	28	29	35	39	32
33	26	30	36	37	27	31	40
48	23	43	21	20	46	18	41
16	50	14	53	52	11	55	9
57	7	3	60	61	6	2	64

Totals = 260.

Fig. 60.

shown in the annexed examples. Fig. 55 illustrates two different forms from which the magic square Fig. 56 is constructed. Fig. 57 shows three different forms which are used in connection with the square in Fig. 58, and in a similar manner Figs. 59 and 60 show four different diagrams and the square derived therefrom. The

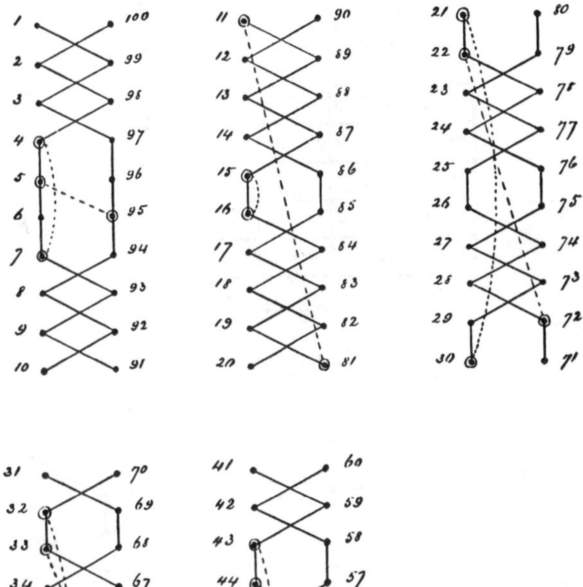

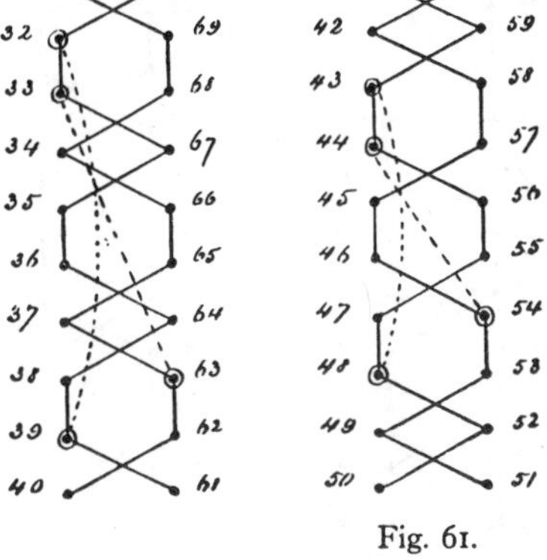

Fig. 61.

foregoing examples are sufficient to illustrate the immense number of different 8×8 magic squares that may be constructed by the aid of various diagrams.

We now come to the magic square of 10×10, and applying the comparative method to the last examples, it will be easy to expand the three diagrams of the 6×6 square (Fig. 49) into five diagrams that are required for the construction of a series of

1	99	3	97	96	5	94	8	92	10
90	12	88	14	86	85	17	83	19	11
80	79	23	77	25	26	74	28	22	71
31	69	68	34	66	65	37	33	62	40
60	42	58	57	45	46	44	53	49	51
50	52	43	47	55	56	54	48	59	41
61	32	38	64	36	35	67	63	39	70
21	29	73	27	75	76	24	78	72	30
20	82	18	84	15	16	87	13	89	81
91	9	93	4	6	95	7	98	2	100

Totals = 505.

Fig. 62.

10×10 squares. These five diagrams are shown in Fig. 61, and in Fig. 62 we have the magic square which is made by copying the

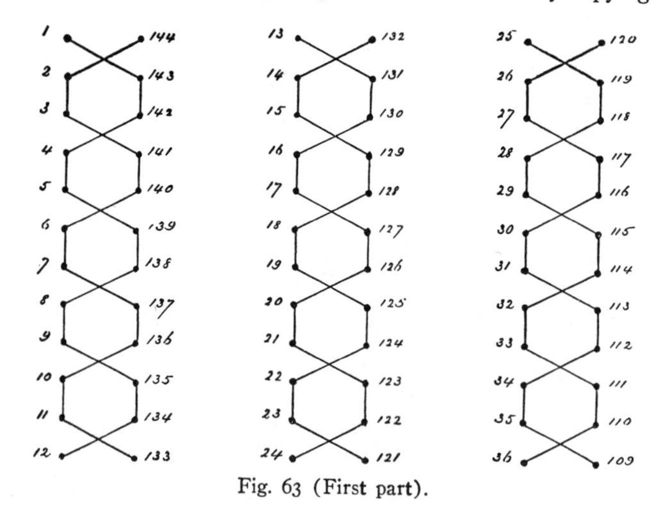

Fig. 63 (First part).

numbers from 1 to 100 in diagrammatic order into the cells of a 10×10 square.

It will be unnecessary to proceed further with the construction of other 10 × 10 squares, for the reader will recognize the striking resemblance between the diagrams of the 6 × 6 and the 10 ×

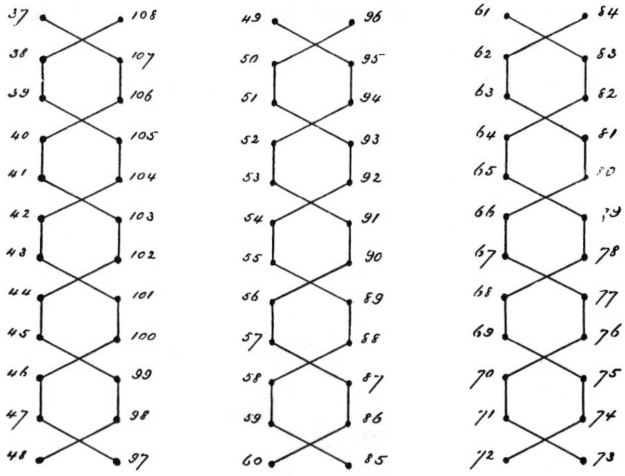

Fig. 63 (Second part).

1	143	142	4	5	139	138	8	9	135	134	12
132	14	15	129	128	18	19	125	124	22	23	121
120	26	27	117	116	30	31	113	112	34	35	109
37	107	106	40	41	103	102	44	45	99	98	48
49	95	94	52	53	91	90	56	57	87	86	60
84	62	63	81	80	66	67	77	76	70	71	73
72	74	75	69	68	78	79	65	64	82	83	61
85	59	58	88	89	55	54	92	93	51	50	96
97	47	46	100	101	43	42	104	105	39	38	108
36	110	111	33	32	114	115	29	28	118	119	25
24	122	123	21	20	126	127	17	16	130	131	13
133	11	10	136	137	7	6	140	141	3	2	144

Totals = 870.

Fig. 64.

10 squares, especially in connection with their respective irregularities.

It will also be seen that the same methods which were used for varying the 6 × 6 diagrams, are equally applicable to the 10 × 10 diagrams, so that an almost infinite variety of changes may be rung on them, from which a corresponding number of 10 × 10 squares may be derived, each of which will be different but will resemble the series of 6 × 6 squares in their curious and characteristic imperfections.

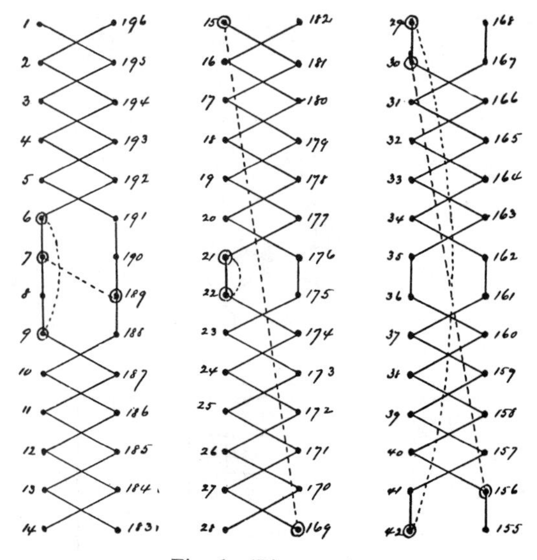

Fig. 65 (First part).

We have thus far studied the construction of even magic squares up to and including that of 10 × 10, and it is worthy of remark that when one-half the number of cells in one side of an even magic square is an even number the square can be made associated, but when it is an uneven number it is impossible to build a fully associated square with a straight arithmetical series. The difficulty can however be easily overcome by using a suitable number series. As this subject is fully treated in Cahpter XI under the heading, "Notes on the Construction of Magic Squares of Orders in which n is of the General Form $4p+2$," it is not discussed here.

Fig. 63 shows a series of diagrams from which the 12 × 12

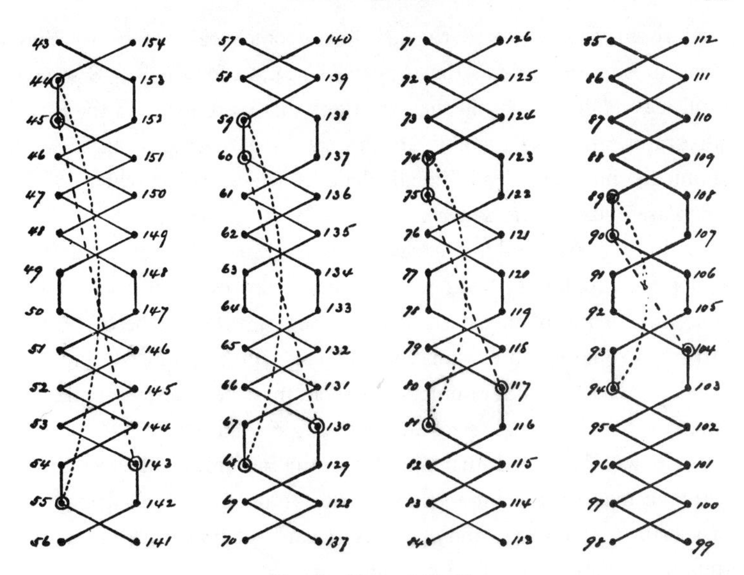

Fig. 65 (Second Part).

1	195	3	193	5	191	190	7	188	10	186	12	184	14
182	16	180	18	178	20	176	175	23	173	25	171	27	15
168	167	31	165	33	163	35	36	160	38	158	40	30	155
43	153	152	46	150	48	148	147	51	145	53	45	142	56
140	58	138	137	61	135	63	64	132	66	60	129	69	127
71	125	73	123	122	76	120	119	79	75	116	82	114	84
112	86	110	88	108	107	91	92	90	103	95	101	97	99
98	100	96	102	89	93	105	106	104	94	109	87	111	85
113	83	115	74	80	118	78	77	121	117	81	124	72	126
70	128	59	67	131	65	133	134	62	136	130	68	139	57
141	44	54	144	52	146	50	49	149	47	151	143	55	154
29	41	157	39	159	37	161	162	34	164	32	166	156	42
28	170	26	172	24	174	21	22	177	19	179	17	181	169
183	13	185	11	187	6	8	189	9	192	4	194	2	196

Fig. 66.

square in Fig. 64 is derived. The geometrical design of these diagrams is the same as that shown in Fig. 52 for the 8 × 8 square, and it is manifest that all the variations that were made in the 8 × 8 diagrams are also possible in the 12 × 12 diagrams, besides an immense number of additional changes which are allowed by the increased size of the square.

In Fig. 65 we have a series of diagrams illustrating the development of the 14 × 14 magic square shown in Fig. 66. These diagrams being plainly derived from the diagrams of the 6 × 6 and 10 × 10 squares, no explanation of them will be required, and it is evident that the diagrammatic method may be readily applied to the construction of all sizes of even magic squares.

It will be noted that the foregoing diagrams illustrate in a graphic manner the interesting results attained by the harmonious association of figures, and they also clearly demonstrate the almost infinite variety of possible combinations.

1			4
	2	3	
	2	3	
1			4

Fig. 67.

1	3	2	4
4	2	3	1
4	2	3	1
1	3	2	4

Fig. 68.

1	4	4	1
3	2	2	3
2	3	3	2
4	1	1	4

Fig. 69.

THE CONSTRUCTION OF EVEN MAGIC SQUARES BY DE LA HIRE'S METHOD.

An associated magic square of 4 × 4 may be constructed as follows:

1. Fill the corner diagonal columns of a 4 × 4 square with the numbers 1 to 4 in arithmetical sequence, starting from the upper and lower left hand corners (Fig. 67).
2. Fill the remaining empty cells with the missing numbers of the series 1 to 4 so that the sum of every perpendicular and horizontal column equals 10 (Fig. 68).

3. Construct another 4×4 square, having all numbers in the same positions relatively to each other as in the last square, but reversing the direction of all horizontal and perpendicular columns (Fig. 69).
4. Form the root square Fig. 70 from Fig. 69 by substituting root numbers for primary numbers, and then add the numbers in this root square to similarly located numbers in the primary square Fig. 68. The result will be the associated square of 4×4 shown in Fig. 72.

By making the root square Fig. 71 from the primary square Fig. 68 and adding the numbers therein to similarly located numbers

PRIMARY NUMBERS	ROOT NUMBERS
1	0
2	4
3	8
4	12

0	12	12	0
8	4	4	8
4	8	8	4
12	0	0	12

Fig. 70.

0	8	4	12
12	4	8	0
12	4	8	0
0	8	4	12

Fig. 71.

1	15	14	4
12	6	7	9
8	10	11	5
13	3	2	16

Fig. 72.

1	12	8	13
15	6	10	3
14	7	11	2
4	9	5	16

Fig. 73.

in the primary square Fig. 69, the same magic square of 4×4 will be produced, but with all horizontal and perpendicular columns reversed in direction as shown in Fig. 73.

The magic square of 6×6 shown in Figure 46 and also a large number of variations of same may be readily constructed by the De la Hire method, and the easiest way to explain the process will be to analyze the above mentioned square into the necessary primary and root squares, using the primary numbers 1 to 6 with their respective root numbers as follows:

Primary numbers ..... 1, 2, 3, 4, 5, 6.
Root numbers .... 0, 6, 12, 18, 24, 30.

The cells of two 6×6 squares may be respectively filled with primary and root numbers by analyzing the contents of each cell in Fig. 46. Commencing at the left-hand cell in the upper row, we note that this cell contains 1. In order to produce this number by the addition of a primary number to a root number it is evident that 0 and 1 must be selected and written into their respective cells. The second number in the top row of Fig. 46 being 35, the root number 30 must be written in the second cell of the root square and the primary number 5 in the second cell of the primary square, and so on throughout all the cells, the finished squares being shown in Figs. 74 and 75.

Another primary square may now be derived from the root square Fig. 74 by writing into the various cells of the former the

1	35	34	3	32	6
30	8	28	27	11	7
24	23	15	16	14	19
13	17	21	22	20	18
12	26	9	10	29	25
31	2	4	33	5	36

Fig. 46 (Dup.)

0	30	30	0	30	0
24	6	24	24	6	6
18	18	12	12	12	18
12	12	18	18	18	12
6	24	6	6	24	24
30	0	0	30	0	30

Fig. 74.

primary numbers that correspond to the root numbers of the latter. This second primary square is shown in Fig. 76. It will be seen that the numbers in Fig. 76 occupy the same relative positions to each other as the numbers of the first primary square (Fig. 75), but the direction of all columns is changed from horizontal to perpendicular, and vice versa.

To distinguish and identify the two primary squares which are used in these operations, the first one (in this case Fig. 75) will in future be termed the A primary square, and the second one (in this case Fig. 76) the B primary square.

It is evident that the magic square of 6×6 shown in Fig. 46 may now be reconstructed by adding the cell numbers in Fig. 74

to the similarly placed cell numbers in Fig. 75. Having thus inversely traced the development of the magic square from its A and B primary and root squares, it will be useful to note some of the general characteristics of even primary squares, and also to study the rules which govern their construction, as these rules will be found instructive in assisting the student to work out an almost endless variety of even magic squares of all dimensions.

1. Referring to the 6 × 6 A primary square shown in Fig. 75, it will be noted that the two corner diagonal columns contain the numbers 1 to 6 in arithmetical order, starting respectively from the upper and lower left hand corner cells, and that the diagonal columns of the B primary square in Fig. 76 also contain the same numbers in arithmetical order but starting

1	5	4	3	2	6
6	2	4	3	5	1
6	5	3	4	2	1
1	5	3	4	2	6
6	2	3	4	5	1
1	2	4	3	5	6

Fig. 75.

1	6	6	1	6	1
5	2	5	5	2	2
4	4	3	3	3	4
3	3	4	4	4	3
2	5	2	2	5	5
6	1	1	6	1	6

Fig. 76.

from the two upper corner cells. The numbers in the two corner diagonal columns are subject to many arrangements which differ from the above but it will be unnecessary to consider them in the present article.

2. The numbers in the A primary square Fig. 75 have the same relative arrangement as those in the B primary square Fig. 76, but the horizontal columns in one square form the perpendicular columns in the other and vice versa. This is a general but not a universal relationship between A and B primary squares.

3. The sum of the series 1 to 6 is 21 and the sum of every column in both A and B 6 × 6 primary squares must also be 21.

4. The sum of every column in a 6×6 root square must be 90, and under these conditions it follows that the sum of every column of a 6×6 magic square which is formed by the combination of a primary square with a root square must be 111 ($21 + 90 = 111$).
5. With the necessary changes in numbers the above rules hold good for all sizes of A and B primary squares and root squares of this class.

We may now proceed to show how a variety of 6×6 magic squares can be produced by different combinations of numbers in

a	1	2	4	3	5	6
b	1	5	4	3	2	6
c	1	5	3	4	2	6
d	6	5	3	4	2	1
e	6	2	3	4	5	1
f	6	2	4	3	5	1

Fig. 77.

1st line	1					6	*a*, *b*, or *c*.
2nd "		2			5		*a*, *e*, or *f*.
3rd "			3	4			*c*, *d*, or *e*.
4th "			3	4			*c*, *d*, or *e*.
5th "		2			5		*a*, *e*, or *f*.
6th "	1					6	*a*, *b*, or *c*.

Fig. 78.

primary and root squares. The six horizontal columns in Fig. 75 show some of the combinations of numbers from 1 to 6 that can be used in 6×6 A primary squares, and the positions of these columns or rows of figures relatively to each other may be changed so as to produce a vast variety of squares which will naturally lead to the development of a corresponding number of 6×6 magic squares.

In order to illustrate this in a systematic manner the different rows of figures in Fig. 75 may be rearranged and identified by letters as given in Fig. 77.

Fig. 78 shows the sequence of numbers in the diagonal columns of these 6×6 A primary squares, and as this arrangement cannot be changed in this series, the various horizontal columns or rows in Fig. 77 must be selected accordingly. The small letters at the right

No. 1.	No. 2.	No. 3.	No. 4.	No. 5.	No. 6.
a	*a*	*b*	*b*	*c*	*c*
f	*e*	*f*	*e*	*a*	*f*
c	*d*	*c*	*d*	*d*	*e*
d	*c*	*d*	*c*	*e*	*d*
e	*f*	*e*	*f*	*f*	*a*
b	*b*	*a*	*a*	*b*	*b*

Fig. 79.

of Fig. 78 indicate the different horizontal columns that may be used for the respective lines in the square; thus either *a*, *b*, or *c* column in Fig. 77 may be used for the first and sixth lines, *a*, *e*, or *f* for the second and fifth, and *c*, *d*, or *e* for the third and fourth lines, but neither *b*, *c*, or *d* can be used in the second or fifth lines, and so forth.

Six different combinations of columns are given in Fig, 79, from which twelve different 6×6 magic squares may be constructed. Taking column No. 1 as an example, Fig. 80 shows an

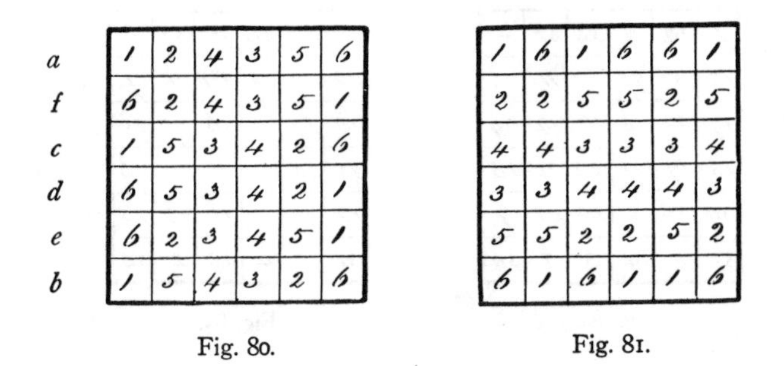

a	1	2	4	3	5	6
f	6	2	4	3	5	1
c	1	5	3	4	2	6
d	6	5	3	4	2	1
e	6	2	3	4	5	1
b	1	5	4	3	2	6

Fig. 80.

1	6	1	6	6	1
2	2	5	5	2	5
4	4	3	3	3	4
3	3	4	4	4	3
5	5	2	2	5	2
6	1	6	1	1	6

Fig. 81.

A primary square made from the combination *a*, *f*, *c*, *d*, *e*, *b*, and Fig. 81 is the B primary square formed by reversing the direction of the horizontal and perpendicular columns of Fig. 80. The root square Fig. 82 is then made from Fig. 81 and the 6×6 magic square in Fig. 84 is the result of adding the cell numbers of Fig. 82 to the corresponding cell numbers in Fig. 80.

The above operation may be varied by reversing the horizontal columns of the root square Fig. 82 right and left as shown in Fig. 83 and then forming the magic square given in Fig. 85. In this way two different magic squares may be derived from each combination.

0	30	0	30	30	0
6	6	24	24	6	24
18	18	12	12	12	18
12	12	18	18	18	12
24	24	6	6	24	6
30	0	30	0	0	30

Fig. 82.

0	30	30	0	30	0
24	6	24	24	6	6
18	12	12	12	18	18
12	18	18	18	12	12
6	24	6	6	24	24
30	0	0	30	0	30

Fig. 83.

It will be noted that all the 6 × 6 magic squares that are constructed by these rules are similar in their general characteristics to the 6 × 6 squares which are built up by the diagrammatic system.

Associated 8×8 magic squares may be constructed in great variety by the method now under consideration, and the different com-

1	32	4	33	35	6
12	8	28	27	11	25
19	23	15	16	14	24
18	17	21	22	20	13
30	26	9	10	29	7
31	5	34	3	2	36

Fig. 84.

1	32	34	3	35	6
30	8	28	27	11	7
19	17	15	16	20	24
18	23	21	22	14	13
12	26	9	10	29	25
31	5	4	33	2	36

Fig. 85.

binations of numbers from 1 to 8 given in Fig. 86 will be found useful for laying out a large number of A primary squares.

Fig. 87 shows the fixed numbers in the diagonal columns of these 8 × 8 A primary squares, and also designates by letters the specific rows of figures which may be used for the different horizontal columns. Thus the row marked *a* in Fig. 86 may be used for the first, fourth, fifth, and eighth horizontal columns but cannot be employed for the second, third, sixth or seventh columns, and so forth.

Fig. 88 suggests half a dozen combinations which will form

1	7	6	4	5	3	2	8	*a*
1	2	6	4	5	3	7	8	*b*
1	2	6	5	4	3	7	8	*c*
1	7	3	4	5	6	2	8	*d*
1	7	3	5	4	6	2	8	*e*
8	2	3	5	4	6	7	1	*aa*
8	7	3	5	4	6	2	1	*bb*
8	7	3	4	5	6	2	1	*cc*
8	2	6	5	4	3	7	1	*dd*
8	2	6	4	5	3	7	1	*ee*

Fig. 86.

as many primary squares, and it is evident that the number of possible variations is very large. It will suffice to develop the first and third of the series in Fig. 88 as examples.

1st line	1							8	*a, b, c, d,* or *e.*
2nd "		2					7		*b, c, aa, dd,* or *ee.*
3rd "			3			6			*d, e, aa,* or *cc.*
4th "				4	5				*a, b, d, cc,* or *ee.*
5th "				4	5				*a, b, d, cc,* or *ee.*
6th "			3			6			*d, e, aa,* or *cc.*
7th "		2					7		*b, c, aa, dd,* or *ee.*
8th "	1							8	*a, b, c, d,* or *e.*

Fig. 87.

Fig. 89 is the A primary square developed from column No. 1 in Fig. 88, and Fig. 90 is the B primary square made by reversing the direction of all horizontal and perpendicular columns of Fig. 89. Substituting root numbers for the primary numbers in Fig. 90, and adding these root numbers to the primary numbers in Fig. 89 gives the regular magic square of 8 × 8 shown in Fig. 91. The latter will be found identical with the square which may be written out directly from diagrams in Fig. 52.

No. 1.	No. 2.	No. 3.	No. 4.	No. 5.	No. 6.
a	*b*	*c*	*d*	*e*	*a*
aa	*b*	*c*	*dd*	*ee*	*b*
aa	*d*	*cc*	*e*	*e*	*e*
a	*b*	*cc*	*d*	*ee*	*d*
a	*b*	*cc*	*d*	*ee*	*d*
aa	*d*	*cc*	*e*	*e*	*e*
aa	*b*	*c*	*dd*	*ee*	*b*
a	*b*	*c*	*d*	*e*	*a*

Fig. 88.

Fig. 92 shows an A primary square produced from column No. 3 in Fig. 88. The B primary square Fig. 93 being made in the regular way by reversing the direction of the columns in Fig. 92.

Primary numbers .. 1, 2, 3, 4, 5, 6, 7, 8.
Root numbers ..... 0, 8, 16, 24, 32, 40, 48, 56.

1	7	6	4	5	3	2	8	*a*
8	2	3	5	4	6	7	1	*aa*
8	2	3	5	4	6	7	1	*aa*
1	7	6	4	5	3	2	8	*a*
1	7	6	4	5	3	2	8	*a*
8	2	3	5	4	6	7	1	*aa*
8	2	3	5	4	6	7	1	*aa*
1	7	6	4	5	3	2	8	*a*

Fig. 89.

1	8	8	1	1	8	8	1
7	2	2	7	7	2	2	7
6	3	3	6	6	3	3	6
4	5	5	4	4	5	5	4
5	4	4	5	5	4	4	5
3	6	6	3	3	6	6	3
2	7	7	2	2	7	7	2
8	1	1	8	8	1	1	8

Fig. 90.

The associated magic square of 8 × 8 in Fig 94 is developed from these two primary squares as in the last example, and it will be found similar to the square which may be formed directly from diagram No. 2 in Fig. 54.

1	63	62	4	5	59	58	8
56	10	11	53	52	14	15	49
48	18	19	45	44	22	23	41
25	39	38	28	29	35	34	32
33	31	30	36	37	27	26	40
24	42	43	21	20	46	47	17
16	50	51	13	12	54	55	9
57	7	6	60	61	3	2	64

Totals = 260.

Fig. 91.

1	2	6	5	4	3	7	8	c
1	2	6	5	4	3	7	8	c
8	7	3	4	5	6	2	1	cc
8	7	3	4	5	6	2	1	cc
8	7	3	4	5	6	2	1	cc
8	7	3	4	5	6	2	1	cc
1	2	6	5	4	3	7	8	c
1	2	6	5	4	3	7	8	c

Fig. 92.

1	1	8	8	8	8	1	1
2	2	7	7	7	7	2	2
6	6	3	3	3	3	6	6
5	5	4	4	4	4	5	5
4	4	5	5	5	5	4	4
3	3	6	6	6	6	3	3
7	7	2	2	2	2	7	7
8	8	1	1	1	1	8	8

Fig. 93.

1	2	62	61	60	59	7	8
9	10	54	53	52	51	15	16
48	47	19	20	21	22	42	41
40	39	27	28	29	30	34	33
32	31	35	36	37	38	26	25
24	23	43	44	45	46	18	17
49	50	14	13	12	11	55	56
57	58	6	5	4	3	63	64

Fig. 94.

1	7	62	60	61	59	2	8
16	10	51	53	52	54	15	9
48	42	19	21	20	22	47	41
33	39	30	28	29	27	24	40
25	31	38	36	37	35	26	32
24	18	43	45	44	46	23	17
56	50	11	13	12	14	55	49
57	63	6	4	5	3	58	64

Fig. 95.

Fig. 95 shows another 8 × 8 magic square which is constructed by combining the A primary square in Fig. 89 with the B primary square in Fig. 93 after changing the latter to a root square in the manner before described. This magic square may also be directly constructed from diagram No. 4 in Fig. 54.

It is evident that an almost unlimited number of different 8 × 8 magic squares may be made by the foregoing methods, and their application to the formation of other and larger squares is so obvious that it will be unnecessarry to present any further examples.

COMPOSITE MAGIC SQUARES.

These squares may be described as a series of small magic squares arranged quadratically in magic square order.

The 9 × 9 square shown in Fig. 96 is the smallest of this class that can be constructed and it consists of nine 3 × 3 sub-squares arranged in the same order as the numerals 1 to 9 inclusive in the 3 × 3 square shown in Fig. 1. The first sub-square occupies the

71	64	69	8	1	6	53	46	51
66	68	70	3	5	7	48	50	52
67	72	65	4	9	2	49	54	47
26	19	24	44	37	42	62	55	60
21	23	25	39	41	43	57	59	61
22	27	20	40	45	38	58	63	56
35	28	33	80	73	78	17	10	15
30	32	34	75	77	79	12	14	16
31	36	29	76	81	74	13	18	11

Totals = 369.

Fig. 96.

middle section of the first horizontal row of sub-squares, and it contains the numbers 1 to 9 inclusive arranged in regular magic square order being a duplicate of Fig. 1. The second sub-square is located in the right hand lower corner of the third horizontal row of sub-squares and it contains the numbers 10 to 18 inclusive arranged in magic square order, and so on to the last sub-square which occupies the middle section of the third horizontal row of

47	58	69	80	1	12	23	34	45
57	68	79	9	11	22	33	44	46
67	78	8	10	21	32	43	54	56
77	7	18	20	31	42	53	55	66
6	17	19	30	41	52	63	65	76
16	27	29	40	51	62	64	75	5
26	28	39	50	61	72	74	4	15
36	38	49	60	71	73	3	14	25
37	48	59	70	81	2	13	24	35

Totals = 369.

Fig. 97.

113	127	126	116	1	15	14	4	81	95	94	84
124	118	119	121	12	6	7	9	92	86	87	89
120	122	123	117	8	10	11	5	88	90	91	85
125	115	114	128	13	3	2	16	93	83	82	96
33	47	46	36	65	79	78	68	97	111	110	100
44	38	39	41	76	70	71	73	108	102	103	105
40	42	43	37	72	74	75	69	104	106	107	101
45	35	34	48	77	67	66	80	109	99	98	112
49	63	62	52	129	143	142	132	17	31	30	20
60	54	55	57	140	134	135	137	28	22	23	25
56	58	59	53	136	138	139	133	24	26	27	21
61	51	50	64	141	131	130	144	29	19	18	32

Totals = 870.

Fig. 98.

sub-squares, and which contains the numbers 73 to 81 inclusive.

This peculiar arrangement of the numbers 1 to 81 inclusive forms a magic square in which the characteristics of the ordinary

9×9 square are multiplied to a remarkable extent, for whereas in the latter square (Fig. 97) there are only twenty columns which sum up to 369, in the compound square of 9×9 there are an immense number of combination columns which yield this amount. This is evident from the fact that there are eight columns in the first sub-square which yield the number 15; also eight columns in the middle sub-square which yield the number 123—and eight columns in the last sub-square which sum up to the number 231—and $15 + 123 + 231 = 369$.

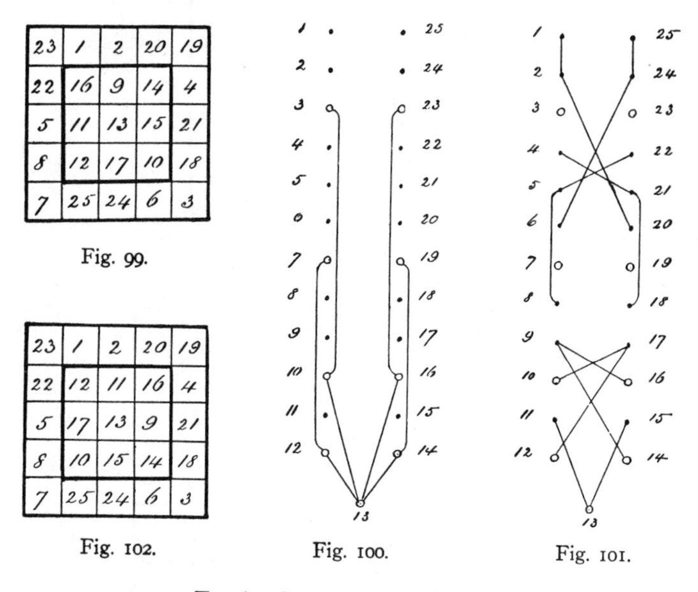

23	1	2	20	19
22	16	9	14	4
5	11	13	15	21
8	12	17	10	18
7	25	24	6	3

Fig. 99.

23	1	2	20	19
22	12	11	16	4
5	17	13	9	21
8	10	15	14	18
7	25	24	6	3

Fig. 102.

Fig. 100.

Fig. 101.

Totals of 3×3 squares $= 39$.
Totals of 5×5 squares $= 65$.

The 15×15 comes next in order and this may be constructed with twenty-five 3×3's or nine 5×5's, and so on in the larger sizes of these squares.

The next larger square of this class is that of 16×16 which can only be built with sixteen sub-squares of 4×4. Next comes the 18×18 compound square which may be constructed with thirty-six sub-squares of 3×3 or with nine sub-squares of 6×6, and so on indefinitely with larger and larger compound squares.

CONCENTRIC MAGIC SQUARES.

Beginning with a small central magic square it is possible to arrange one or more panels of numbers concentrically around it so that after the addition of each panel, the enlarged square will still retain magic qualifications.

Either a 3×3 or a 4×4 magic square may be used as a nucleus, and the square will obviously remain either odd or even, according to its beginning, irrespective of the number of panels which may be successively added to it. The center square will

19	2	20	1	23
4	16	9	14	22
18	11	13	15	8
21	12	17	10	5
3	24	6	25	7

Fig. 103.

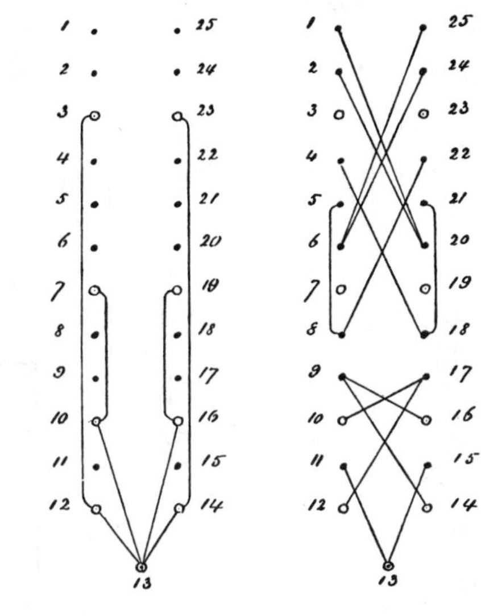

Fig. 104.

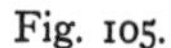

Fig. 105.

Totals of 3×3 square $= 39$.
Totals of 5×5 square $= 65$.

naturally be associated, but after one or more panels have been added the enlarged square will no longer be associated, because the peculiar features of its construction will not permit the sum of every pair of diametrically opposite numbers to equal the sum of the first and last numbers of the series used. The sum of every horizontal and perpendicular column and of the two corner diagonal columns will, however, be the same amount.

The smallest concentric square that can be constructed is that of 5×5, an example of which is illustrated in Fig. 99.

The center square of 3×3 begins with 9 and continues, with increments of 1, up to 17, the center number being 13 in accordance with the general rule for a 5×5 square made with the series of

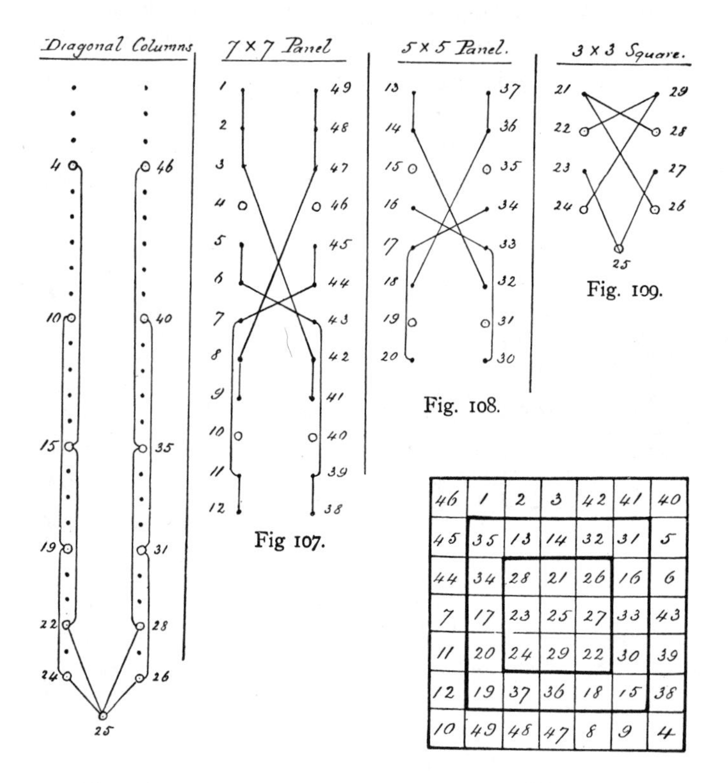

46	1	2	3	42	41	40
45	35	13	14	32	31	5
44	34	28	21	26	16	6
7	17	23	25	27	33	43
11	20	24	29	22	30	39
12	19	37	36	18	15	38
10	49	48	47	8	9	4

Fig. 106. Fig 107. Fig. 108. Fig. 109. Fig. 110.

Totals of 3×3 square $= 75$
Totals of 5×5 square $= 125$
Totals of 7×7 square $= 175$

numbers 1 to 25. The development of the two corner diagonal columns is given in diagram Fig. 100, the numbers for these columns being indicated by small circles. The proper sequence of the other twelve numbers in the panels is shown in Fig. 101. The relative positions of the nine numbers in the central 3 × 3 square cannot be changed, but the entire square may be inverted or turned one quarter, one half, or three quarters around, so as to vary the

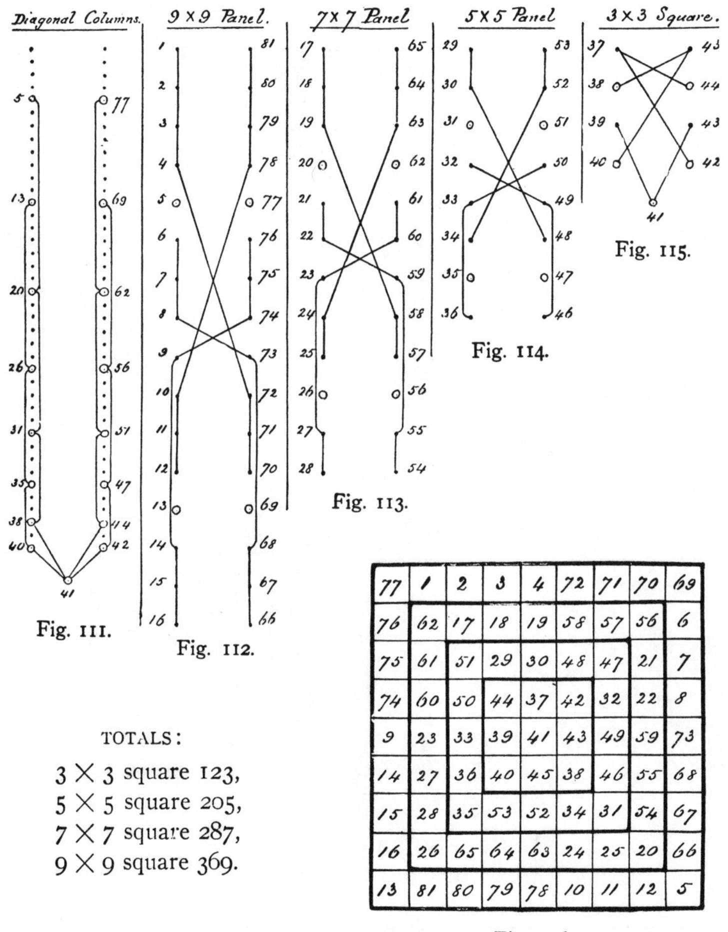

Fig. 111. Fig. 112. Fig. 113. Fig. 114. Fig. 115.

TOTALS:

3 × 3 square 123,
5 × 5 square 205,
7 × 7 square 287,
9 × 9 square 369.

77	1	2	3	4	72	71	70	69
76	62	17	18	19	58	57	56	6
75	61	51	29	30	48	47	21	7
74	60	50	44	37	42	32	22	8
9	23	33	39	41	43	49	59	73
14	27	36	40	45	38	46	55	68
15	28	35	53	52	34	31	54	67
16	26	65	64	63	24	25	20	66
13	81	80	79	78	10	11	12	5

Fig. 116.

position of the numbers in it relatively to the surrounding panel numbers. Fig. 102 shows a 5 × 5 concentric square in which the panel numbers occupy the same cells as in Fig. 99, but the central

3 × 3 square is turned around one quarter of a revolution to the right.

Several variations may also be made in the location of the panel numbers, an example being given in Figs. 103, 104, and 105. Many

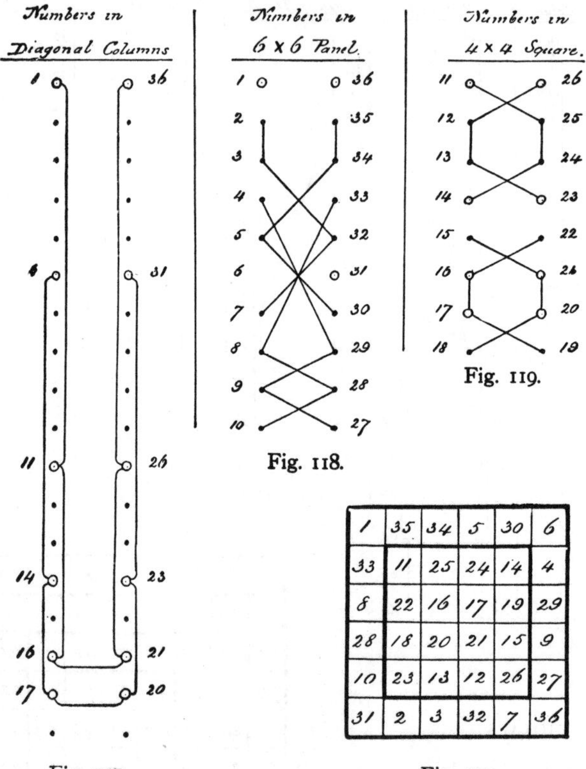

Fig. 118.

Fig. 119.

1	35	34	5	30	6
33	11	25	24	14	4
8	22	16	17	19	29
28	18	20	21	15	9
10	23	13	12	26	27
31	2	3	32	7	36

Fig. 117.

Fig. 120.

Totals of 4 × 4 square = 74.
Totals of 6 × 6 square = 111.

other changes in the relative positions of the panel numbers are selfevident.

One of many variations of the 7 × 7 concentric magic square is shown in Fig. 110. The 3 × 3 central square in this example is started with 21 and finished with 29 in order to comply with the general rule that 25 must occupy the center cell in a 7 × 7 square that includes the series of numbers 1 to 49. The numbers for the two corner diagonal columns are indicated in their proper order by small circles in Fig. 106, and the arrangement of the panel numbers is given in Figs. 107, 108, and 109. As a final example of an

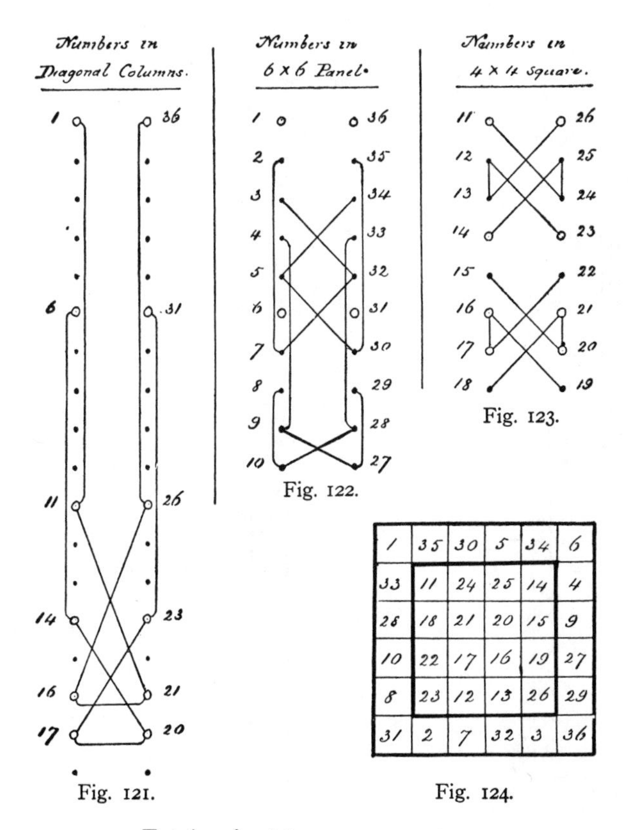

Fig. 121.

Fig. 122.

Fig. 123.

1	35	30	5	34	6
33	11	24	25	14	4
28	18	21	20	15	9
10	22	17	16	19	27
8	23	12	13	26	29
31	2	7	32	3	36

Fig. 124.

Totals of 4 × 4 square = 74.
Totals of 6 × 6 square = 111.

odd concentric square Fig. 116 shows one of 9 × 9, its development being given in Figs. 111, 112, 113, 114, and 115.

All these diagrams are simple and obvious expansions of those shown in Figs. 100 and 101 in connection with the 5 × 5 concentric square, and they and their numerous variations may be expanded indefinitely and used for the construction of larger odd magic squares of this class.

The smallest even concentric magic square is that of 6×6, of

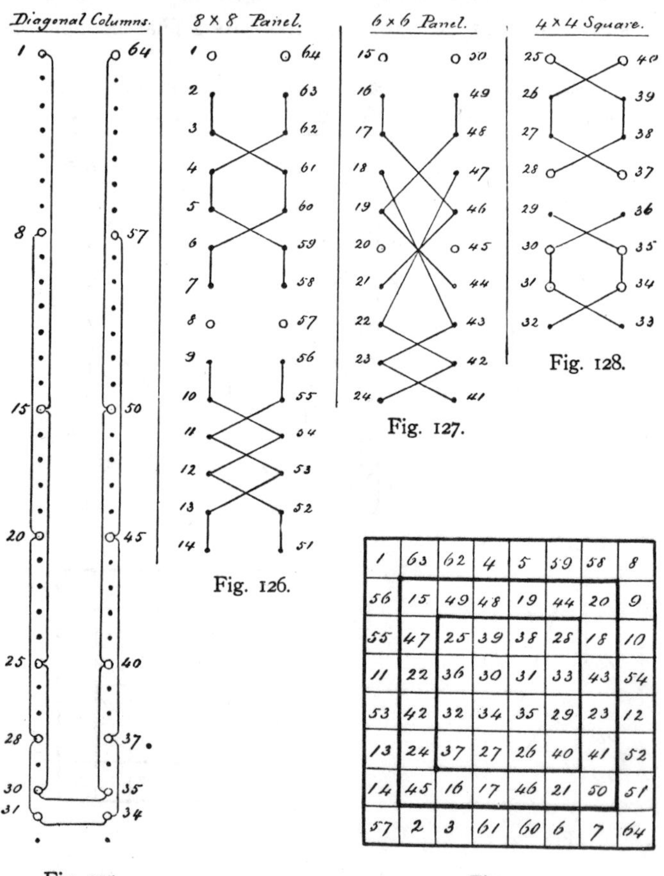

Fig. 125. Fig. 126. Fig. 127. Fig. 128.

1	63	62	4	5	59	58	8
56	15	49	48	19	44	20	9
55	47	25	39	38	28	18	10
11	22	36	30	31	33	43	54
53	42	32	34	35	29	23	12
13	24	37	27	26	40	41	52
14	45	16	17	46	21	50	51
57	2	3	61	60	6	7	64

Fig. 129.

Totals of 4×4 square $= 130$.
Totals of 6×6 square $= 195$.
Totals of 8×8 square $= 260$.

which Fig. 120 is an example. The development of this square may be traced in the diagrams given in Figs. 117, 118, and 119. The center square of 4×4 is associated, but after the panel is added the enlarged square ceases to be so, as already noted. Figs. 121, 122, 123, and 124 illustrate another example of this square with diagrams of development.

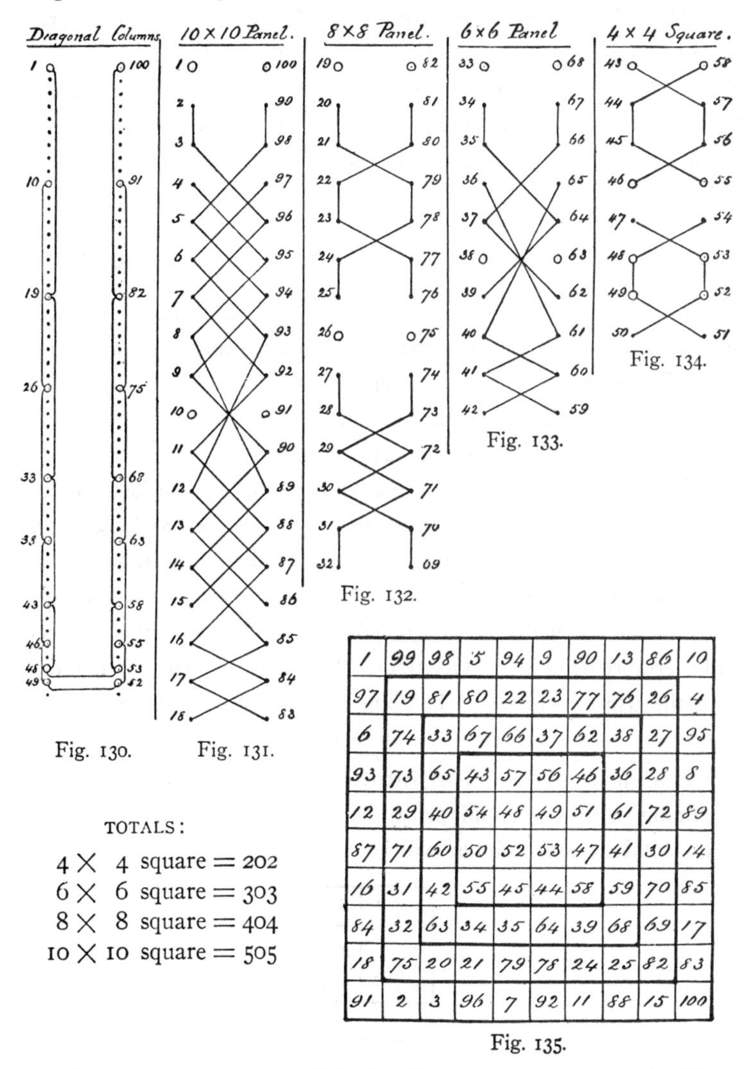

Fig. 130. Fig. 131. Fig. 132. Fig. 133. Fig. 134.

TOTALS:

4 × 4 square = 202
6 × 6 square = 303
8 × 8 square = 404
10 × 10 square = 505

1	99	98	5	94	9	90	13	86	10
97	19	81	80	22	23	77	76	26	4
6	74	33	67	66	37	62	38	27	95
93	73	65	43	57	56	46	36	28	8
12	29	40	54	48	49	51	61	72	89
87	71	60	50	52	53	47	41	30	14
16	31	42	55	45	44	58	59	70	85
84	32	63	34	35	64	39	68	69	17
18	75	20	21	79	78	24	25	82	83
91	2	3	96	7	92	11	88	15	100

Fig. 135.

A concentric square of 8 × 8 with diagrams are given in Figs. 125, 126, 127, 128, and 129, and one of 10 × 10 in Figs. 130, 131, 132, 133, 134, and 135. It will be seen that all these larger squares have been developed in a very easy manner from successive expansions of the diagrams used for the 6×6 square in Figs. 117, 118, and 119.

The rules governing the formation of concentric magic squares have been hitherto considered somewhat difficult, but by the aid of diagrams, their construction in great variety and of any size has been reduced to an operation of extreme simplicity, involving only the necessary patience to construct the diagrams and copy the numbers.

GENERAL NOTES ON THE CONSTRUCTION OF MAGIC SQUARES.

There are two variables that govern the summation of magic squares formed of numbers that follow each other with equal increments throughout the series, viz.:

1. The Initial, or starting number.
2. The Increment, or increasing number.

When these two variables are known, the summations can be easily determined, or when either of these variables and the summation are known, the other variable can be readily derived.

The most interesting problem in this connection is the construction of squares with predetermined summations, and this subject will therefore be first considered, assuming that the reader is familiar with the usual methods of building odd and even squares.

* * *

If a square of 3×3 is constructed in the usual manner, that is, beginning with unity and proceeding with regular increments of 1, the total of each column will be 15.

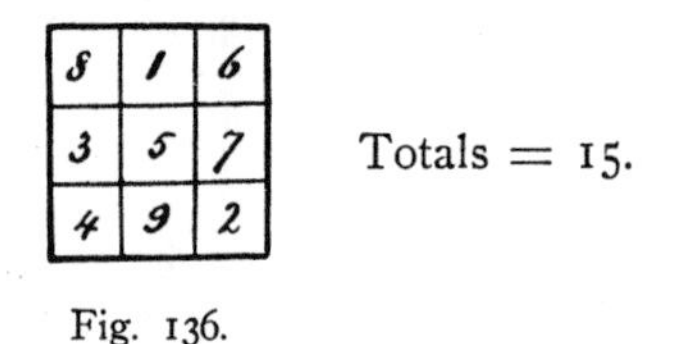

Fig. 136.

If 2 is used as the initial number instead of 1 and the square is again constructed with regular increments of 1, the total of each column will be 18.

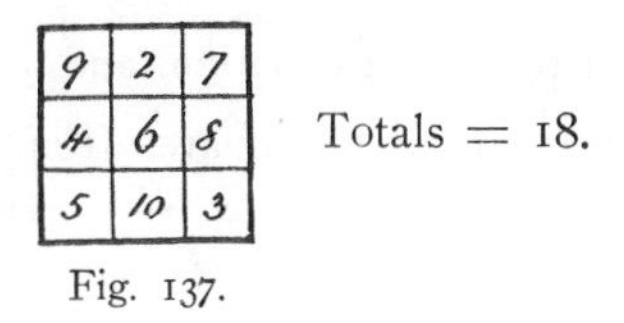

Fig. 137.

If 2 is still used as the initial number and the square is once more constructed with regular increments of 2 instead of 1, the total of each column will be 30.

16	2	12
6	10	14
8	18	4

Total = 30.

Fig. 138.

It therefore follows that there must be initial numbers, the use of which with given increments will entail summations of any predetermined amount, and there must also be increments, the use of which with given initial numbers, will likewise produce predetermined summations.

These initial numbers and increments may readily be determined by a simple form of equation which will establish a connection between them and the summation numbers.

Let:

$A =$ initial number,

$\beta =$ increment,

$n =$ number of cells in one side of square,

$S =$ summation.

Then, if $A = 1$ and $\beta = 1$

$$\frac{n}{2}(n^2 + 1) = S.$$

If A and β are more or less than unity, the following general formula may be used:

$$An + \beta\frac{n}{2}(n^2 - 1) = S.$$

It will be found convenient to substitute a constant, (K) for

$\frac{n}{2}(n^2 - 1)$ in the above equation, and a table of these constants is therefore appended for all squares from 3×3 to 12×12.

Squares:	Const. = K
3×3	12
4×4	30
5×5	60
6×6	105
7×7	168
8×8	252
9×9	360
10×10	495
11×11	660
12×12	858

When using the above constants the equation will be:

$$An + \beta K = S.$$

EXAMPLES.

What initial number is required for the square of 3×3, with 1 as the increment, to produce 1903 as the summation?

Transposing the last equation:

$$\frac{S - \beta K}{n} = A,$$

or

$$\frac{1903 - (1 \times 12)}{3} = 630\,^1/_3 = \text{Initial No.}$$

$637\frac{1}{3}$	$630\frac{1}{3}$	$635\frac{1}{3}$
$632\frac{1}{3}$	$634\frac{1}{3}$	$636\frac{1}{3}$
$633\frac{1}{3}$	$638\frac{1}{3}$	$631\frac{1}{3}$

Totals = 1903.

Fig. 130.

We will now apply the same equation to a square of 4×4, in which case:

$$\frac{1903 - (1 \times 30)}{4} = 468\tfrac{1}{4} = \text{Initial No.}$$

$468\frac{1}{4}$	$482\frac{1}{4}$	$481\frac{1}{4}$	$471\frac{1}{4}$
$479\frac{1}{4}$	$473\frac{1}{4}$	$474\frac{1}{4}$	$476\frac{1}{4}$
$475\frac{1}{4}$	$477\frac{1}{4}$	$478\frac{1}{4}$	$472\frac{1}{4}$
$480\frac{1}{4}$	$470\frac{1}{4}$	$469\frac{1}{4}$	$483\frac{1}{4}$

Totals = 1903.

Fig. 140.

Also to a square of 5 × 5,

$$\frac{1903 - (1 \times 60)}{5} = 368.6 = \text{Initial No.}$$

384.6	391.6	368.6	375.6	382.6
390.6	372.6	374.6	381.6	383.6
371.6	373.6	380.6	387.6	389.6
377.6	379.6	386.6	388.6	370.6
378.6	385.6	392.6	369.6	376.6

Totals = 1903.

Fig. 141.

And for a square of 6 × 6.

$$\frac{1903 - (1 \times 105)}{6} = 299\,{}^{2}/_{3} = \text{Initial No.}$$

$299\frac{2}{3}$	$333\frac{2}{3}$	$332\frac{2}{3}$	$301\frac{2}{3}$	$330\frac{2}{3}$	$304\frac{2}{3}$
$328\frac{2}{3}$	$306\frac{2}{3}$	$326\frac{2}{3}$	$325\frac{2}{3}$	$309\frac{2}{3}$	$305\frac{2}{3}$
$322\frac{2}{3}$	$321\frac{2}{3}$	$313\frac{2}{3}$	$314\frac{2}{3}$	$312\frac{2}{3}$	$317\frac{2}{3}$
$311\frac{2}{3}$	$315\frac{2}{3}$	$319\frac{2}{3}$	$320\frac{2}{3}$	$318\frac{2}{3}$	$316\frac{2}{3}$
$310\frac{2}{3}$	$324\frac{2}{3}$	$307\frac{2}{3}$	$308\frac{2}{3}$	$327\frac{2}{3}$	$323\frac{2}{3}$
$329\frac{2}{3}$	$300\frac{2}{3}$	$302\frac{2}{3}$	$331\frac{2}{3}$	$303\frac{2}{3}$	$334\frac{2}{3}$

Totals = 1903.

Fig. 142.

The preceding examples illustrate the construction of squares built up with progressive increments of 1, but the operation may be varied by using increments that are greater or less than unity.

EXAMPLES.

What initial number must be used in a square of 3 × 3, with increments of 3, to produce a summation of 1903?

Applying the equation given on page 56, but making $\beta = 3$ instead of 1, we have:

$$\frac{1903 - (3 \times 12)}{3} = 622\,{}^1/_3.$$

$622\,{}^1/_3$ is therefore the initial number and by using this in a 3×3 square with progressive increments of 3, the desired results are obtained.

643 1/3	622 1/3	637 1/3
628 1/3	634 1/3	640 1/3
631 1/3	646 1/3	625 1/3

Totals = 1903.

Fig. 143.

To find the initial number with increments of 10.

$$\frac{1903 - (10 \times 12)}{3} = 594\,{}^1/_3 = \text{Initial No.}$$

664 1/3	594 1/3	644 1/3
614 1/3	634 1/3	654 1/3
624 1/3	674 1/3	604 1/3

Totals = 1903.

Fig. 144.

Or to find the initial number with increments of ${}^1/_3$.

$$\frac{1903 - ({}^1/_3 \times 12)}{3} = 633 = \text{Initial No.}$$

635 1/3	633	634 2/3
633 2/3	634 1/3	635
634	635 2/3	633 1/3

Totals = 1903.

Fig. 145.

These examples being sufficient to illustrate the rule, we will pass on another step and show how to build squares with predetermined summations, using any desired initial numbers, with proper increments.

EXAMPLES.

What increment number must be used in a square of 3×3, wherein 1 is the initial number and 1903 the desired summation?

Referring to equation on page 56 and transposing, we have

$$\frac{S - An}{K} = \beta = \text{increment, or}$$

$$\frac{1903 - (1 \times 3)}{12} = 158\,{}^1/_3 = \text{Increment.}$$

Starting therefore with unity and building up the square with successive increments of $158\,{}^1/_3$, we obtain the desired result.

$1109\frac{1}{3}$	1	$792\frac{2}{3}$
$317\frac{2}{3}$	$634\frac{1}{3}$	951
476	$1267\frac{2}{3}$	$159\frac{1}{3}$

Totals = 1903.

Fig. 146.

When it is desired to start with any number larger or smaller than unity, the numbers in the equation can be modified accordingly. Thus if 4 is selected as an initial number, the equation will be:

$$\frac{1903 - (4 \times 3)}{12} = 157\,{}^7/_{12} = \text{Increment.}$$

$1107\frac{1}{12}$	4	$791\frac{11}{12}$
$319\frac{2}{12}$	$634\frac{4}{12}$	$949\frac{6}{12}$
$476\frac{9}{12}$	$1264\frac{8}{12}$	$161\frac{7}{12}$

Totals = 1903.

Fig. 147.

With an initial number of $^1/_3$.

$$\frac{1903 - (^1/_3 \times 3)}{12} = 158\tfrac{1}{2} = \text{Increment.}$$

$1109\frac{10}{12}$	$\frac{4}{12}$	$792\frac{10}{12}$
$317\frac{4}{12}$	$634\frac{4}{12}$	$951\frac{4}{12}$
$475\frac{10}{12}$	$1268\frac{4}{12}$	$158\frac{10}{12}$

Totals = 1903.

Fig. 148.

It is thus demonstrated that any initial number may be used providing (in a square of 3×3) it is less than one-third of the summation. In a square of 4×4 it must be less than one-fourth of the summation, and so on.

To illustrate an extreme case, we will select 634 as an initial number in a 3×3 square and find the increment which will result in a summation of 1903.

$$\frac{1903 - (634 \times 3)}{12} = {}^{1}/_{12} = \text{Increment.}$$

$634\frac{7}{12}$	634	$634\frac{5}{12}$
$634\frac{2}{12}$	$634\frac{4}{12}$	$634\frac{6}{12}$
$634\frac{3}{12}$	$634\frac{8}{12}$	$634\frac{1}{12}$

Totals = 1903.

Fig. 149.

Having now considered the formation of magic squares with predetermined summations by the use of proper initial numbers and increments, it only remains to show that the summation of any square may be found, when the initial number and the increment are given, by the application of the equation shown on page 56, viz.:

$$\text{A}n + \beta\text{K} = \text{S}.$$

EXAMPLES.

Find the summation of a square of 3×3 using 5 as the initial number, and 7 as the increment.

$$(5 \times 3) + (7 \times 12) = 99 = \text{Summation.}$$

54	5	40
19	33	47
26	61	12

Totals = 99.

Fig. 150.

What will be the summation of a square of 4×4 using 9 as an initial number and 11 as an increment?

$$(9 \times 4) + (11 \times 30) = 366 = \text{Summation.}$$

9	163	152	42
130	64	75	97
86	108	119	53
141	31	20	174

Totals = 366.

Fig. 151.

The preceding equations may also be used for the construction of magic squares involving zero and minus quantities, as illustrated in the following examples.

What will be the summation of a square of 3 × 3, using 10 as the initial number with —2 increments?

$$(10 \times 3) + (-2 \times 12) = 6 = \text{Summation.}$$

-4	10	0
6	2	-2
4	-6	8

Totals = 6.

Fig. 152.

What initial number must be used in a square of 3 × 3 with increments of —3 to produce a summation of 3?

$$\frac{3-(-3 \times 12)}{3} = 13 = \text{Initial No.}$$

-8	13	-2
7	1	-5
4	-11	10

Totals = 3.

Fig. 153.

What initial number is required for a 3 × 3 square, with increments of 1, to produce a summation of 0?

$$\frac{0-(1 \times 12)}{3} = -4 = \text{Initial No.}$$

3	-4	1
-2	0	2
-1	4	-3

Totals = 0.

Fig. 154.

What increment must be used in a square of 3×3 wherein 12 is the initial number and — 12 the required summation?

$$\frac{-12-(12 \times 3)}{12} = -4 = \text{Increment.}$$

−16	12	−8
4	−4	−12
0	−20	8

Totals = — 12.

Fig. 155.

What increment must be used in a square of 4×4 wherein 48 is the initial number and 42 the summation?

$$\frac{42-(48 \times 4)}{30} = -5 = \text{Increment.}$$

48	−22	−17	33
−7	23	18	8
13	3	−2	28
−12	38	43	−27

Totals = 42.

Fig. 156.

The foregoing rules have been applied to examples in squares of small size only for the sake of brevity and simplicity, but the principles explained can evidently be expanded to any extent that may be desired.

Numbers following each other with uniform increments have been used throughout this article in the construction of magic squares, in order to illustrate their formation according to certain rules in a simple manner. It has however been shown by various writers that the series of numbers used in the construction of every magic square is divided by the breakmoves into n groups of n numbers per group (n representing the number of cells in one side of the square), and that the numbers in these groups do not *necessarily* follow each other in regular order with equal increments, but under certain well defined rules they may be arranged in a great variety of irregular sequences and still produce perfect magic squares.

Referring to Fig. 40 as an example, many different 5×5 squares may be formed by varying the sequence of the five groups, and also by changing the arrangement of the numbers in each group.

Instead of writing the five diagonal columns in Fig. 40 with the numbers 1 to 25 in arithmetical order thus:

a.	1	2	3	4	5
b.	6	7	8	9	10
c.	11	12	13	14	15
d.	16	17	18	19	20
e.	21	22	23	24	25

they may be arranged in the order *b e c a d,* which will develop the 5×5 square shown in Fig. 17.

Other variations may be made by re-arranging the consecutive numbers in each group, as for example thus:

a.	1	4	3	2	5
b.	6	9	8	7	10
c.	11	14	13	12	15
d.	16	19	18	17	20
e.	21	24	23	22	25

The foregoing may be considered as only suggestive of many ways of grouping numbers by which magic squares may be produced in great variety, which however will be generally found to follow regular constructive rules, providing that these rules are applied to series of numbers arranged in similar consecutive order.

CHAPTER II.

MAGIC CUBES.

THE curious and interesting characteristics of magic squares may be developed in figures of three dimensions constituting magic cubes.

Cubes of odd numbers may be constructed by direct and continuous process, and cubes of even numbers may be built up by the aid of geometrical diagrams. In each case the constructive methods resemble those which were previously explained in connection with odd and even magic squares.

As the cube is a figure of three dimensions it is naturally more difficult to construct in magic formation than the square (which has only two dimensions) because the interrelations between the various numbers are more complext than those in a square and not so easily adjusted one with the other to sum the magic constants.

THE ESSENTIAL CHARACTERISTICS OF MAGIC CUBES.

A magic cube consists of a series of numbers so arranged in cubical form that each row of numbers running parallel with any of its edges, and also each of its four great diagonals shall sum the same amount. Any cubical arrangement of numbers that fulfils these conditions may be properly termed a magic cube. As in the case of magic squares, various interesting but non-essential features may be added to these requisites, and in this way many different kinds of magic cubes may be constructed. In the present chapter, however, associated or regular magic cubes will be principally described.

ASSOCIATED OR REGULAR MAGIC CUBES OF ODD NUMBERS.

The smallest magic cube is naturally $3 \times 3 \times 3$.

Fig. 157 shows one of these cubes, and in columns I, II and III, Fig. 158, there are given the nine different squares which it contains. In this cube there are twenty-seven straight columns, two diagonal columns in each of the three middle squares, and four diagonal columns connecting the eight corners of the cube, making in all thirty-seven columns each of which sums up to 42. The center number is also 14 or $(n^3 + 1)/2$ and the sum of any pair of diametrically opposite numbers is 28 or $n^3 + 1$.

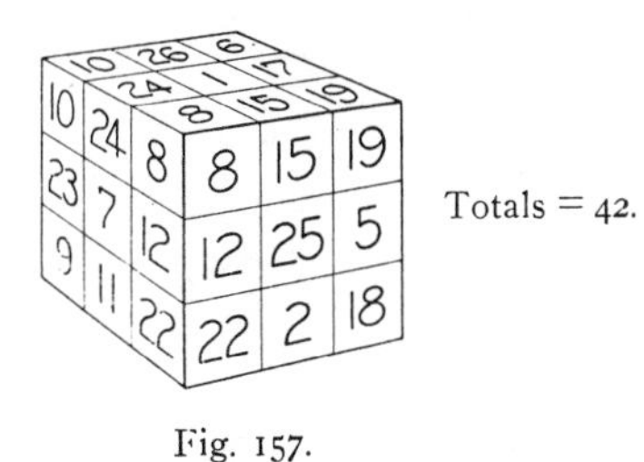

Fig. 157.

In describing the direct method of building odd magic squares, many forms of regular advance moves were explained, including right and left diagonal sequence, and various so-called "knight's moves." It was also shown that the order of regular advance was periodically broken by other well-defined spacings which were termed "breakmoves." In building odd magic squares, only one form of breakmove was employed in each square, but in the construction of odd magic cubes, two kinds are required in each cube which for distinction may be termed n and n^2 breakmoves respectively. In magic cubes which commence with unity and proceed with increments of 1, the n^2 breakmoves occur between each multiple of n^2 and the next following number, which in a $3 \times 3 \times 3$ cube brings them between 9 and 10, 18 and 19, and also between the last and first numbers of the series, 27 and 1. The n breakmoves are made between all other multiples of *n*, which in the above case brings them between 3 and 4, 6 and 7, 12 and 13, 15 and 16, 21 and 22, and 24 and 25. With this explanation the rules for building the magic cube shown in Fig. 1 may now be formulated, and for convenience of observation and construction, the cube is divided horizontally into three sections or layers, each section being shown separately in Column 1, Fig. 158.

It may be mentioned that when a move is to be continued *upward* from the top square it is carried around to the bottom square,

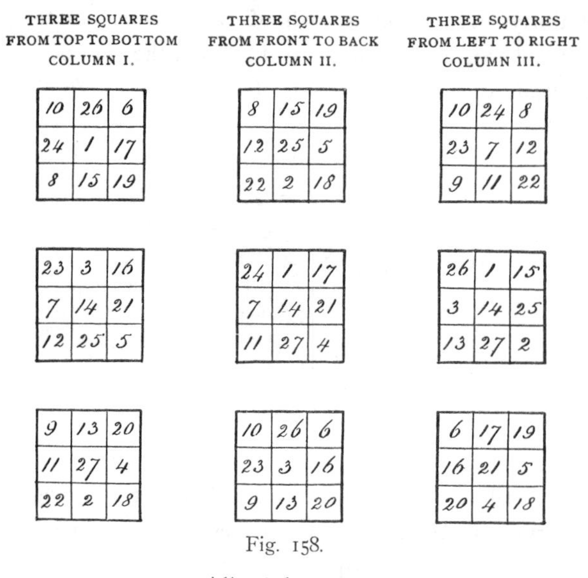

Fig. 158.

All totals = 42.

and when a move is to be made *downward* from the bottom square, it is carried around to the top square, the conception being similar to that of the horizontal cylinder used in connection with odd magic squares.

Commencing with 1 in the center cell of the top square, the cells in the three squares are filled with consecutive numbers up to 27 in accordance with the following directions:

Advance move. One cell down in next square up (from last entry).

n breakmove. One cell in downward right-hand diagonal in next square down (from last entry).

n^2 breakmove. Same cell in next square down (from last entry).

If it is desired to build this cube from the three vertical squares from front to back of Fig. 157, as shown in Column II, Fig. 158, the directions will then be as follows: commencing with 1 in the middle cell of the upper row of numbers in the middle square,

Advance move. One cell up in next square up.

n breakmove. One cell in downward right-hand diagonal in next square up.

n^2 breakmove. Next cell down in same square.

TABLE I.

	A	B	C		A	B	C		A	B	C
1	1	1	1	10	2	1	1	19	3	1	1
2	1	1	2	11	2	1	2	20	3	1	2
3	1	1	3	12	2	1	3	21	3	1	3
4	1	2	1	13	2	2	1	22	3	2	1
5	1	2	2	14	2	2	2	23	3	2	2
6	1	2	3	15	2	2	3	24	3	2	3
7	1	3	1	16	2	3	1	25	3	3	1
8	1	3	2	17	2	3	2	26	3	3	2
9	1	3	3	18	2	3	3	27	3	3	3

Fig. 159.

Finally, the same cube may be constructed from the three vertical squares running from left to right side of Fig. 157, as shown in Column III, Fig. 158 commencing, as in the last example, with 1 in the middle cell of the upper row of numbers in the middle square, and proceeding as follows:

Advance move. Three consecutive cells in upward right-hand diagonal in same square (as last entry).

n breakmove. One cell in downward right-hand diagonal in next square down.

n^2 breakmove. One cell down in same square (as last entry).

Five variations may be derived from this cube in the simple way illustrated in Table I on the preceding page.

Assign three-figure values to the numbers 1 to 27 inclusive in terms of 1, 2, 3 as given in Table I, Fig. 159, and change the numbers in the three squares in Column I, Fig. 158, to their corresponding three-figure values, thus producing the square shown in Fig. 160. It is evident that if the arrangement of numbers in the three squares in Column I were unknown, they could be readily produced from Fig. 160 by the translation of the three-figure values into regular numbers in accordance with Table I, but more than

	A	B	C	A	B	C	A	B	C	
Top Square	2	1	1	3	3	2	1	2	3	1st Line
	3	2	3	1	1	1	2	3	2	2nd "
	1	3	2	2	2	3	3	1	1	3rd "
Middle Square	3	2	2	1	1	3	2	3	1	1st Line
	1	3	1	2	2	2	3	1	3	2nd "
	2	1	3	3	3	1	1	2	2	3rd "
Bottom Square	1	3	3	2	2	1	3	1	2	1st Line
	2	1	2	3	3	3	1	2	1	2nd "
	3	2	1	1	1	2	2	3	3	3rd "

Fig. 160.

this can be accomplished. The letters A, B, C, in Table I indicate the normal order of the numerals 1, 2, 3, but by changing this order other triplets of 3×3 squares can be made which will differ more or less from the original models in the arrangement of their cell numbers, but which will retain their general magic characteristics. The changes which may be rung on A, B, C, are naturally six, as follows:

A. B. C.	C. B. A.
B. C. A.	B. A. C.
C. A. B.	A. C. B.

The combination of 1, 2, 3 being given in normal order in the original cube, the five cubes formed from the other combinations are shown in Figs. 161-165.

These magic cubes may also be constructed by the direct method in accordance with the annexed directions.

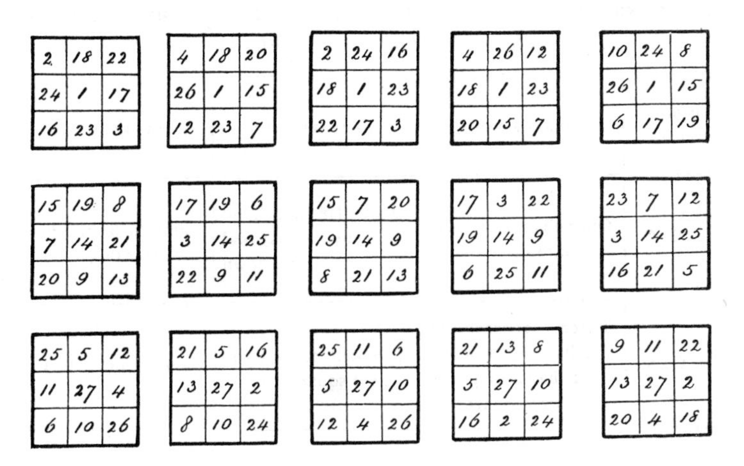

2	18	22
24	1	17
16	23	3

15	19	8
7	14	21
20	9	13

25	5	12
11	27	4
6	10	26

4	18	20
26	1	15
12	23	7

17	19	6
3	14	25
22	9	11

21	5	16
13	27	2
8	10	24

2	24	16
18	1	23
22	17	3

15	7	20
19	14	9
8	21	13

25	11	6
5	27	10
12	4	26

4	26	12
18	1	23
20	15	7

17	3	22
19	14	9
6	25	11

21	13	8
5	27	10
16	2	24

10	24	8
26	1	15
6	17	19

23	7	12
3	14	25
16	21	5

9	11	22
13	27	2
20	4	18

FIG. 161 (B.C.A.) FIG. 162. (C.A.B.) FIG. 163. (C.B.A.) FIG. 164. (B.A.C.) FIG. 165. (A.C.B.)

Fig. 166 is an example of another $3 \times 3 \times 3$ cube in which the first number occupies a corner cell, and the last number fills the diametrically opposite corner cell, the middle number coming in

TOP SQUARE.

1	17	24
15	19	8
26	6	10

MIDDLE SQUARE.

23	3	16
7	14	21
12	25	5

BOTTOM SQUARE.

18	22	2
20	9	13
4	11	27

Fig. 166.

the center cell in accordance with the rule. Fig. 167 shows this cube with the numbers changed to their three-figure values from which five variations of Fig. 166 may be derived, or they may be constructed directly by the directions which are marked with the changes of A. B. C. for convenient reference.

The analysis of the numbers in Fig. 157 and Fig. 166 into their three-figure values in terms of 1, 2, 3, as shown in Figs. 160 and 167, makes clear the curious mathematical order of their arrangement which is not apparent on the face of the regular numbers as

DIRECTIONS FOR CONSTRUCTING THE 3×3×3 MAGIC CUBE SHOWN IN FIG. 157 AND FIVE VARIATIONS OF THE SAME.

COMBINATION	ADVANCE MOVES	n BREAKMOVES	n^2 BREAKMOVES
A. B. C.	One cell down in next square up	One cell in right-hand downward diagonal in next square down	Same cell in next square down
B. C. A.	Three consecutive cells in upward left-hand diagonal in same square	One cell to left in next square up	Same as in A. B. C.
C. A. B.	One cell to right in next square up	One cell up in next square up	Same as in A. B. C.
C. B. A.	Same as in B. C. A.	Same as in C. A. B.	Same as in A. B. C.
B. A. C.	Same as in A. B. C.	Same as in B. C. A.	Same as in A. B. C.
A. C. B.	Same as in C. A. B.	Same as in A. B. C.	Same as in A. B. C.

they appear in the various cells of the cubes. For example, it may be seen that in every subsquare in Figs. 160 and 167 (corresponding to horizontal columns in the cubes) the numbers 1, 2, 3 are each repeated three times. Also in every horizontal and perpendicular column there is the same triple repetition. Furthermore, all the diagonal columns in the cubes which sum up to 42, if followed into their analyses in Figs. 160 and 167 will also be found to carry similar repetitions. A brief study of these figures will also disclose other curious mathematical qualities pertaining to their intrinsic symmetrical arrangement.

The next odd magic cube in order is $5 \times 5 \times 5$, and Fig. 168 shows one of its many possible variations. For convenience, it is divided into five horizontal sections or layers, forming five 5×5 squares from the top to the bottom of the cube.

Commencing with 1 in the first cell of the middle horizontal

	A	B	C	A	B.	C	A	B	C	
Top Square	1	1	1	2	3	2	3	2	3	1st Line
	2	2	3	3	1	1	1	3	2	2nd "
	3	3	2	1	2	3	2	1	1	3rd "
Middle Square	3	2	2	1	1	3	2	3	1	1st Line
	1	3	1	2	2	2	3	1	3	2nd "
	2	1	3	3	3	1	1	2	2	3rd "
Bottom Square.	2	3	3	3	2	1	1	1	2	1st Line
	3	1	2	1	3	3	2	2	1	2nd "
	1	2	1	2	1	2	3	3	3	3rd "

Fig. 167.

column in the third square, this cube may be constructed by filling in the various cells with consecutive numbers up to 125 in accordance with the following directions:

Advance moves. One cell up in next square down.
n breakmove. Two cells to the left and one cell down (knight's move) in same square as the last entry.
n^2 breakmove. One cell to right in same square as last entry.

This cube exhibits some interesting qualifications. Examining first the five horizontal squares from the top to the bottom of the cube as shown in Fig. 168, there are:

50 straight columns summing up to......... 315
10 corner diagonal columns summing up to.. 315
40 sub-diagonal columns summing up to.... 315
Total 100 columns having the same summation.

DIRECTIONS FOR CONSTRUCTING THE 3×3×3 MAGIC CUBE SHOWN IN FIG. 166 AND FIVE VARIATIONS OF THE SAME.

COMBINATIONS	ADVANCE MOVES	n BREAKMOVES	n^2 BREAKMOVES
A. B. C.	One cell to left in next square up	One cell in upward left-hand diagonal in next square down	One cell in downward right-hand diagonal in next square down
B. C. A.	Three consecutive cells in upward left-hand diagonal in same square	One cell in upward right-hand diagonal in next square up	Same as in A. B. C.
C. A. B.	One cell up in next square up	One cell in downward left-hand diagonal in next square up	Same as in A. B. C.
C. B. A.	Same as in B. C. A.	Same as in C. A. B.	Same as in A. B. C.
B. A. C.	Same as in A. B. C.	Same as in B. C. A.	Same as in A. B. C.
A. C. B.	Same as in C. A, B.	Same as in A. B. C.	Same as in A B. C.

In the five vertical squares from front to back of this cube there are:

50 straight columns summing up to......... 315
6 corner diagonal columns summing up to .. 315
20 sub-diagonal columns summing up to.... 315
Total 76 columns having the same summation.

In the five vertical squares from right to left of cube, there are, as in the last case, 76 columns which all sum up to 315. In the complete cube there are also four great diagonals and also a number of broken diagonals that sum up to 315.

1.

67	98	104	10	36
110	11	42	73	79
48	54	85	111	17
86	117	23	29	60
4	35	61	92	123

TOP SQUARE.

3.

50	51	82	113	19
88	119	25	26	57
1	32	63	94	125
69	100	101	7	38
107	13	44	75	76

5.

3	34	65	91	122
66	97	103	9	40
109	15	41	72	78
47	53	84	115	16
90	116	22	28	59

BOTTOM SQUARE

2.

106	12	43	74	80
49	55	81	112	18
87	118	24	30	56
5	31	62	93	124
68	99	105	6	37

4.

89	120	21	27	58
2	33	64	95	121
70	96	102	8	39
108	14	45	71	77
46	52	83	114	20

Fig. 168.

A table similar to Fig. 159 may be laid out giving three-figure values for the numbers in $5 \times 5 \times 5$ cubes from 1 to 125, and by changing the numbers in Fig. 168 to these three-figure values, a square similar to Fig. 160 will be produced from which five variations of Fig. 168 may be derived. Similar results, however, can be obtained with less work by means of a table of numbers constructed as shown in Fig. 169. (Table II.)

The three-figure values of cell numbers in $5 \times 5 \times 5$ magic cubes are found from this table as follows:

Select the root-number which is nearest to the cell-number, but *below it in value.* Then write down

1. The section number in which the root-number is found,
2. The primary number over the root-number,
3. The difference between the root-number and the cell-number.

Three figures will thus be determined which will represent the required three-figure value of the cell-number.

Examples. The first number in the first row of the upper square in Fig. 168 is 67. The nearest root-number to this and below it in value is 65 in section 3 under the primary number 4 and the

TABLE II.

Primary Nos.	1	2	3	4	5	Section 1
Root Nos	0	5	10	15	20	
Primary Nos.	1	2	3	4	5	Section 2
Root Nos	25	30	35	40	45	
Primary Nos.	1	2	3	4	5	Section 3
Root Nos.	50	55	60	65	70	
Primary Nos.	1	2	3	4	5	Section 4
Root Nos.	75	80	85	90	95	
Primary Nos.	1	2	3	4	5	Section 5
Root Nos.	100	105	110	115	120	

Fig. 169.

difference between the root-number and the cell-number is 2. The three-number value of 67 is therefore 3. 4. 2. Again, the fourth number in the same row is 10. The nearest root-number but *below it in value* is 5 in section 1 under the primary number 2, and the difference between the root-number and the cell-number is 5. The three-figure value of 10 is therefore 1. 2. 5. By these simple operations the three-figure values of all the cell-numbers in the $5 \times 5 \times 5$ cube in Fig. 168 may be quickly determined, and by the system of transposition previously explained, five variations of this cube may be constructed.

The shorter method of building these $5 \times 5 \times 5$ cubes by the direct process of filling the different cells in regular order with consecutive numbers may, however, be considered by some to be preferable to the more roundabout way. (See directions in the following table.)

DIRECTIONS FOR CONSTRUCTING THE $5 \times 5 \times 5$ MAGIC CUBE SHOWN IN FIG. 168 AND FIVE VARIATIONS OF THE SAME.

COMBINATIONS	ADVANCE MOVES	n BREAKMOVES	n^2 BREAKMOVES
A. B. C.	One cell up in next square down	Two cells to left and one down in same square as last entry	One cell to right in same square as last entry
B. C. A.	Two cells to left and one up for five consecutive numbers in same square	Two cells in upward left hand diagonal in next square down	Same as in A. B. C.
C. A. B.	Two cells in left hand downward diagonal in next square up	One cell in right-hand downward diagonal in next square up	Same as in A. B. C.
C. B. A,	Same as in B, C. A.	Same as in C. A. B.	Same as in A. B. C.
B. A. C.	Same as in A. B. C.	Same as in B. C. A.	Same as in A. B. C.
A. C. B.	Same as in C. A. B.	Same as in A. B. C.	Same as in A. B.C.

Fig. 170 is another example of a $5 \times 5 \times 5$ magic cube which is commenced in the upper left-hand corner of the top square, and finished in the lower right-hand corner of the bottom square, the middle number of the series (63) appearing in the center cell of the cube according to rule.

Odd magic cubes may be commenced in various cells other than those shown in the preceding pages, and they may be built up with an almost infinite number of variations. It would, however, be only superfluous and tiresome to amplify the subject further, as the examples already submitted cover the important points of construction, and may readily be applied to further extensions.

1.

1	82	38	119	75
74	5	81	37	118
117	73	4	85	36
40	116	72	3	84
83	39	120	71	2

TOP SQUARE.

3.

65	16	97	28	109
108	64	20	96	27
26	107	63	19	100
99	30	106	62	18
17	98	29	110	61

5.

124	55	6	87	43
42	123	54	10	86
90	41	122	53	9
8	89	45	121	52
51	7	88	44	125

BOTTOM SQUARE.

2.

33	114	70	21	77
76	32	113	69	25
24	80	31	112	68
67	23	79	35	111
115	66	22	78	34

4.

92	48	104	60	11
15	91	47	103	59
58	14	95	46	102
101	57	13	94	50
49	105	56	12	93

Fig. 170.

Any sizes of odd magic cubes larger than $5 \times 5 \times 5$ may be constructed by the directions which govern the formation of $3 \times 3 \times 3$ and $5 \times 5 \times 5$ cubes.

ASSOCIATED OR REGULAR MAGIC CUBES OF EVEN NUMBERS.

Magic cubes of even numbers may be built by the aid of geometric diagrams, similar to those illustrated in the preceding chapter, which describes the construction of even magic squares.

Fig. 171 shows one of the many possible arrangements of a $4 \times 4 \times 4$ cube, the diagram of which is given in Fig. 172.

There are fifty-two columns in this cube which sum up to 130, viz., sixteen vertical columns from the top of the cube to the

DIRECTIONS FOR CONSTRUCTING THE $5 \times 5 \times 5$ MAGIC CUBE SHOWN IN FIG. 170 AND FIVE VARIATIONS OF THE SAME.

COMBINATIONS	ADVANCE MOVES	n BREAKMOVES	n^2 BREAKMOVES
A. B. C.	Five consecutive cells in upward left hand diagonal in next square up	One cell in upward right-hand diagonal in next square up	One cell in downward right-hand diagonal in next square down
B. C. A.	Two cells down in second square down	One cell in downward left-hand diagonal in second square down	Same as in A. B. C.
C. A. B.	Two cells to right in next square up	Two cells in downward right hand diagonal in next square down	Same as in A. B. C.
C. B. A.	Some as in B. C. A.	Same as in C. A. B.	Same as in A. B. C.
B. A. C.	Same as in A. B. C.	Same as in B. C. A.	Same as in A. B. C.
A. C. B.	Same as in C. A. B.	Same as in A. B. C.	Same as in A. B. C.

bottom, sixteen horizontal columns from the front to the back, sixteen horizontal columns from right to left, and four diagonal columns uniting the four pairs of opposite corners. The sum of any two numbers, which are diametrically opposite to each other and equidistant from the center of the cube also equals 65 or $n^3 + 1$.

Another feature of this cube is that the sum of the four numbers in each of the forty-eight sub-squares of 2×2 is 130.

It has been shown in the chapter on "Magic Squares" that the

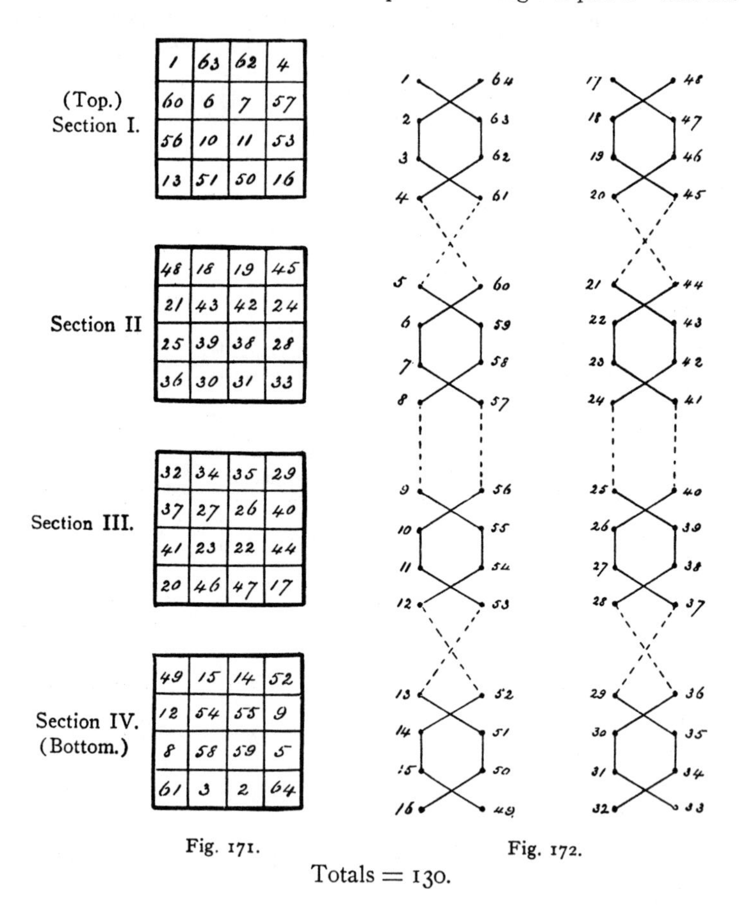

Fig. 171.

Fig. 172.

Totals = 130.

square of 4×4 could be formed by writing the numbers 1 to 16 in arithmetical order, then leaving the numbers in the two corner diagonals unchanged, but changing all the other numbers to their complements with 17 or $n^2 + 1$. It will be noted in the magic cube of $4 \times 4 \times 4$, given in Fig. 171, that in the first and last of the four sections (I and IV) this rule also holds good. In the two middle sections (II and III) the rule is reversed; the numbers in the two corner diagonals being complements with 65 or $n^3 + 1$, and all the other numbers in arithmetical order.

Fig. 173 shows four squares or sections of a cube, with the numbers 1 to 64 written in arithmetical order. Those numbers that occupy corresponding cells in Fig. 171 are enclosed within circles. If all the other numbers in Fig. 173 are changed to their complements with 65, the total arrangement of numbers will then be the same as in Fig. 171.

In his interesting and instructive chapter entitled "Reflections on Magic Squares"* Dr. Paul Carus gives a novel and ingenious analysis of even squares in different "orders" of numbering, these orders being termed respectively *o, ro, i* and *ri*. It is shown that the two magic squares of 4×4 (in the chapter referred to) con-

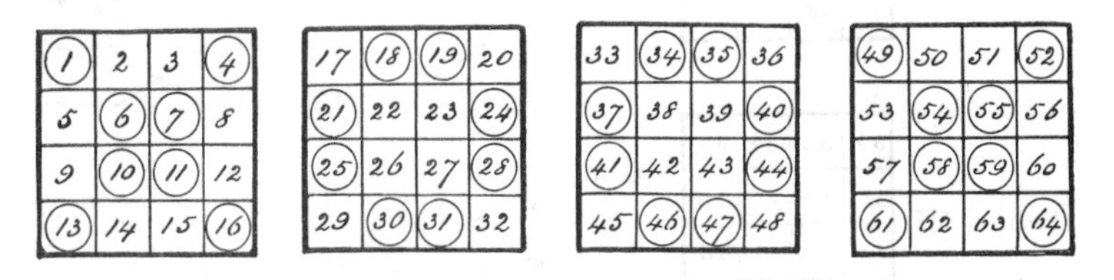

Fig. 173.

sist only of *o* and *ro* numbers; *ro* numbers being in fact the complements of *o* numbers with $n^2 + 1$. This rule also obtains in the magic cube of $4 \times 4 \times 4$ given in Fig. 171. The four sections of this cube may in fact be filled out by writing the *o* numbers, in arithmetical order in the cells of the two corner diagonal columns of sections I and IV, and in all the cells of sections II and III, excepting those of the two corner diagonal columns, and then writing the *ro* numbers, also in arithmetical order, in the remaining empty cells of the four sections.

Fig. 171 may be considered as typical of all magic cubes of $4 \times 4 \times 4$ and their multiples, of this class, but a great many variations may be effected by simple transpositions. For example, Fig.

* See p. 113 ff.

174 is a $4 \times 4 \times 4$ cube which is constructed by writing the four numbers that are contained in the 2×2 sub-squares (Fig. 171) in a straight line, and there are many other possible transpositions which will change the relative order of the numbers, without destroying the magic characteristics of the cube.

Section I. (Top.)

1	63	60	6
62	4	7	57
56	10	13	51
11	53	50	16

Section II.

48	18	21	43
19	45	42	24
25	39	36	30
38	28	31	33

Section III.

32	34	37	27
35	29	26	40
41	23	20	46
22	44	47	17

Section IV. (Bottom.)

49	15	12	54
14	52	55	9
8	58	61	3
59	5	2	64

Fig. 174.

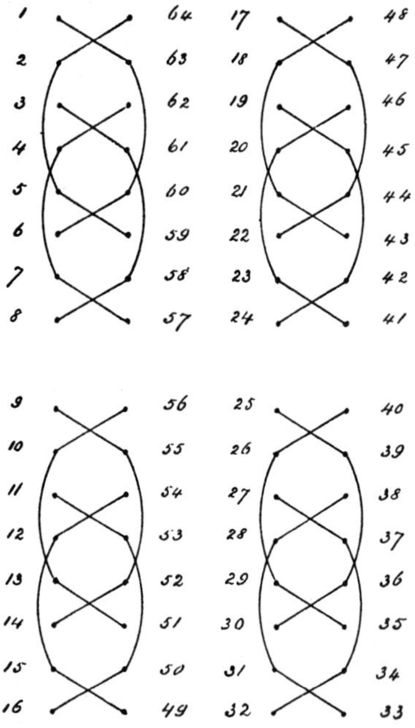

1	64	17	48
2	63	18	47
3	62	19	46
4	61	20	45
5	60	21	44
6	59	22	43
7	58	23	42
8	57	24	41
9	56	25	40
10	55	26	39
11	54	27	38
12	53	28	37
13	52	29	36
14	51	30	35
15	50	31	34
16	49	32	33

Fig. 175.

Totals = 130.

The arrangement of the numbers in Fig. 174 follows the diagrammatic order shown in Fig. 175.

The next even magic cube is $6 \times 6 \times 6$, but as Chapter IX of this book has been devoted to a description of these cubes they will be passed over here.

The $8 \times 8 \times 8$ magic cube follows next in order. Fig. 176 shows this cube divided, for convenience, into eight horizontal layers or sections, and Fig. 177 gives the diagrammatic order of the numbers in the first and eighth sections, the intermediate sections being built from similar diagrams, numbered in arithmetical order.

1	511	510	4	5	507	506	8
504	10	11	501	500	14	15	497
496	18	19	493	492	22	23	489
25	487	486	28	29	483	482	32
33	479	478	36	37	475	474	40
472	42	43	469	468	46	47	465
464	50	51	461	460	54	55	457
57	455	454	60	61	451	450	64

Section I.

384	130	131	381	380	134	135	377
137	375	374	140	141	371	370	144
145	367	366	148	149	363	362	152
360	154	155	357	356	158	159	353
352	162	163	349	348	166	167	345
169	343	342	172	173	339	338	176
177	335	334	180	181	331	330	184
328	186	187	325	324	190	191	321

Section III.

448	66	67	445	444	70	71	441
73	439	438	76	77	435	434	80
81	431	430	84	85	427	426	88
424	90	91	421	420	94	95	417
416	98	99	413	412	102	103	409
105	407	406	108	109	403	402	112
113	399	398	116	117	395	394	120
392	122	123	389	388	126	127	385

Section II.

193	319	318	196	197	315	314	200
312	202	203	309	308	206	207	305
304	210	211	301	300	214	215	297
217	295	294	220	221	291	290	224
225	287	286	228	229	283	282	232
280	234	235	277	276	238	239	273
272	242	243	269	268	246	247	265
249	263	262	252	253	259	258	256

Section IV.

(First Part.)
Fig. 176.

It will be seen from these diagrams that the $8 \times 8 \times 8$ magic cube is simply an expansion of the $4 \times 4 \times 4$ cube, just as the 8×8 magic square is an expansion of the 4×4 square. In like manner all the diagrams which were given for different arrangements of 8×8 magic squares may also be employed in the construction of $8 \times 8 \times 8$ magic cubes.

An examination of Fig. 176 will show that, like the $4 \times 4 \times 4$ cube in Fig. 171 it is built up of *o* and *ro* numbers exclusively. In sections I, IV, V, and VIII, the cells in the corner diagonal columns, and in certain other cells which are placed in definite geometrical relations thereto, contain *o* numbers, while all the other cells con-

257	255	254	260	261	251	250	264
248	266	267	245	244	270	271	241
240	274	275	237	236	278	279	233
281	231	230	284	285	227	226	288
289	223	222	292	293	219	218	296
216	298	299	213	212	302	303	209
208	306	307	205	204	310	311	201
313	199	198	316	317	195	194	320

Section V.

128	386	387	125	124	390	391	121
393	119	118	396	397	115	114	400
401	111	110	404	405	107	106	408
104	410	411	101	100	414	415	97
96	418	419	93	92	422	423	89
425	87	86	428	429	83	82	432
433	79	78	436	437	75	74	440
72	442	443	69	68	446	447	65

Section VII.

192	322	323	189	188	326	327	185
329	183	182	332	333	179	178	336
337	175	174	340	341	171	170	344
168	346	347	165	164	350	351	161
160	354	355	157	156	358	359	153
361	151	150	364	365	147	146	368
369	143	142	372	373	139	138	376
136	378	379	133	132	382	383	129

Section VI.

449	63	62	452	453	59	58	456
56	458	459	53	52	462	463	49
48	466	467	45	44	470	471	41
473	39	38	476	477	35	34	480
481	31	30	484	485	27	26	488
24	490	491	21	20	494	495	17
16	498	499	13	12	502	503	9
505	7	6	508	509	3	2	512

Section VIII.

(Second Part.)
Fig. 176.

tain *ro* numbers. In sections II, III, VI, and VII, the relative positions of the *o* and *ro* numbers are reversed.

By noting the symmetrical disposition of these two orders of numbers in the different sections, the cube may be readily constructed without the aid of any geometrical diagrams. Fig. 178 shows sections I and II of Fig. 176 filled with *o* and *ro* symbols

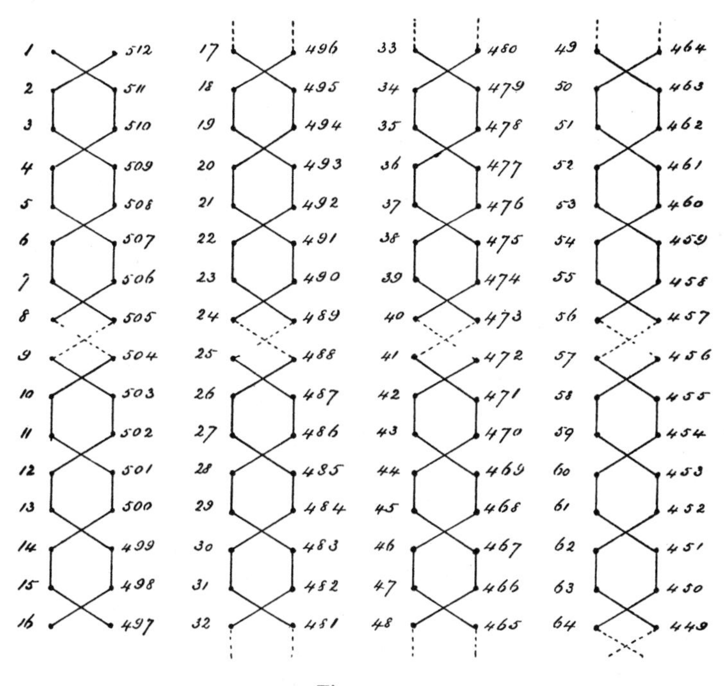

Fig. 177.

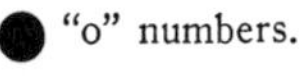
"o" numbers. "ro" numbers.

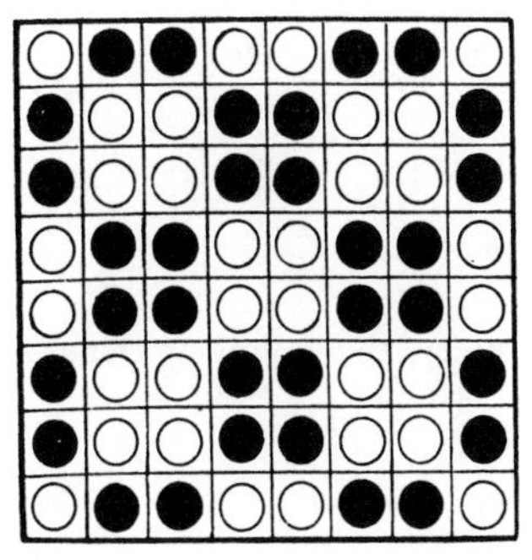

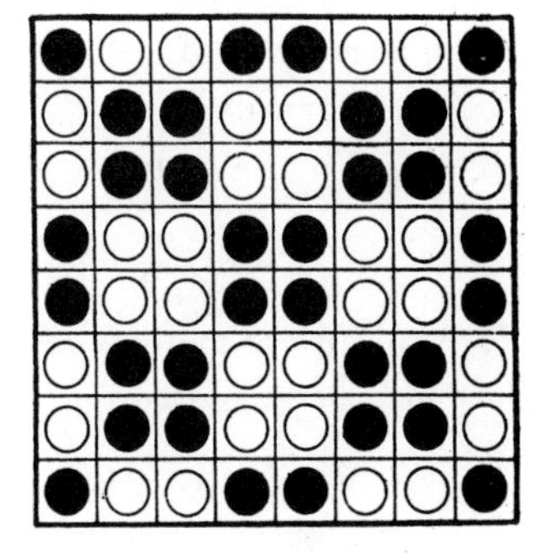

Fig. 178.

without regard to numerical values, and the relative symmetrical arrangement of the two orders is therein plainly illustrated. This clear and lucid analysis, for which we are indebted to Dr. Carus, reduces the formation of a rather complicated numerical structure to an operation of the utmost simplicity.

In this cube there are 192 straight columns, and 4 great diagonals (which unite the eight corners of the cube) each of which sums up to 2052; also 384 half columns and the same number of 2×2 sub-squares each of which has the summation of 1026. It will also be seen that the sum of any two numbers, which are located in cells diametrically opposite to each other and equidistant from the center of the cube, is 513 or $n^3 + 1$.

GENERAL NOTES ON MAGIC CUBES.

Magic cubes may be constructed having any desired summations by using suitable initial numbers with given increments, or by applying proper increments to given initial numbers.

* * *

The formula for determining the summations of magic cubes is similar to that which was given in connection with magic squares and may be expressed as follows:

Let:

$A =$ initial number,

$\beta =$ increment,

$n =$ number of cells in each column of cube,

$S =$ summation;

then if $A = 1$ and $\beta = 1$:—

$$\frac{n}{2}(n^3 + 1) = S.$$

If A and β are more or less than unity, the following general formula may be employed:

$$An + \beta\frac{n}{2}(n^3 - 1) = S.$$

To shorten the above equation, $\frac{n}{2}(n^3 - 1)$ may be expressed as a constant (K) for each size of cube as follows:

Cubes.	Const. = K.
$3 \times 3 \times 3$	39
$4 \times 4 \times 4$	126
$5 \times 5 \times 5$	310
$6 \times 6 \times 6$	645
$7 \times 7 \times 7$	1197
$8 \times 8 \times 8$	2044
$9 \times 9 \times 9$	3276
$10 \times 10 \times 10$	4995

When using the above constants the equation will be:

(1) $$An + \beta K = S,$$

or:

(2) $$\frac{S - An}{K} = \beta,$$

or:

(3) $$\frac{S - \beta K}{n} = A.$$

EXAMPLES.

What increment number is required for the cube of $3 \times 3 \times 3$ with an initial number of 10 to produce summations of 108?

Expressing equation (2) in figure values:

$$\frac{108 - (10 \times 3)}{39} = 2$$

28	60	20
56	10	42
24	38	46

54	14	40
22	36	50
32	58	18

26	34	48
30	62	16
52	12	44

Fig. 179. S = 108.

What increments should be used in a cube of $4 \times 4 \times 4$ to produce summations of 704 if the initial number is 50?

$$\frac{704 - (50 \times 4)}{126} = 4.$$

50	298	294	62
286	70	74	274
270	86	90	258
98	250	246	110

Section I (Top).

238	118	122	226
130	218	214	142
146	202	198	158
190	166	170	178

Section II.

174	182	186	162
194	154	150	206
210	138	134	222
126	230	234	114

Section III.

242	106	102	254
94	262	266	82
98	278	282	66
290	58	54	302

Section IV (Bottom).

Fig. 180.

Totals = 704.

What initial number must be used with increments of 10 to produce summations of 1906 in a $3 \times 3 \times 3$ cube?

Expressing equation (3) in figure values:

$$\frac{1906 - (10 \times 39)}{3} = 505\,{}^1/_3.$$

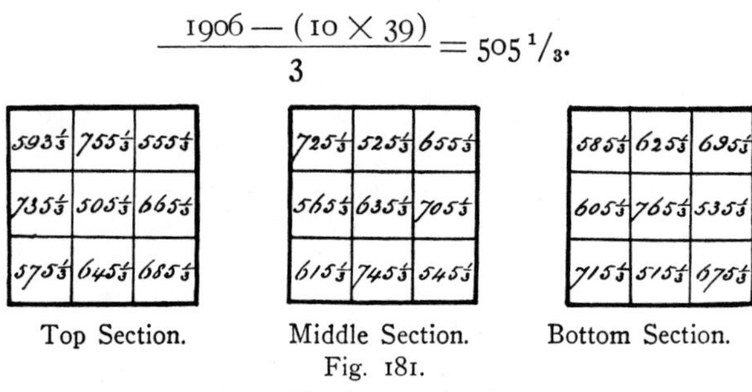

$593\frac{1}{3}$	$755\frac{1}{3}$	$555\frac{1}{3}$
$735\frac{1}{3}$	$505\frac{1}{3}$	$665\frac{1}{3}$
$575\frac{1}{3}$	$645\frac{1}{3}$	$685\frac{1}{3}$

Top Section.

$725\frac{1}{3}$	$525\frac{1}{3}$	$655\frac{1}{3}$
$565\frac{1}{3}$	$635\frac{1}{3}$	$705\frac{1}{3}$
$615\frac{1}{3}$	$745\frac{1}{3}$	$545\frac{1}{3}$

Middle Section.

$585\frac{1}{3}$	$625\frac{1}{3}$	$695\frac{1}{3}$
$605\frac{1}{3}$	$765\frac{1}{3}$	$535\frac{1}{3}$
$715\frac{1}{3}$	$515\frac{1}{3}$	$675\frac{1}{3}$

Bottom Section.

Fig. 181.

Totals = 1906.

What initial number is required for the cube of $5 \times 5 \times 5$, with 4 as increment number, to produce summations of 1906?*

$$\frac{1906 - (4 \times 310)}{5} = 133.2$$

The preceding simple examples will be sufficient to illustrate the formulæ given, and may suggest other problems to those who are interested in the subject.

It will be noted that the magic cubes which have been described in this chapter are all in the same general class as the magic squares which formed the subject of the previous chapter.

There are, however, many classes of magic squares and corresponding cubes which differ from these in the general arrange-

* This example was contributed by the late Mr. D. B. Ventres of Deep River, Conn.

ment of numbers and in various other features, while retaining the common characteristic of having similar column values. An example of this differentiation is seen in the interesting "Jaina" square

397.2	521.2	545.2	169.2	273.2
569.2	173.2	297.2	421.2	445.2
321.2	345.2	469.2	573.2	197.2
473.2	597.2	221.2	245.2	369.2
145.2	269.2	373.2	497.2	621.2

Section I (Top).

553.2	177.2	301.2	425.2	449.2
325.2	349.2	453.2	577.2	201.2
477.2	601.2	225.2	249.2	353.2
149.2	253.2	377.2	501.2	625.2
401.2	525.2	549.2	153.2	277.2

Section II.

329.2	333.2	457.2	581.2	205.2
481.2	605.2	229.2	233.2	357.2
133.2	257.2	381.2	505.2	629.2
405.2	529.2	533.2	157.2	281.2
557.2	181.2	305.2	429.2	433.2

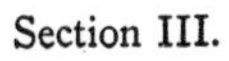
Section III.

485.2	609.2	213.2	237.2	361.2
137.2	261.2	385.2	509.2	613.2
409.2	513.2	537.2	161.2	285.2
561.2	185.2	309.2	413.2	437.2
313.2	337.2	461.2	585.2	209.2

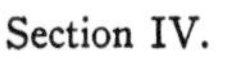
Section IV.

141.2	265.2	389.2	493.2	617.2
393.2	517.2	541.2	165.2	289.2
565.2	189.2	293.2	417.2	441.2
317.2	341.2	465.2	589.2	193.2
489.2	593.2	217.2	241.2	365.2

Section V.
Fig. 182.

described by Dr. Carus in his "Reflections on Magic Squares." Squares of this class can readily be expanded into cubes which will naturally carry with them the peculiar features of the squares.

Another class is illustrated in the "Franklin Squares," and these can also be expanded into cubes constructed on the same general principles.

The subject of magic squares and cubes is indeed inexhaustible and may be indefinitely extended. The philosophical significance of these studies has been so ably set forth by Dr. Carus that the writer considers it unnecessary to add anything in this connection, but he trusts that the present endeavor to popularize these interesting problems may some time lead to useful results.

CHAPTER III.

THE FRANKLIN SQUARES.

THE following letter with squares of 8×8 and 16×16 is copied from "Letters and papers on Philosophical subjects by Benjamin Franklin, LL. D., F.R.S.," a work which was printed in London, England, in 1769.

FROM BENJAMIN FRANKLIN ESQ. OF PHILADELPHIA.
TO PETER COLLINSON ESQ. AT LONDON.

DEAR SIR:—According to your request I now send you the arithmetical curiosity of which this is the history.

Being one day in the country at the house of our common friend, the late learned Mr. Logan, he showed me a folio French book filled with magic squares, wrote, if I forget not by one Mr. Frenicle, in which he said the author had discovered great ingenuity and dexterity in the management of numbers; and though several other foreigners had distinguished themselves in the same way, he did not recollect that any one Englishman had done anything of the kind remarkable.

I said it was perhaps a mark of the good sense of our mathematicians that they would not spend their time in things that were merely *difficiles nugæ,* incapable of any useful application. He answered that many of the arithmetical or mathematical questions publicly proposed in England were equally trifling and useless. Perhaps the considering and answering such questions, I replied, may not be altogether useless if it produces by practice an habitual readiness and exactness in mathematical disquisitions, which readiness may, on many occasions be of real use. In the same way says he, may the making of these squares be of use. I then confessed to him that in my younger days, having once some leisure (which I still think I might have employed more usefully) I had amused myself in making these kind of magic squares, and, at length had acquired such a knack at it, that I could fill the cells of any magic square of reasonable size with a series of numbers as fast as I could write them, disposed in such a manner that the sums of every row, horizontal, perpendicular or diagonal, should be equal; but not being satisfied with these, which I looked on as com-

52	61	4	13	20	29	36	45
14	3	62	51	46	35	30	19
53	60	5	12	21	28	37	44
11	6	59	54	43	38	27	22
55	58	7	10	23	26	39	42
9	8	57	56	41	40	25	24
50	63	2	15	18	31	34	47
16	1	64	49	48	33	32	17

Fig. 183.

mon and easy things, I had imposed on myself more difficult tasks, and succeeded in making other magic squares with a variety of properties, and much more curious. He then showed me several in the same book of an uncommon and more curious kind; but as I thought none of them equal to some I remembered to have made, he desired me to let him see them; and accordingly the next time I visited him, I carried him a square of 8 which I found among my old papers, and which I will now give you with an account of its properties (see Fig. 183). The properties are:

1. That every straight row (horizontal or vertical) of 8 numbers added together, makes 260, and half of each row, half of 260.

2. That the bent row of 8 numbers ascending and descending

diagonally, viz., from 16 ascending to 10 and from 23 descending to 17 and every one of its parallel bent rows of 8 numbers make 260, etc., etc. And lastly the four corner numbers with the four middle numbers

200	217	232	249	8	25	40	57	72	89	104	121	136	153	168	185
58	39	26	7	250	231	218	199	186	167	154	135	122	103	90	71
198	219	230	251	6	27	38	59	70	91	102	123	134	155	166	187
60	37	28	5	252	229	220	197	188	165	156	133	124	101	92	69
201	216	233	248	9	24	41	56	73	88	105	120	137	152	169	184
55	42	23	10	247	234	215	202	183	170	151	138	119	106	87	74
203	214	235	246	11	22	43	54	75	86	107	118	139	150	171	182
53	44	21	12	245	236	213	204	181	172	149	140	117	108	85	76
205	212	237	244	13	20	45	52	77	84	109	116	141	148	173	180
51	46	19	14	243	238	211	206	179	174	147	142	115	110	83	78
207	210	239	242	15	18	47	50	79	82	111	114	143	146	175	178
49	48	17	16	241	240	209	208	177	176	145	144	113	112	81	80
196	221	228	253	4	29	36	61	68	93	100	125	132	157	164	189
62	35	30	3	254	227	222	195	190	163	158	131	126	90	94	67
194	223	226	255	2	31	34	63	66	95	98	127	130	159	162	191
64	33	32	1	256	225	224	193	192	161	160	129	128	97	96	65

Fig. 184.

5	8	9	12
14	15	2	3
11	10	7	6
4	1	16	13

Fig. 185.

make 260. So this magical square seems perfect in its kind, but these are not all its properties, there are 5 other curious ones which at some time I will explain to you.

Mr. Logan then showed me an old arithmetical book in quarto, wrote, I think by one Stifelius, which contained a square of 16 which he said he should imagine to be a work of great labour; but if I forget not, it had only the common properties of making the same sum, viz., 2056 in every row, horizontal, vertical and diagonal. Not willing to be outdone by Mr. Stifelius, even in the size of my square, I went home, and made that evening the following magical square of 16 (see Fig. 184) which besides having all the properties of the foregoing square of 8, i. e., it would make 2056 in all the same rows and diagonals, had this added, that a four-square hole being cut in a piece of paper of such a size as to take in and show through it just 16 of the little squares, when laid on the greater

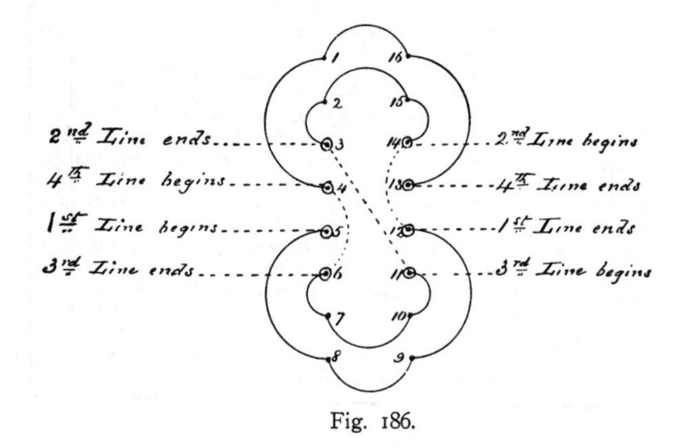

Fig. 186.

square, the sum of the 16 numbers so appearing through the hole, wherever it was placed on the greater square should likewise make 2056. This I sent to our friend the next morning, who after some days sent it back in a letter with these words:

> "I return to thee thy astonishing
> "or most stupendous piece
> "of the magical square in which"....

—but the compliment is too extravagant and therefore, for his sake, as well as my own I ought not to repeat it. Nor is it necessary, for I make no question but you will readily allow the square of 16 to be the most magically magical of any magic square ever made by any magician.

I am etc. B. F.

It will be seen that the squares shown in Figures 183 and 184 are not perfect according to the rules for magic squares previously

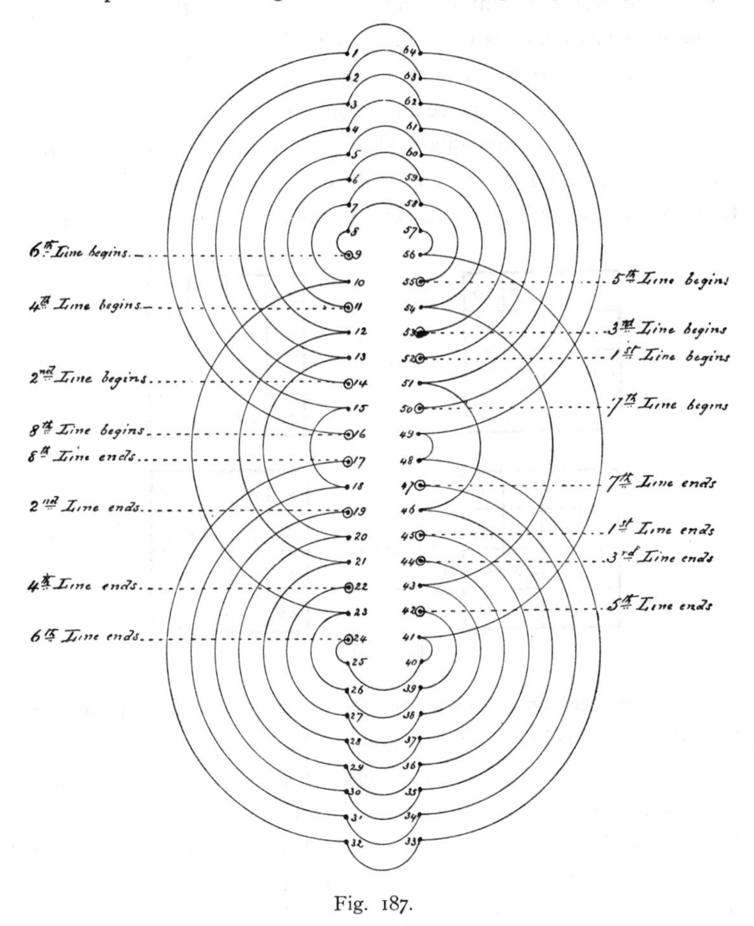

Fig. 187.

given, but the interesting feature of their *bent diagonal columns* calls for more than passing notice. In order to facilitate the study of their construction, a 4 × 4 square is given in Fig. 185 which presents similar characteristics.

The dotted lines in this square indicate four bent diagonal columns, each of which has a total of 34; three of these columns being intact within the square and one being broken. Four bent diagonal columns may be formed from each of the four sides of the square, but only twelve of these sixteen columns have the proper totals. Adding to these the eight straight columns, we find that this square contains twenty columns with summations of 34. The 4 × 4 "Jaina" square contains sixteen columns which sum up to 34 while the ordinary 4 × 4 magic square may contain only twelve.

The 8 × 8 Franklin square (Fig. 183) contains forty-eight columns which sum up to 260, viz., eight horizontal, eight vertical, sixteen bent horizontal diagonals, and sixteen bent vertical diagonals,

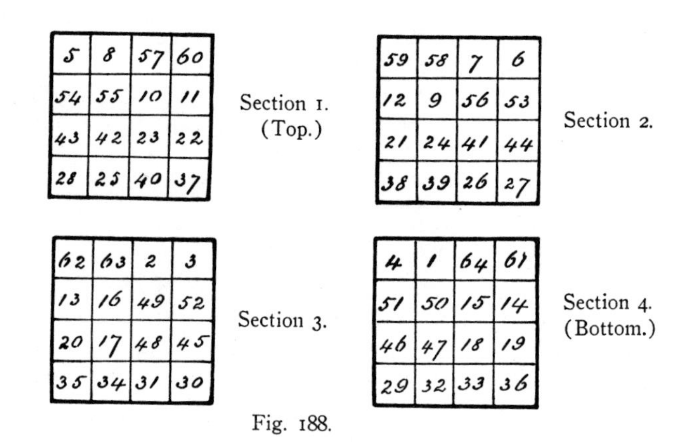

Section 1. (Top.)

5	8	57	60
54	55	10	11
43	42	23	22
28	25	40	37

Section 2.

59	58	7	6
12	9	56	53
21	24	41	44
38	39	26	27

Section 3.

62	63	2	3
13	16	49	52
20	17	48	45
35	34	31	30

Section 4. (Bottom.)

4	1	64	61
51	50	15	14
46	47	18	19
29	32	33	36

Fig. 188.

whereas the pandiagonal associated 8 × 8 magic square may contain only thirty-two columns and diagonals of the same summation.

In addition to the other characteristics mentioned by Franklin in his letter concerning his 8 × 8 magic square it may be stated that the sum of the numbers in any 2 × 2 sub-square contained therein is 130, and that the sum of any four numbers that are arranged diametrically equidistant from the center of the square also equals 130.

In regard to his 16 × 16 square, Franklin states in his letter that the sum of the numbers in any 4 × 4 sub-square contained therein is 2056. The sub-division may indeed be carried still further, for it will be observed that the sum of the numbers in any 2 × 2 sub-square is 514, and there are also other curious features which a little study will disclose.

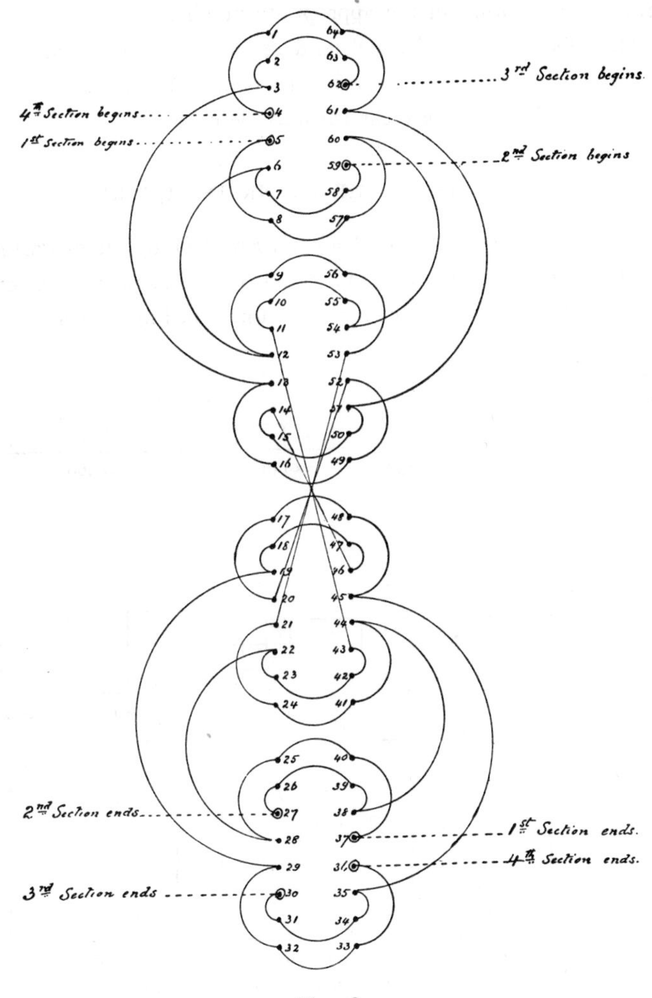

Fig. 189.

The Franklin Squares possess a unique and peculiar symmetry in the arrangement of their numbers which is not clearly observable on their faces, but which is brought out very strikingly in their geometrical diagrams as given in Figs. 186 and 187, which illustrate respectively the diagrams of the 4×4 and 8×8 squares.

Magic cubes may be readily constructed by expanding these diagrams and writing in the appropriate numbers.

The cube of $4 \times 4 \times 4$ and its diagram are given as examples in Figs. 188 and 189, and it will be observed that the curious characteristics of the square are carried into the cube.

AN ANALYSIS OF THE FRANKLIN SQUARES.

In *The Life and Times of Benjamin Franklin,* by James Parton, (Vol. I, pp. 255-257), there is an account of two magic squares, one 8×8, the other 16×16, which are given here in Figs. 191 and 192.

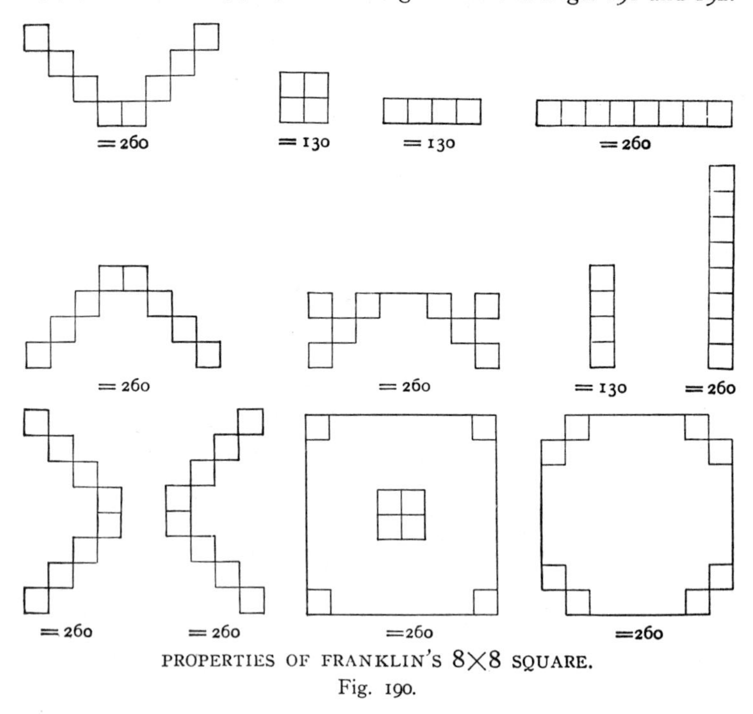

PROPERTIES OF FRANKLIN'S 8×8 SQUARE.
Fig. 190.

Mr. Parton explains the 8×8 square as follows:

"This square, as explained by its contriver, contains astonishing "properties: every straight row (horizontal or vertical) added to-

52	61	4	13	20	29	36	45
14	3	62	51	46	35	30	19
53	60	5	12	21	28	37	44
11	6	59	54	43	38	27	22
55	58	7	10	23	26	39	42
9	8	57	56	41	40	25	24
50	63	2	15	18	31	34	47
16	1	64	49	48	33	32	17

FRANKLIN 8×8 SQUARE.
Fig. 191.

200	217	232	249	8	25	40	57	72	89	104	121	136	153	168	185
58	39	26	7	250	231	218	199	186	167	154	135	122	103	90	71
198	219	230	251	6	27	38	59	70	91	102	123	134	155	166	187
60	37	28	5	252	229	220	197	188	165	156	133	124	101	92	69
201	216	233	248	9	24	41	56	73	88	105	120	137	152	169	184
55	42	23	10	247	234	215	202	183	170	151	138	119	106	87	74
203	214	235	246	11	22	43	54	75	86	107	118	139	150	171	182
53	44	21	12	245	236	213	204	181	172	149	140	117	108	85	76
205	212	237	244	13	20	45	52	77	84	109	116	141	148	173	180
51	46	19	14	243	238	211	206	179	174	147	142	115	110	83	78
207	210	239	242	15	18	47	50	79	82	111	114	143	146	175	178
49	48	17	16	241	240	209	208	177	176	145	144	113	112	81	80
196	221	228	253	4	29	36	61	68	93	100	125	132	157	164	189
62	35	30	3	254	227	222	195	190	163	158	131	126	99	94	67
194	223	226	255	2	31	34	63	66	95	98	127	130	159	162	191
64	33	32	1	256	225	224	193	192	161	160	129	128	97	96	65

FRANKLIN 16×16 SQUARE.
Fig. 192.

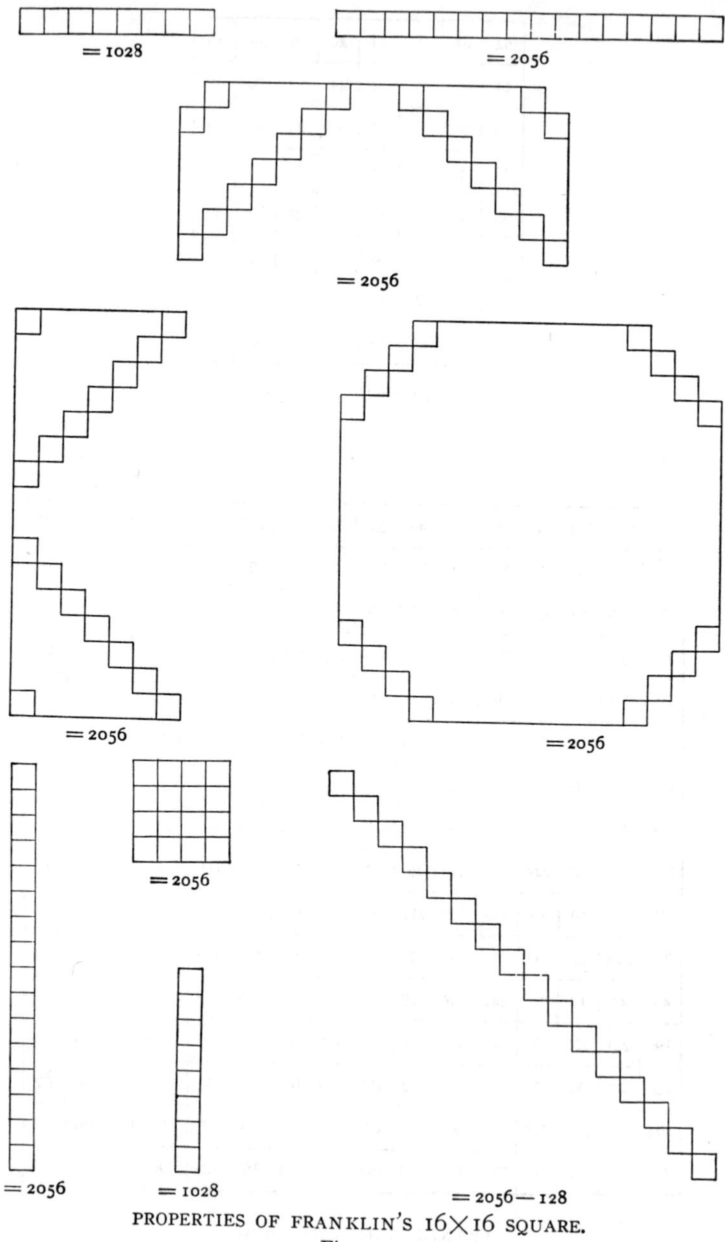

PROPERTIES OF FRANKLIN'S 16×16 SQUARE.
Fig. 193.

"gether makes 260, and each half row half 260. The bent row of "eight numbers ascending and descending diagonally, viz., from 16 "ascending to 10, and from 23 descending to 17, and every one of "its parallel bent rows of eight numbers, makes 260. Also, the bent

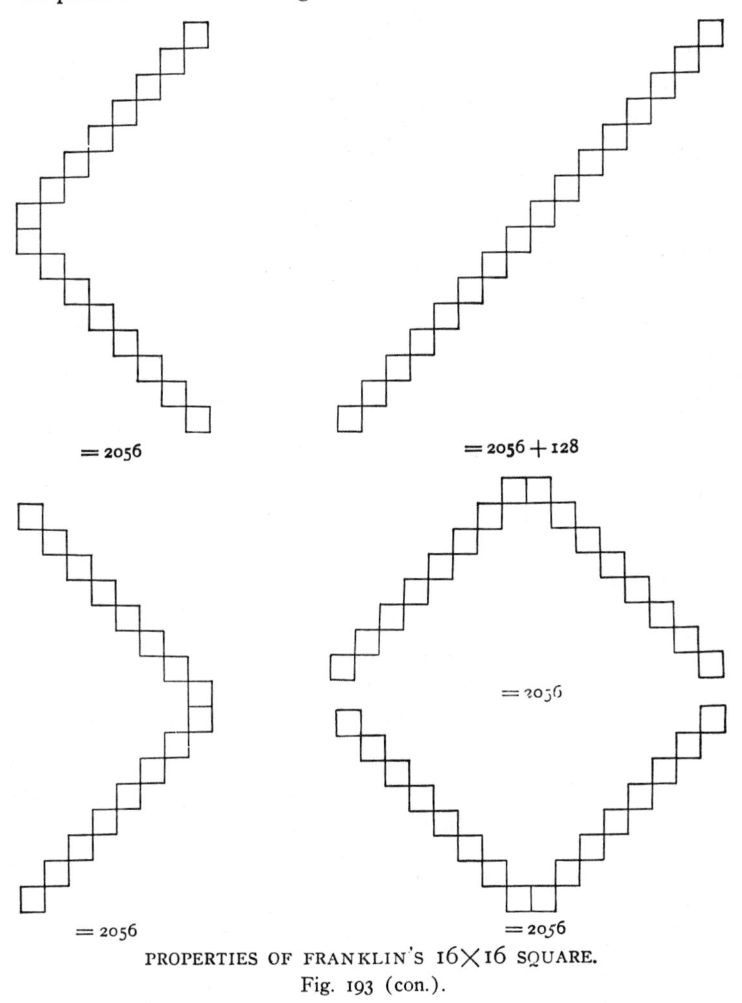

PROPERTIES OF FRANKLIN'S 16×16 SQUARE.
Fig. 193 (con.).

"row from 52 descending to 54, and from 43 ascending to 45, and "every one of its parallel bent rows of eight numbers, makes 260. "Also, the bent row from 45 to 43, descending to the left, and from

"23 to 17, descending to the right, and every one of its parallel bent "rows of eight numbers, makes 260. Also, the bent row from 52 "to 54, descending to the right, and from 10 to 16, descending to "the left, and every one of its parallel bent rows of eight numbers, "makes 260. Also, the parallel bent rows next to the above-men-"tioned, which are shortened to three numbers ascending and three "descending, etc., as from 53 to 4 ascending and from 29 to 44 "descending, make, with the two corner numbers, 260. Also, the two "numbers, 14, 61, ascending, and 36, 19, descending, with the lower "four numbers situated like them, viz., 50, 1, descending, and 32, 47, "ascending, makes 260. And, lastly, the four corner numbers, with "the four middle numbers, make 260.

"But even these are not all the properties of this marvelous "square. Its contriver declared that it has 'five other curious ones,' "which he does not explain; but which the ingenious reader may "discover if he can."

These remarkable characteristics which Mr. Parton enumerates are illustrated graphically in the accompanying diagrams in which the relative position of the cells containing the numbers which make up the number 260, is indicated by the relation of the small hollow squares (Fig. 190).

Franklin's 16×16 square is constructed upon the same principle as the smaller, and Mr. Parton continues:

"Nor was this the most wonderful of Franklin's magical "squares. He made one of sixteen cells in each row, which besides "possessing the properties of the squares given above (the amount, "however added, being always 2056), had also this most remark-"able peculiarity: a square hole being cut in a piece of paper of such "a size as to take in and show through it just sixteen of the little "squares, when laid on the greater square, the sum of sixteen num-"bers, so appearing through the hole, wherever it was placed on the "greater square, should likewise make 2056."

The additional peculiarity which Mr. Parton notes of the 16×16 square is no more remarkable than the corresponding fact which is true of the smaller square, that the sum of the numbers in any

	1	2	3	4	5	6	7	8	
A	1	2	3	4	5	6	7	8	A
B	9	10	11	12	13	14	15	16	B
C	17	18	19	20	21	22	23	24	C
D	25	26	27	28	29	30	31	32	D
E	33	34	35	36	37	38	39	40	E
F	41	42	43	44	45	46	47	48	F
G	49	50	51	52	53	54	55	56	G
H	57	58	59	60	61	62	63	64	H
	1	2	3	4	5	6	7	8	

Fig. 194. THE PLAN OF CONSTRUCTION.

A_1	B_8	C	D	E	F	G	H
H_7	G_2	F	E	D	C	B	A
3	6						
5	4						
5	4						
3	6						
7	2						
1	8						

Fig. 195. First Step.
KEY TO THE SCHEME OF SIMPLE ALTERNATION.

A_1	B_8	C_1	D_8	E_1	F_8	G_1	H_8
H_7	G_2	F_7	E_2	D_7	C_2	B_7	A_2
A_3	B_6	C_3	D_6	E_3	F_6	G_3	H_6
H_5	G_4	F_5	E_4	D_5	C_4	B_5	A_4
A_5	B_4	C_5	D_4	E_5	F_4	G_5	H_4
H_3	G_6	F_3	E_6	D_3	C_6	B_3	A_6
A_7	B_2	C_7	D_2	E_7	F_2	G_7	H_2
H_1	G_8	F_1	E_8	D_1	C_8	B_1	A_8

Fig. 196. Second Step.
COMPLETED SCHEME OF SIMPLE ALTERNATION.

1	16	17	32	33	48	49	64
63	50	47	34	31	18	15	2
3	14	19	30	35	46	51	62
61	52	45	36	29	20	13	4
5	12	21	28	37	44	53	60
59	54	43	38	27	22	11	6
7	10	23	26	39	42	55	58
57	56	41	40	25	24	9	8

Fig. 197. Third Step.
8×8 MAGIC SQUARE CONSTRUCTED BY SIMPLE ALTERNATION.

2×2 combination of its cells yields 130. The properties of the larger square are also graphically represented here (Fig. 193).

A clue to the construction of these squares may be found as follows:

We write down the numbers in numerical order and call the cells after the precedent of the chess-board, with two sets of symbols, letters and numbers. We call this "the plan of construction" (Fig. 194).

Before we construct the general scheme of Franklin's square we will build up another magic square, a little less complex in principle, which will be preparatory work for more complicated squares. We will simply intermix the ordinary series of numbers according to a definite rule alternately reversing the letters so that the odd rows are in alphabetical order and the even ones reversed. In order to distribute the numbers in a regular fashion so that no combination of letter and number would occur twice, we start with 1 in the upper left-hand corner and pass consecutively downwards, alternating between the first and second cells in the successive rows, thence ascending by the same method of simple alternation from 1 in the lower left-hand corner. We have now the key to a scheme for the distribution of numbers in an 8×8 magic square. It is the first step in the construction of the Franklin 8×8 magic square, and we call it "the key to the scheme of simple alternation" (Fig. 195).

It goes without saying that the effect would be the same if we begin in the same way in the right-hand corners,—only we must beware of a distribution that would occasion repetitions.

To complete the scheme we have to repeat the letters, alternately inverting their order row after row, and the first two given figures must be repeated throughout every row, as they are started. The top and bottom rows will read 1, 8; 1, 8; 1, 8; 1, 8. The second row from the top and also from the bottom will be 7, 2; 7, 2; 7, 2: 7, 2. The third row from the top and bottom will be 3, 6; 3, 6; 3, 6; 3, 6; and the two center rows 5, 4; 5, 4; 5, 4; 5, 4. In every line the sum of two consecutive figures yields 9. This is the second step, yielding the completed scheme of simple alternation (Fig. 196).

The square is now produced by substituting for the letter and figure combinations, the corresponding figures according to the consecutive arrangement in the plan of construction (Fig. 197).

Trying the results we find that all horizontal rows sum up to 260, while the vertical rows are alternately $260-4$, and $260+4$. The diagonal from the upper right to the lower left corner yields a sum of $260+32$, while the other diagonal from the left upper corner descending to the right lower corner makes $260-32$. The upper halves of the two diagonals yield 260, and also the sum of the lower halves, and the sum total of both diagonals is accordingly 520 or 2×260. The sum of the two left-hand half diagonals results in $260-16$, and the sum of the two half diagonals to the right-hand side makes $260+16$. The sum of the four central cells plus the four extreme corner cells yields also 260.

Considering the fact that the figures 1 to 8 of our scheme run up and down in alternate succession, we naturally have an arrangement of figures in which sets of two belong together. This binate peculiarity is evidenced in the result just stated, that the rows yield sums which are the same with an alternate addition and subtraction of an equal amount. So we have a symmetry which is astonishing and might be deemed magical, if it were not a matter of intrinsic necessity.

We represent these peculiarities in the adjoined diagrams (Fig. 198) which, however, by no means exhaust all the possibilities.

We must bear in mind that these magic squares are to be regarded as continuous; that is to say, they are as if their opposite sides in either direction passed over into one another as if they were joined both ways in the shape of a cylinder. In other words when we cross the boundary of the square on the right hand, the first row of cells outside to the right has to be regarded as identical with the first row of cells on the left; and in the same way the uppermost or first horizontal row of cells corresponds to the first row of cells below the bottom row. This remarkable property of the square will bring out some additional peculiarities which mathematicians may easily derive according to general principles; especially what was stated of the sum of the lower and upper half-

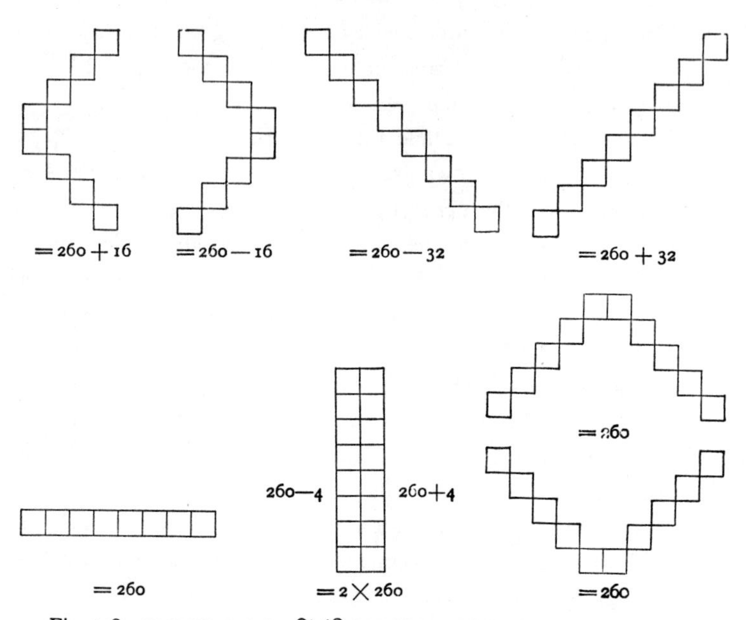

Fig. 198. PROPERTIES OF 8×8 SQUARE BY SIMPLE ALTERNATION.

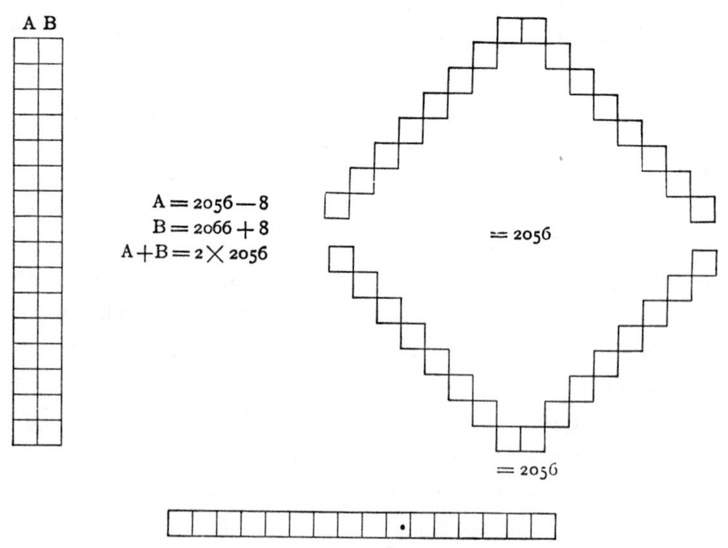

Fig. 199. PROPERTIES OF 16×16 SQUARE BY SIMPLE ALTERNATION.

diagonal of any bent series of cells running staircase fashion either upward or downward to the center, and hence proceeding in the opposite way to the other side.

The magic square constructed according to the method of sim-

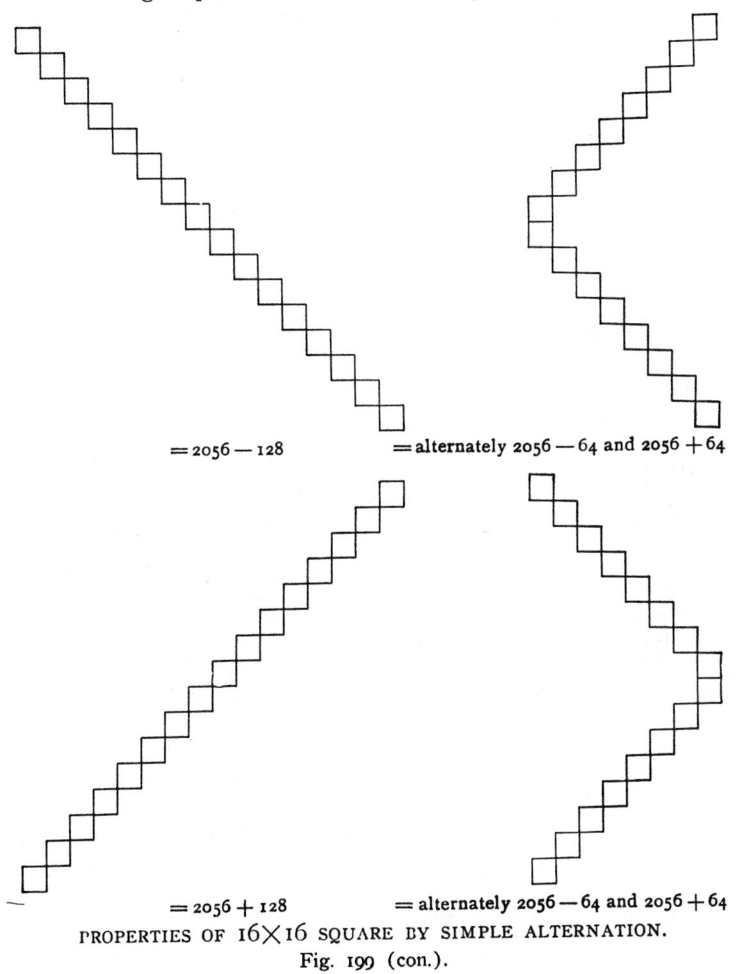

PROPERTIES OF 16×16 SQUARE BY SIMPLE ALTERNATION.
Fig. 199 (con.).

ple alternation of figures is not, however, the square of Benjamin Franklin, but we can easily transform the former into the latter by slight modifications.

We notice that in certain features the sum total of the bent

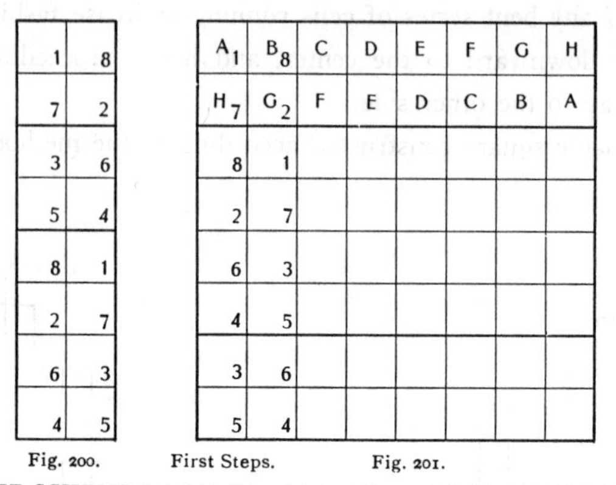

Fig. 200. First Steps. Fig. 201.

KEY TO THE SCHEME OF ALTERNATION WITH BINATE TRANSPOSITION.

A_1	B_8	C_1	D_8	E_1	F_8	G_1	H_8
H_7	G_2	F_7	E_2	D_7	C_2	B_7	A_2
A_8	B_1	C_8	D_1	E_8	F_1	G_8	H_1
H_2	G_7	F_2	E_7	D_2	C_7	B_2	A_7
A_6	B_3	C_6	D_3	E_6	F_3	G_6	H_3
H_4	G_5	F_4	E_5	D_4	C_5	B_4	A_5
A_3	B_6	C_3	D_6	E_3	F_6	G_3	H_6
H_5	G_4	F_5	E_4	D_5	C_4	B_5	A_4

Fig. 202. Second Step.
SCHEME OF ALTERNATION WITH BINATE TRANSPOSITION.

1	16	17	32	33	48	49	64
63	50	47	34	31	18	15	2
8	9	24	25	40	41	56	57
58	55	42	39	26	23	10	7
6	11	22	27	38	43	54	59
60	53	44	37	28	21	12	5
3	14	19	30	35	46	51	62
61	52	45	36	29	20	13	4

Fig. 203. Third Step.
SQUARE CONSTRUCTED BY ALTERNATION WITH BINATE TRANSPOSITION

G_4	H_5	A_4	B_5	C_4	D_5	E_4	F_5
B_6	A_3	H_6	G_3	F_6	E_3	D_6	C_3
G_5	H_4	A_5	B_4	C_5	D_4	E_5	F_4
B_3	A_6	H_3	G_6	F_3	E_6	D_3	C_6
G_7	H_2	A_7	B_2	C_7	D_2	E_7	F_2
B_1	A_8	H_1	G_8	F_1	E_8	D_1	C_8
G_2	H_7	A_2	B_7	C_2	D_7	E_2	F_7
B_8	A_1	H_8	G_1	F_8	E_1	D_8	C_1

Fig. 204. SCHEME OF FRANKLIN'S 8×8 SQUARE.

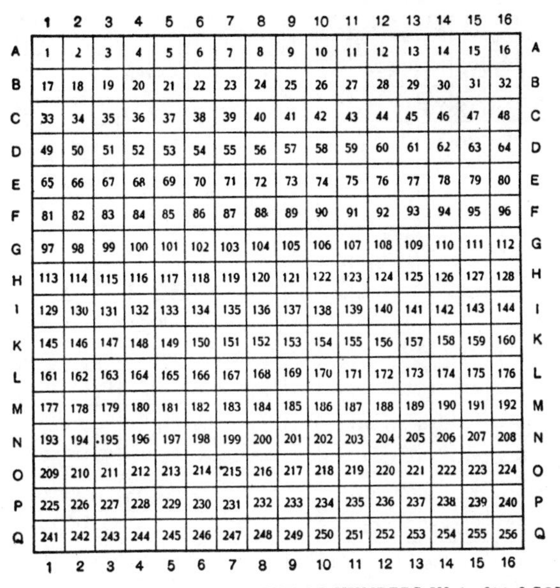

	1	2	3	4	5	6	7	8	9	10	11	12	13	14	15	16	
A	1	2	3	4	5	6	7	8	9	10	11	12	13	14	15	16	A
B	17	18	19	20	21	22	23	24	25	26	27	28	29	30	31	32	B
C	33	34	35	36	37	38	39	40	41	42	43	44	45	46	47	48	C
D	49	50	51	52	53	54	55	56	57	58	59	60	61	62	63	64	D
E	65	66	67	68	69	70	71	72	73	74	75	76	77	78	79	80	E
F	81	82	83	84	85	86	87	88	89	90	91	92	93	94	95	96	F
G	97	98	99	100	101	102	103	104	105	106	107	108	109	110	111	112	G
H	113	114	115	116	117	118	119	120	121	122	123	124	125	126	127	128	H
I	129	130	131	132	133	134	135	136	137	138	139	140	141	142	143	144	I
K	145	146	147	148	149	150	151	152	153	154	155	156	157	158	159	160	K
L	161	162	163	164	165	166	167	168	169	170	171	172	173	174	175	176	L
M	177	178	179	180	181	182	183	184	185	186	187	188	189	190	191	192	M
N	193	194	195	196	197	198	199	200	201	202	203	204	205	206	207	208	N
O	209	210	211	212	213	214	215	216	217	218	219	220	221	222	223	224	O
P	225	226	227	228	229	230	231	232	233	234	235	236	237	238	239	240	P
Q	241	242	243	244	245	246	247	248	249	250	251	252	253	254	255	256	Q
	1	2	3	4	5	6	7	8	9	10	11	12	13	14	15	16	

Fig. 205. CONSECUTIVE ARRANGEMENT OF NUMBERS IN A 16×16 SQUARE

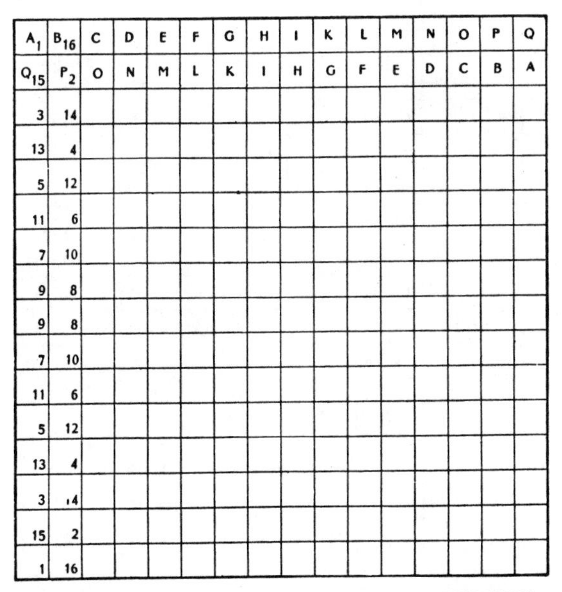

A_1	B_{16}	C	D	E	F	G	H	I	K	L	M	N	O	P	Q
Q_{15}	P_2	O	N	M	L	K	I	H	G	F	E	D	C	B	A
3	14														
13	4														
5	12														
11	6														
7	10														
9	8														
9	8														
7	10														
11	6														
5	12														
13	4														
3	14														
15	2														
1	16														

Fig. 206. KEY TO THE SCHEME OF SIMPLE ALTERNATION.

A_1	B_{16}	C_1	D_{16}	E_1	F_{16}	G_1	H_{16}	I_1	K_{16}	L_1	M_{16}	N_1	O_{16}	P_1	Q_{16}
Q_{15}	P_2	O_{15}	N_2	M_{15}	L_2	K_{15}	I_2	H_{15}	G_2	F_{15}	E_2	D_{15}	C_2	B_{15}	A_2
A_3	B_{14}	C_3	D_{14}	E_3	F_{14}	G_3	H_{14}	I_3	K_{14}	L_3	M_{14}	N_3	O_{14}	P_3	Q_{14}
Q_{13}	P_4	O_{13}	N_4	M_{13}	L_4	K_{13}	I_4	H_{13}	G_4	F_{13}	E_4	D_{13}	C_4	B_{13}	A_4
A_5	B_{12}	C_5	D_{12}	E_5	F_{12}	G_5	H_{12}	I_5	K_{12}	L_5	M_{12}	N_5	O_{12}	P_5	Q_{12}
Q_{11}	P_6	O_{11}	N_6	M_{11}	L_6	K_{11}	I_6	H_{11}	G_6	F_{11}	E_6	D_{11}	C_6	B_{11}	A_6
A_7	B_{10}	C_7	D_{10}	E_7	F_{10}	G_7	H_{10}	I_7	K_{10}	L_7	M_{10}	N_7	O_{10}	P_7	Q_{10}
Q_9	P_8	O_9	N_8	M_9	L_8	K_9	I_8	H_9	G_8	F_9	E_8	D_9	C_8	B_9	A_8
A_9	B_8	C_9	D_8	E_9	F_8	G_9	H_8	I_9	K_8	L_9	M_8	N_9	O_8	P_9	Q_8
Q_7	P_{10}	O_7	N_{10}	M_7	L_{10}	K_7	I_{10}	H_7	G_{10}	F_7	E_{10}	D_7	C_{10}	B_7	A_{10}
A_{11}	B_6	C_{11}	D_6	E_{11}	F_6	G_{11}	H_6	I_{11}	K_6	L_{11}	M_6	N_{11}	O_6	P_{11}	Q_6
Q_5	P_{12}	O_5	N_{12}	M_5	L_{12}	K_5	I_{12}	H_5	G_{12}	F_5	E_{12}	D_5	C_{12}	B_5	A_{12}
A_{13}	B_4	C_{13}	D_4	E_{13}	F_4	G_{13}	H_4	I_{13}	K_4	L_{13}	M_4	N_{13}	O_4	P_{13}	Q_4
Q_3	P_{14}	O_3	N_{14}	M_3	L_{14}	K_3	I_{14}	H_3	G_{14}	F_3	E_{14}	D_3	C_{14}	B_3	A_{14}
A_{15}	B_2	C_{15}	D_2	E_{15}	F_2	G_{15}	H_2	I_{15}	K_2	L_{15}	M_2	N_{15}	O_2	P_{15}	Q_2
Q_1	P_{16}	O_1	N_{16}	M_1	L_{16}	K_1	I_{16}	H_1	G_{16}	F_1	E_{16}	D_1	C_{16}	B_1	A_{16}

Fig. 207. SCHEME OF SIMPLE ALTERNATION.

1	32	33	64	65	96	97	128	129	160	161	192	193	224	225	256
255	226	223	194	191	162	159	130	127	98	95	66	63	34	31	2
3	30	35	62	67	94	99	126	131	158	163	190	195	222	227	254
253	228	221	196	189	164	157	132	125	100	93	68	61	36	29	4
5	28	37	60	69	92	101	124	133	156	165	188	197	220	229	252
251	230	219	198	187	166	155	134	123	102	91	70	59	38	27	6
7	26	39	58	71	90	103	122	135	154	167	186	199	218	231	250
249	232	217	200	185	168	153	136	121	104	89	72	57	40	25	8
9	24	41	56	73	88	105	120	137	152	169	184	201	216	233	248
247	234	215	202	183	170	151	138	119	106	87	74	55	42	23	10
11	22	43	54	75	86	107	118	139	150	171	182	203	214	235	246
245	236	213	204	181	172	149	140	117	108	85	76	53	44	21	12
13	20	45	52	77	84	109	116	141	148	173	180	205	212	237	244
243	238	211	206	179	174	147	142	115	110	83	78	51	46	19	14
15	18	47	50	79	82	111	114	143	146	175	178	207	210	239	242
241	240	209	208	177	176	145	144	113	112	81	80	49	48	17	16

Fig. 208. 16×16 MAGIC SQUARE CONSTRUCTED BY SIMPLE ALTERNATION.

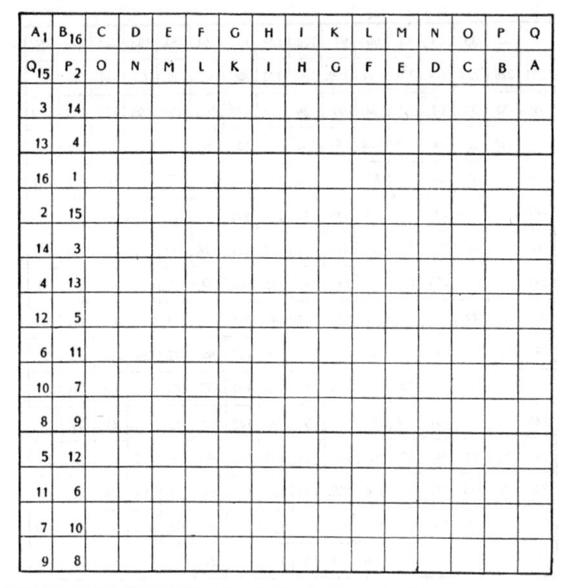

A_1	B_{16}	C	D	E	F	G	H	I	K	L	M	N	O	P	Q
Q_{15}	P_2	O	N	M	L	K	I	H	G	F	E	D	C	B	A
3	14														
13	4														
16	1														
2	15														
14	3														
4	13														
12	5														
6	11														
10	7														
8	9														
5	12														
11	6														
7	10														
9	8														

ig. 209. KEY TO THE SCHEME OF ALTERNATION WITH QUATERNATE TRANSPOSITION.

A_1	B_{16}	C_1	D_{16}	E_1	F_{16}	G_1	H_{16}	I_1	K_{16}	L_1	M_{16}	N_1	O_{16}	P_1	Q_{16}
Q_{15}	P_2	O_{15}	N_2	M_{15}	L_2	K_{15}	I_2	H_{15}	G_2	F_{15}	E_2	D_{15}	C_2	B_{15}	A_2
A_3	B_{14}	C_3	D_{14}	E_3	F_{14}	G_3	H_{14}	I_3	K_{14}	L_3	M_{14}	N_3	O_{14}	P_3	Q_{14}
Q_{13}	P_4	O_{13}	N_4	M_{13}	L_4	K_{13}	I_4	H_{13}	G_4	F_{13}	E_4	D_{13}	C_4	B_{13}	A_4
A_{16}	B_1	C_{16}	D_1	E_{16}	F_1	G_{16}	H_1	I_{16}	K_1	L_{16}	M_1	N_{16}	O_1	P_{16}	Q_1
Q_2	P_{15}	O_2	N_{15}	M_2	L_{15}	K_2	I_{15}	H_2	G_{15}	F_2	E_{15}	D_2	C_{15}	B_2	A_{15}
A_{14}	B_3	C_{14}	D_3	E_{14}	F_3	G_{14}	H_3	I_{14}	K_3	L_{14}	M_3	N_{14}	O_3	P_{14}	Q_3
Q_4	P_{13}	O_4	N_{13}	M_4	L_{13}	K_4	I_{13}	H_4	G_{13}	F_4	E_{13}	D_4	C_{13}	B_4	A_{13}
A_{12}	B_5	C_{12}	D_5	E_{12}	F_5	G_{12}	H_5	I_{12}	K_5	L_{12}	M_5	N_{12}	O_5	P_{12}	Q_5
Q_6	P_{11}	O_6	N_{11}	M_6	L_{11}	K_6	I_{11}	H_6	G_{11}	F_6	E_{11}	D_6	C_{11}	B_6	A_{11}
A_{10}	B_7	C_{10}	D_7	E_{10}	F_7	G_{10}	H_7	I_{10}	K_7	L_{10}	M_7	N_{10}	O_7	P_{10}	Q_7
Q_8	P_9	O_8	N_9	M_8	L_9	K_8	I_9	H_8	G_9	F_8	E_9	D_8	C_9	B_8	A_9
A_5	B_{12}	C_5	D_{12}	E_5	F_{12}	G_5	H_{12}	I_5	K_{12}	L_5	M_{12}	N_5	O_{12}	P_5	Q_{12}
Q_{11}	P_6	O_{11}	N_6	M_{11}	L_6	K_{11}	I_6	H_{11}	G_6	F_{11}	E_6	D_{11}	C_6	B_{11}	A_6
A_7	B_{10}	C_7	D_{10}	E_7	F_{10}	G_7	H_{10}	I_7	K_{10}	L_7	M_{10}	N_7	O_{10}	P_7	Q_{10}
Q_9	P_8	O_9	N_8	M_9	L_8	K_9	I_8	H_9	G_8	F_9	E_8	D_9	C_8	B_9	A_8

Fig. 210. SCHEME OF ALTERNATION WITH QUATERNATE TRANSPOSITION.

1	32	33	64	65	96	97	128	129	160	161	192	193	224	225	256
255	226	223	194	191	162	159	130	127	98	95	66	63	34	31	2
3	30	35	62	67	94	99	126	131	158	163	190	195	222	227	254
253	228	221	196	189	164	157	132	125	100	93	68	61	36	29	4
16	17	48	49	80	81	112	113	144	145	176	177	208	209	240	241
242	239	210	207	178	175	146	143	114	111	82	79	50	47	18	15
14	19	46	51	78	83	110	115	142	147	174	179	206	211	238	243
244	237	212	205	180	173	148	141	116	109	84	77	52	45	20	13
12	21	44	53	76	85	108	117	140	149	172	181	204	213	236	245
246	235	214	203	182	171	150	139	118	107	86	75	54	43	22	11
10	23	42	55	74	87	106	119	138	151	170	183	202	215	234	247
248	233	216	201	184	169	152	137	120	105	88	73	56	41	24	9
5	28	37	60	69	92	101	124	133	156	165	188	197	220	229	252
251	230	219	198	187	166	155	134	123	102	91	70	59	38	27	6
7	26	39	58	71	90	103	122	135	154	167	186	199	218	231	250
249	232	217	200	185	168	153	136	121	104	89	72	57	40	25	8

Fig. 211. A SQUARE CONSTRUCTED BY ALTERNATION WITH QUATERNATE TRANSPOSITION.

N_8	O_9	P_8	Q_9	A_8	B_9	C_8	D_9	E_8	F_9	G_8	H_9	I_8	K_9	L_8	M_9
D_{10}	C_7	B_{10}	A_7	Q_{10}	P_7	O_{10}	N_7	M_{10}	L_7	K_{10}	I_7	H_{10}	G_7	F_{10}	E_7
N_6	O_{11}	P_6	Q_{11}	A_6	B_{11}	C_6	D_{11}	E_6	F_{11}	G_6	H_{11}	I_6	K_{11}	L_6	M_{11}
D_{12}	C_5	B_{12}	A_5	Q_{12}	P_5	O_{12}	N_5	M_{12}	L_5	K_{12}	I_5	H_{12}	G_5	F_{12}	E_5
N_9	O_8	P_9	Q_8	A_9	B_8	C_9	D_8	E_9	F_8	G_9	H_8	I_9	K_8	L_9	M_8
D_7	C_{10}	B_7	A_{10}	Q_7	P_{10}	O_7	N_{10}	M_7	L_{10}	K_7	I_{10}	H_7	G_{10}	F_7	E_{10}
N_{11}	O_6	P_{11}	Q_6	A_{11}	B_6	C_{11}	D_6	E_{11}	F_6	G_{11}	H_6	I_{11}	K_6	L_{11}	M_6
D_5	C_{12}	B_5	A_{12}	Q_5	P_{12}	O_5	N_{12}	M_5	L_{12}	K_5	I_{12}	H_5	G_{12}	F_5	E_{12}
N_{13}	O_4	P_{13}	Q_4	A_{13}	B_4	C_{13}	D_4	E_{13}	F_4	G_{13}	H_4	I_{13}	K_4	L_{13}	M_4
D_3	C_{14}	B_3	A_{14}	Q_3	P_{14}	O_3	N_{14}	M_3	L_{14}	K_3	I_{14}	H_3	G_{14}	F_3	E_{14}
N_{15}	O_2	P_{15}	Q_2	A_{15}	B_2	C_{15}	D_2	E_{15}	F_2	G_{15}	H_2	I_{15}	K_2	L_{15}	M_2
D_1	C_{16}	B_1	A_{16}	Q_1	P_{16}	O_1	N_{16}	M_1	L_{16}	K_1	I_{16}	H_1	G_{16}	F_1	E_{16}
N_4	O_{13}	P_4	Q_{13}	A_4	B_{13}	C_4	D_{13}	E_4	F_{13}	G_4	H_{13}	I_4	K_{13}	L_4	M_{13}
D_{14}	C_3	B_{14}	A_3	Q_{14}	P_3	O_{14}	N_3	M_{14}	L_3	K_{14}	I_3	H_{14}	G_3	F_{14}	E_3
N_2	O_{15}	P_2	Q_{15}	A_2	B_{15}	C_2	D_{15}	E_2	F_{15}	G_2	H_{15}	I_2	K_{15}	L_2	M_{15}
D_{16}	C_1	B_{16}	A_1	Q_{16}	P_1	O_{16}	N_1	M_{16}	L_1	K_{16}	I_1	H_{16}	G_1	F_{16}	E_1

Fig. 212. SCHEME OF FRANKLIN'S 16×16 SQUARE.

diagonals represents regularities which counterbalance one another on the right- and the left-hand side. In order to offset these results we have to shift the figures of our scheme.

We take the diagram which forms the key to the scheme of our distribution by simple alternation (Fig. 195), and cutting it in the middle, turn the lower half upside down, giving the first two rows as seen in Fig. 200 in which the heavy lines indicate the cutting. Cutting then the upper half in two (i. e., in binate sections), and transposing the second quarter to the bottom, we have the key to the entire arrangement of figures; in which the alternation starts as in the scheme for simple alternation but skips the four center rows passing from 2 in the second cell of the second row to 3 in the first cell of the seventh, and from 4 in the second cell of the eighth passing to 5 in the first cell, and thence upwards in similar alternation, again passing over the four central rows to the second and ending with 8 in the second cell of the first row. Then the same alternation is produced in the four center rows. It is obvious that this can not start in the first cell as that would duplicate the first row, so we start with 1 in the second cell passing down uninterruptedly to 4 and ascending as before from 5 to 8.

A closer examination will show that the rows are binate, which means in sets of two. The four inner numbers, 3, 4, 5, 6 and the two outer sets of two numbers each, 1, 2 and 7, 8, are brought together thus imparting to the whole square a binate character (Fig. 202).

We are now provided with a key to build up a magic square after the pattern of Franklin. We have simply to complete it in the same way as our last square repeating the letters with their order alternately reversed as before, and repeating the figures in each line.

When we insert their figure values we have a square which is not the same as Franklin's, but possesses in principle the same qualities (Fig. 203).

To make our 8×8 square of binate transposition into the Franklin square we must first take its obverse square; that is to say, we preserve exactly the same order but holding the paper with the figures toward the light we read them off from the obverse side, and then take the mirror picture of the result, holding the mirror on either horizontal side. So far we have still our square with the peculiarities of our scheme, but which lacks one of the incidental characteristics of Franklin's square. We must notice that he makes four cells in both horizontal and vertical directions sum up to 130 which property is necessarily limited only to two sets of four cells in each row. If we write down the sum of $1+2+3+4+5+6+7+8=2\times18$, we will find that the middle set $3+4+5+6$ is equal to the rest consisting of the sum of two extremes, $1+4$, and $7+8$. In this way we cut out in our scheme (Fig. 202), the rows represented by the letters C, D, E, F in either order and accordingly we can shift either of the two first or two last vertical rows to the other side. Franklin did the former, thus beginning his square with G_4 in the left upper corner as in Fig. 204. We have indicated this division by heavier lines in both schemes.

The greater square of Franklin, which is 16×16, is made after the same fashion, and the adjoined diagrams (Figs. 205-212) will sufficiently explain its construction.

We do not know the method employed by Franklin; we possess only the result, but it is not probable that he derived his square according to the scheme employed here.

Our 16×16 square is not exactly the same as the square of Franklin, but it belongs to the same class. Our method gives the key to the construction, and it is understood that the system here represented will allow us to construct many more squares by simply pushing the square beyond its limits into the opposite row which by this move has to be transferred.

There is the same relation between Franklin's 16×16 square and our square constructed by alternation with quaternate transposition, that exists between the corresponding 8×8 squares.

P. C.

CHAPTER IV.

REFLECTIONS ON MAGIC SQUARES.

MATHEMATICS, especially in the field where it touches philosophy, has always been my foible, and so Mr. W. S. Andrews's article on "Magic Squares" tempted me to seek a graphic key to the interrelation among their figures which should reveal at a glance the mystery of their construction.

THE ORDER OF FIGURES.

In odd magic squares, 3×3, 5×5, 7×7, etc., there is no difficulty whatever, as Mr. Andrews's diagrams show at a glance (Fig. 213). The consecutive figures run up slantingly in the form

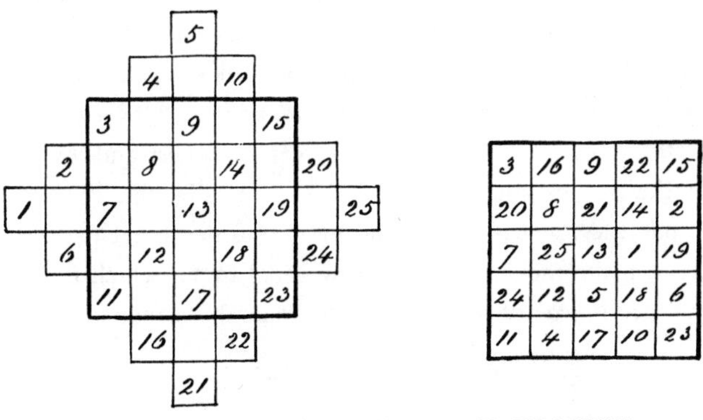

Fig. 213. A SPECIMEN OF 5×5 MAGIC SQUARE.

of a staircase, so as to let the next higher figure pass over into the next higher or lower cell of the next row, and those figures that according to this method would fall outside of the square, revert into it as if the magic square were for the time (at the moment of crossing its boundary) connected with its opposite side into the shape of a cylinder. This cannot be done at once with both its two opposite vertical and its two opposite horizontal sides, but the process is easily represented in the plane by having the magic square extended on all its sides, and on passing its limits on one side we must treat the extension as if we had entered into the magic square on the side opposite to where we left it. If we now transfer the figures to their respective places in the inside square, they are shoved over in a way which by a regular transposition will counteract their regular increase of counting and so equalize the sums of entire rows.

The case is somewhat more complicated with even magic squares, and a suggestion which I propose to offer here, pertains to their formation. Mr. Andrews begins their discussion by stating that "in regard to regular or associated magic squares it is not only necessary that each row, column and corner diagonal shall sum the same amount, but also that the pairs of numbers which sum $n^2 + 1$ must occupy cells which are located diametrically equidistant from the center of the square."

The smallest magic square of even numbers is, of course, 4×4; and he points out that if we write the figures in their regular order in a 4×4 square, those standing on the diagonal lines can remain in their places, while the rest are to be reversed so as to replace every figure by its complementary to 17 (i. e., 2 by 15, 3 by 14, 5 by 12, 9 by 8) the number 17 being the sum of the highest and lowest numbers of the magic square (i. e., $n^2 + 1$). It is by this reversal of figures that the inequalities of the natural order are equalized again, so as to make the sum of each row equal to 34, which is one fourth of the sum total of all figures, the general formula being

$$\frac{1 + 2 + 3 + 4 + \ldots . n}{n} = \frac{n}{2}(n^2 + 1).$$

We will now try to find out more about the relation which the magic square arrangement bears to the normal sequence of figures.

For each corner there are two ways, one horizontal and one vertical, in which figures can be written in the normal sequence; accordingly there are altogether eight possible arrangements, from which we select one as fundamental, and regard all others as mere variations, produced by inverting and reversing the order.

As the fundamental arrangement we choose the ordinary way of writing from the left to the right, proceeding in parallel lines downward. We call this "the original order" or *o*. Its reverse proceeds from the lower right-hand corner toward the left, and line by line upward, thus beginning the series where the ordinary arrengement ends, and ending where it started, as reflected on the ground glass of a camera. We call this order "the reversed original," or simply *ro*.

Another order is produced by following the Hebrew and Arabic mode of writing: we begin in the upper right-hand corner, proceeding to the left, and then continue in the same way line by line downward. This, the inverse direction to the original way, we call briefly *i* or "mirror" order.

The reverse order of *i*, starting in the lower left corner, proceeding to the right, and line by line upward, we call *ri*, or "lake" order. Further on we shall have occasion to present these four orders by the following symbols: *o* by ●; *ro* by ❁; *i* by ✠; *ri* by +.

1	2	3	4	5	6
7	8	9	10	11	12
13	14	15	16	17	18
19	20	21	22	23	24
25	26	27	28	29	30
31	32	33	34	35	36

ORDER *o* (●), ORIGINAL.

6	5	4	3	2	1
12	11	10	9	8	7
18	17	16	15	14	13
24	23	22	21	20	19
30	29	28	27	26	25
36	35	34	33	32	31

ORDER *i* (✠), MIRROR.

31	32	33	34	35	36
25	26	27	28	29	30
19	20	21	22	23	24
13	14	15	16	17	18
7	8	9	10	11	12
1	2	3	4	5	6

ORDER *ri* (+), LAKE.

36	35	34	33	32	31
30	29	28	27	26	25
24	23	22	21	20	19
18	17	16	15	14	13
12	11	10	9	8	7
6	5	4	3	2	1

ORDER *ro* (❁), CAMERA.

Fig. 214.

1	15	14	4
12	6	7	9
8	10	11	5
13	3	2	16

16	2	3	13
5	11	10	8
9	7	6	12
4	14	15	1

1	63	59	4	5	62	58	8
56	10	54	13	12	51	15	49
24	47	19	45	44	22	42	17
25	34	38	28	29	35	39	32
33	26	30	36	37	27	31	40
48	23	43	21	20	46	18	41
16	50	14	53	52	11	55	9
57	7	3	60	61	6	2	64

1	143	142	4	5	139	138	8	9	135	134	12
132	14	15	129	128	18	19	125	124	22	23	121
120	26	27	117	116	30	31	113	112	34	35	109
37	107	106	40	41	103	102	44	45	99	98	48
49	95	94	52	53	91	90	56	57	87	86	60
84	62	63	81	80	66	67	77	76	70	71	73
72	74	75	69	68	78	79	65	64	82	83	61
85	59	58	88	89	55	54	92	93	51	50	96
97	47	46	100	101	43	42	104	105	39	38	108
36	110	111	33	32	114	115	29	28	118	119	25
24	122	123	21	20	126	127	17	16	130	131	13
133	11	10	136	137	7	6	140	141	3	2	144

Fig. 215. EVEN SQUARES IN MULTIPLES OF FOUR.*

* These squares, 4×4 and its multiples, consist of *o* and *ro* orders only, and it will be sufficient to write out the two 4×4 squares, which show how *o* and *ro* are mutually interchangeable.

o	*ro*	*ro*	*o*
ro	*o*	*o*	*ro*
ro	*o*	*o*	*ro*
o	*ro*	*ro*	*o*

ro	*o*	*o*	*ro*
o	*ro*	*ro*	*o*
o	*ro*	*ro*	*o*
ro	*o*	*o*	*ro*

It will be noticed that *i* is the vertical mirror picture of *o* and *ro* of *ri*, and *vice versa.* Further if the mirror is placed upon one of the horizontal lines, *ri* is the mirror picture of *o* as well as *ro* of *i* and *vice versa.*

There are four more arrangements. There is the Chinese way of writing downward in vertical columns as well as its inversion, and the reversed order of both. This method originated by the use of bamboo strips as writing material in China, and we may utilize the two vowel sounds of the word "bamboo" (viz., *a* and *u*) to name the left and the right downward order, *a* the left and *u* the right, the reverse of the right *ru* and of the left *ra,* but for our present purpose there will be no occasion to use them.

Now we must bear in mind that magic squares originate from the ordinary and normal consecutive arrangement by such transpositions as will counteract the regular increase of value in the normally progressive series of figures; and these transpositions depend upon the location of the several cells. All transpositions in the cells of even magic squares are brought about by the substitution of figures of the *ro, i,* and *ri* order for the original figures of the ordinary or *o* order, and the symmetry which dominates these changes becomes apparent in the diagrams, which present at a glance the order to which each cell in a magic square belongs.

Numbers of the same order are grouped not unlike the Chladni acoustic figures, and it seems to me that the origin of the regularity of both the magic figures and this phenomenon of acoustics, is due to an analogous law of symmetry.

The dominance of one order *o, ro, i,* or *ri,* in each cell of an even magic square, is simply due to a definite method of their selection from the four different orders of counting. Never can a figure appear in a cell where it does not belong by right of some regular order, either *o, ro, i,* or *ri.*

The magic square of 4 × 4, consists only of *o* and *ro* figures, and the same rule applies to the simplest construction of even squares of multiples of four, such as 8 × 8, and 12 × 12.

There are several ways of constructing a magic square of 6 × 6. Our first sample consists of 12 *o,* 12 *ro,* 6 *ri,* and 6 *i* figures. The

12 *o* hold the diagonal lines. The 12 *ro* go parallel with one of these diagonals, and stand in such positions that if the whole magic square were diagonally turned upon itself, they would exactly cover the 6 *i,* and 6 *ri* figures. And again the 6 *i* and 6 *ri* also hold toward each other places in the same way corresponding to one another; if the magic square were turned upon itself around the other diagonal, each *ri* figure would cover one of the *i* order.

1	35	4	33	32	6
12	8	28	27	11	25
24	17	15	16	20	19
13	23	21	22	14	18
30	26	9	10	29	7
31	2	34	3	5	36

1	5	33	34	32	6
30	8	28	9	11	25
18	23	15	16	20	19
24	14	21	22	17	13
7	26	10	27	29	12
31	35	4	3	2	36

Fig. 216. 6×6 EVEN SQUARES.

If we compare the magic squares with the sand-covered glass plates which Chladni used, and think of every cell as equally filled with the four figures that would fall upon it according to the normal sequence of *o, ro, i,* and *ri*; and further if we compare their change into a magic square to a musical note harmonizing whole rows into equal sums, we would find (if by some magic process the different values of the several figures would mechanically be turned up so

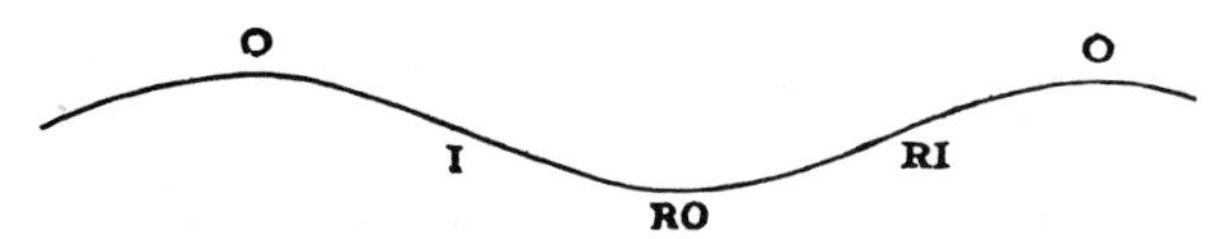

as to be evenly balanced in rows) that they would present geometrically harmonious designs as much as the Chladni acoustic figures.

The progressive transformations of *o, ro, i,* and *ri,* by mirroring, are not unlike the air waves of notes in which *o* represents the crest of the wave, *ro* the trough, *i* and *ri* the nodes.

In placing the mirror at right angles progressively from *o* to *i,* from *i* to *ro,* from *ro* to *ri,* and from *ri* to *o,* we return to the beginning thus completing a whole sweep of the circle.* The re-

* See diagram on page 115.

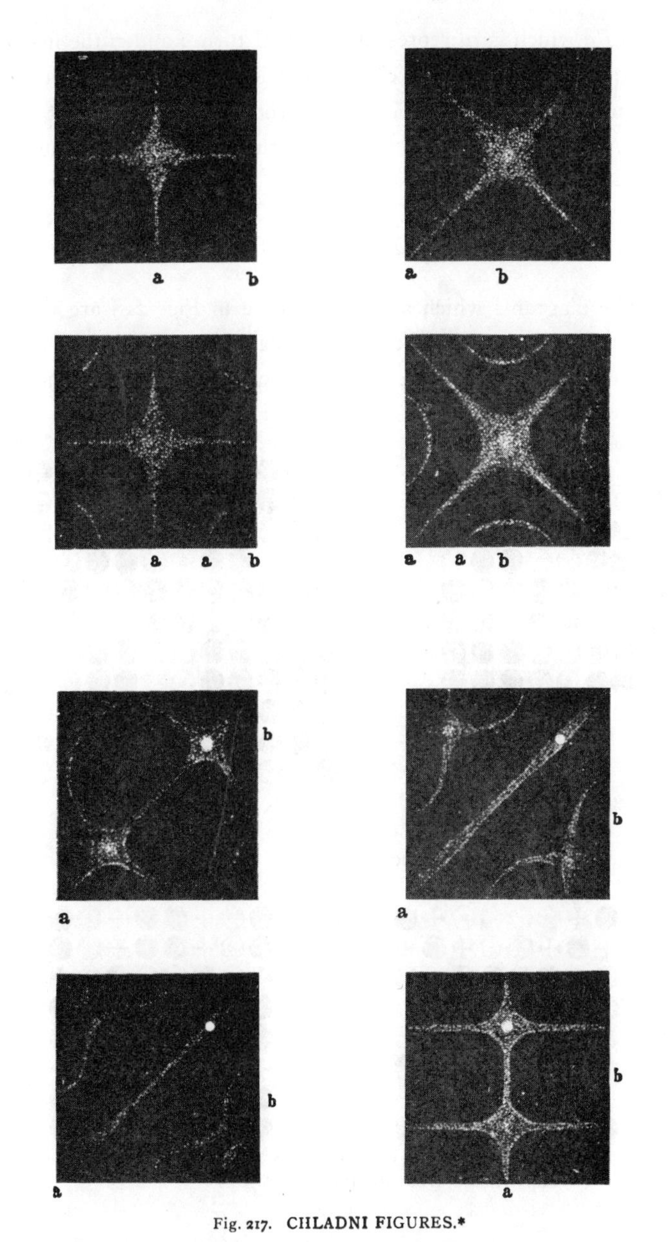

Fig. 217. CHLADNI FIGURES.*

* The letter *a* indicates where the surface is touched with a finger; while *b* marks the place where the bow strikes the glass plate. In the four upper

verse of *o* which is *ro* represents one-half turn, *i* and *ri* the first and third quarter in the whole circuit, and it is natural, therefore, that a symmetry-producing wave should produce a similar effect in the magic square to that of a note upon the sand of a Chladni glass plate.

MAGIC SQUARES IN SYMBOLS.

The diagrams which are offered here in Fig. 218 are the best evidence of their resemblance to the Chladni figures, both exhibiting in their formation, the effect of the law of symmetry. The most

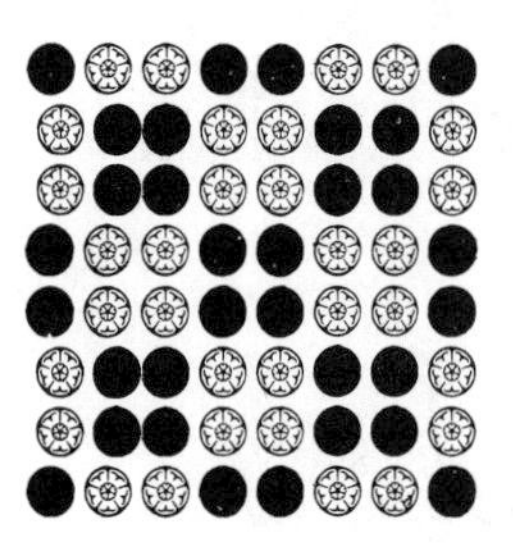

8 × 8. 32 *o* and 32 *ro*. 10 × 10. 72 *o* and 72 *ro*.

SQUARES OF MULTIPLES OF FOUR.

Constructed only of *o* and *ro*.

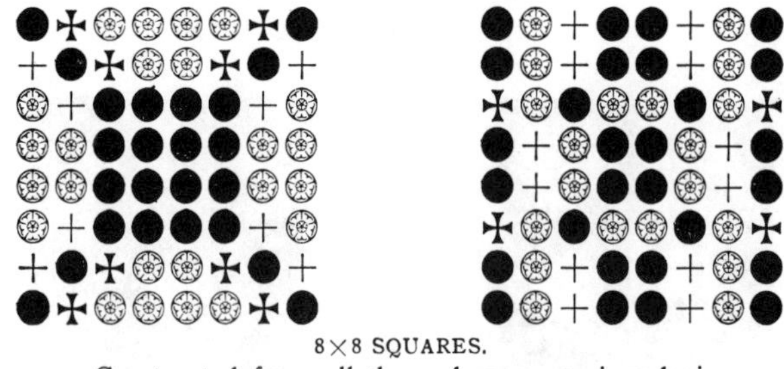

8 × 8 SQUARES.

Constructed from all the orders, *o, ro, i,* and *ri*.

Fig. 218.

diagrams the plate has been fastened in the center, while in the lower ones it has been held tight in an excentric position, indicated by the white dot.

elegant way of rendering the different orders, *i, ri, o,* and *ro,* visible at a glance, would be by printing the cells in four different colors,

ANOTHER 8×8 SQUARE.

It will be noted that in this square the arrangement of the *o* symbols corresponds very closely to the distribution of the sand in the second of the Chladni diagrams. The same may be said of the two following figures, and it is especially true of the first one of the 8×8 squares just preceding.

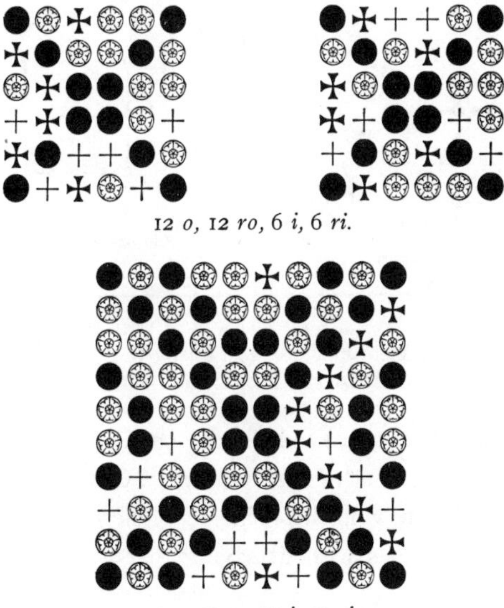

12 *o*, 12 *ro*, 6 *i*, 6 *ri*.

40 *o*, 40 *ro*, 10 *i*, 10 *ri*.

The reader will notice that there is a remarkable resemblance between the symmetry displayed in this figure and in the fourth of the Chladni diagrams.

Fig. 218. (con.). EXAMPLES OF 6×6 AND 10×10 MAGIC SQUARES.

but for proving our case, it will be sufficient to have the four orders represented by four symbols, omitting their figure values, and we here propose to indicate the order of *o* by ●, *ro* by ❁, *i* by ✠, *ri* by +.

THE MAGIC SQUARE IN CHINA.

In the introduction to the Chou edition of the *Yih King,* we find some arithmetical diagrams and among them the *Loh-Shu,* the scroll of the river Loh, which is a mathematical square from 1 to 9, so written that all the odd numbers are expressed by white dots, i. e., yang symbols, the emblem of heaven, while the even numbers

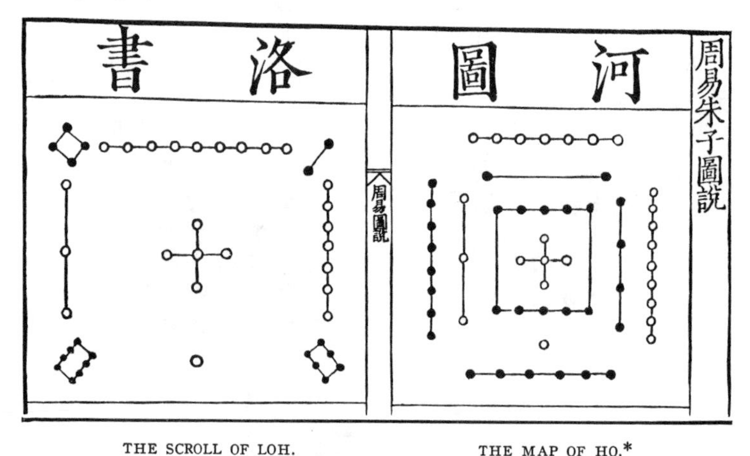

THE SCROLL OF LOH. THE MAP OF HO.*

(According to Ts'ai Yüang-ting.)

Fig. 219. TWO ARITHMETICAL DESIGNS OF ANCIENT CHINA.

are in black dots, i. e., yin symbols, the emblem of earth. The invention of the scroll is attributed to Fuh-Hi, the mythical founder of Chinese civilization, who according to Chinese reports lived 2858-2738 B. C. But it goes without saying that we have to deal here with a reconstruction of an ancient document, and not with the document itself. The scroll of Loh is shown in Fig. 219.

The first unequivocal appearance of the Loh-Shu in the form of a magic square is in the latter part of the posterior Chou dynasty

* The map of Ho properly does not belong here, but we let it stand because it helps to illustrate the spirit of the times when the scroll of Loh was composed in China. The map of Ho contains five groups of odd and even figures, the numbers of heaven and earth respectively. If the former are regarded as positive and the latter as negative, the difference of each group will uniformly yield + 5 or — 5.

(951-1126 A. D.) or the beginning of the Southern Sung dynasty (1127-1333 A. D.). The Loh-Shu is incorporated in the writings of Ts'ai Yüan-Ting who lived from 1135-1198 A. D. (cf. Mayers, *Chinese Reader's Manual,* I, 754a), but similar arithmetical diagrams are traceable as reconstructions of primitive documents among scholars that lived under the reign of Sung Hwei-Tsung, which lasted from 1101-1125 A. D. (See Mayers, *C. R. M.*, p. 57.)

The *Yih King* is unquestionably very ancient and the symbols yang and yin as emblems of heaven and earth are inseparable from its contents. They existed at the time of Confucius (551-479 B. C.), for he wrote several chapters which are called appendices to the *Yih King,* and in them he says (III, I, IX, 49-50. *S. B. E.,* XVI, p. 365.):

"To heaven belongs 1; to earth, 2; to heaven, 3; to earth, 4; to heaven, 5; to earth, 6; to heaven, 7; to earth, 8; to heaven, 9; to earth, 10.

"The numbers belonging to heaven are five, and those belonging to earth are five. The numbers of these two series correspond to each other, and each one has another that may be considered its mate. The heavenly numbers amount to 25, and the earthly to 30. The numbers of heaven and earth together amount to 55. It is by these that the changes and transformations are effected, and the spiritlike agencies kept in movement."

This passage was written about 500 B. C. and is approximately simultaneous with the philosophy of Pythagoras in the Occident, who declares number to be the essence of all things.

One thing is sure, that the magic square among the Chinese cannot have been derived from Europe. It is highly probable, however, that both countries received suggestions and a general impulse from India and perhaps ultimately from Babylonia. But the development of the yang and yin symbols in their numerical and occult significance can be traced back in China to a hoary antiquity so as to render it typically Chinese, and thus it seems strange that the same idea of the odd numbers as belonging to heaven and the even ones to earth appears in ancient Greece.

I owe the following communication to a personal letter from

Professor David Eugene Smith of the Teachers' College of New York:

"There is a Latin aphorism, probably as old as Pythagoras, *Deus imparibus numeris gaudet.* Virgil paraphrases this as follows: *Numero deus impare gaudet.* (Ecl. viii, 75). In the edition I have at hand* there is a footnote which gives the ancient idea of the nature of odd and even numbers, saying:

"...*impar numerus immortalis, quia dividi integer non potest, par numerus mortalis, quia dividi potest; licet Varro dicat Pythagoreos putare imparem numerum habere finem, parem esse infinitum* [a curious idea which I have not seen elsewhere]; *ideo medendi causa multarumque rerum impares numeros servari*: *nam, ut supra dictum est, superi dii impari, inferi pari gaudent.*

"There are several references among the later commentators to the fact that the odd numbers are masculine, divine, heavenly, while the even ones were feminine, mortal, earthly, but I cannot just at this writing place my hands upon them.

"As to the magic square, Professor Fujisawa, at the International Congress of Mathematicians at Paris in 1900, made the assertion that the mathematics derived at an early time from the Chinese (independent of their own native mathematics which was of a somewhat more scientific character), included the study of these squares, going as far as the first 400 numbers. He did not, however, give the dates of these contributions, if indeed they are known."

As to other magic squares, Professor Smith writes in another letter:

"The magic square is found in a work by Abraham ben Ezra in the eleventh century. It is also found in Arabic works of the twelfth century. In 1904, Professor Schilling contributed to the Mathematical Society of Göttingen the fact that Professor Kielhorn had found a Jaina inscription of the twelfth or thirteenth century

---

* P. Virgilii Maronis | Opera, | cum integris commentariis | Servii, Philargyrii, Pierii, | Accedunt | Scaligeri et Lindenbrogii | . . . . . | Pancratius Masvicius | . . . | Tom. I, | . . . | Leonardiae, | . . . | . . cIↃIↃccxvii. |

in the city of Khajuraho, India, a magic square of the notable peculiarity that each sub-square sums to 34."

Fig 220 is the square which Professor Smith encloses.

We must assume that we are confronted in many cases with an independent parallel development, but it appears that suggestions must have gone out over the whole world in most primitive times perhaps from Mesopotamia, the cradle of Babylonian civilization, or later from India, the center of a most brilliant development of scientific and religious thought.

How old the magic square in China may be, is difficult to say. It seems more than probable that its first appearance in the twelfth century is not the time of its invention, but rather the date of a

7	12	1	14
2	13	8	11
16	3	10	5
9	6	15	4

Fig. 220.

recapitulation of former accomplishments, the exact date of which can no longer be determined.

THE JAINA SQUARE.

Professor Kielhorn's Jaina square is not "an associated or regular magic square" according to Mr. Andrews's definition, quoted above. While the sums of all the rows, horizontal, vertical, and diagonal, are equal, the figures equidistant from the center are not equal to $n^2 + 1$, viz., the sum of the first and last numbers of the series. Yet it will be seen that in other respects this square is more regular, for it represents a distribution of the figure values in what might be called absolute equilibrium.

First we must observe that the Jaina square is *continuous*, by which I mean that it may vertically as well as horizontally be turned upon itself and the rule still holds good that wherever we may start four consecutive numbers in whatever direction, backward or forward, upward or downward, in horizontal, vertical, or slanting lines, always yield the same sum, viz. 34, which is $2(n^2+1)$; and so does any small square of 2×2 cells. Since we can not bend the square upon itself at once in two directions, we make the result visible in Fig. 221, by extending the square in each direction by half its own size.

Wherever 4×4 cells are taken out from this extended square, we shall find them satisfying all the conditions of this peculiar kind of magic squares.

The construction of this ancient Jaina equilibrium-square requires another method than we have suggested for Mr. Andrews'

10	5	16	3	10	5	16	3
15	4	9	6	15	4	9	6
1	14	7	12	1	14	7	12
8	11	2	13	8	11	2	13
10	5	16	3	10	5	16	3
15	4	9	6	15	4	9	6
1	14	7	12	1	14	7	12
8	11	2	13	8	11	2	13

Fig. 221.

"associated squares," and the following considerations will afford us the key as shown in Fig. 222.

First we write the numbers down into the cells of the square in their consecutive order and call the four rows in one direction A, B, C, D; in the other direction 1, 2, 3, 4. Our aim is to redistribute them so as to have no two numbers of the same denomination in the same row. In other words, each row must contain one and only one of each of the four letters, and also one and only one of each of the four figures.

We start in the left upper corner and write down in the first horizontal row the letters A, B, C, and D, in their ordinary succession, and in the second horizontal row, the same letters in their inverted order. We do the same with the numbers in the first and second vertical rows. All that remains to be done is to fill out the rest in such a way as not to repeat either a letter or a number. In the first row there are still missing for C and D the numbers 2 and 3, of which 2 must belong to C, for C_3 appears already in the second row and 3 is left for D.

In the second row there are missing 1 and 4, of which 1 must belong to B, because we have B_4 in the first row.

In the first vertical row the letters B and C are missing, of which B must belong to 3, leaving C to 4.

	1	2	3	4
A	1	2	3	4
B	5	6	7	8
C	9	10	11	12
D	13	14	15	16

In Consecutive Order.

A_1	B_4	C	D
D_2	C_3	B	A
3	2		
4	1		

The Start for a Redistribution.

A_1	B_4	C_2	D_3
D_2	C_3	B_1	A_4
B_3	A_2	D_4	C_1
C_4	D_1	A_3	B_2

The Perfected Redistribution.

1	8	10	15
14	11	5	4
7	2	16	9
12	13	3	6

Figure Values of the Square.

Fig. 222.

In the second vertical row A and D are missing for 1 and 2. A_1 and D_2 exist, so A must go to 2, and D to 1.

In the same simple fashion all the columns are filled out, and then the cell names replaced by their figure values, which yields the same kind of magic square as the one communicated by Prof. Smith, with these differences only, that ours starts in the left corner with number 1 and the vertical rows are exchanged with the horizontal ones. It is scarcely necessary to point out the beautiful symmetry in the distribution of the figures which becomes fully apparent when we consider their cell names. Both the letters, A,

B, C, D, and the figures, 1, 2, 3, 4, are harmoniously distributed over the whole square, so as to leave to each small square its distinct individuality, as appears from Fig. 223.

A	B	C	D
D	C	B	A
B	A	D	C
C	D	A	B

1	4	2	3
2	3	1	4
3	2	4	1
4	1	3	2

Fig. 223.

The center square in each case exhibits a cross relation, thus:

C	B
A	D

3	1
2	4

In a similar way each one of the four groups of four cells in each of the corners possesses an arrangement of its own which is symmetrically different from the others.

P. C.

CHAPTER V.

A MATHEMATICAL STUDY OF MAGIC SQUARES.

A NEW ANALYSIS.

MAGIC squares are not simple puzzles to be solved by the old rule of "Try and try again," but are visible results of "order" as applied to numbers. Their construction is therefore governed by laws that are as fixed and immutable as the laws of geometry.

It will be the object of this essay to investigate these laws, and evolve certain rules therefrom. Many rules have been published

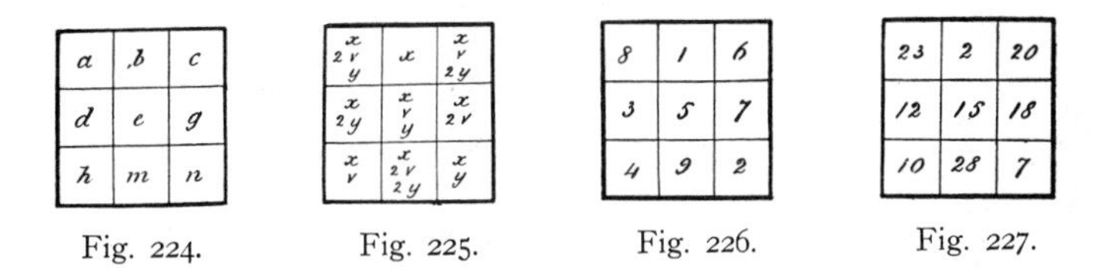

Fig. 224. Fig. 225. Fig. 226. Fig. 227.

by which various magic squares may be constructed, but they do not seem to cover the ground comprehensively.

Let Fig. 224 represent a 3×3 magic square. By inspection we note that:

$$h + c = b + m$$
$$\text{and } h + m = g + c$$
$$\text{therefore } 2h = b + g$$

In this way four equations may be evolved as follows:

$$2h = b + g$$
$$2n = b + d$$
$$2c = d + m$$
$$2a = m + g$$

It will be seen that the first terms of these equations are the quantities which occur in the four corner cells, and therefore that the quantity in each corner cell is a mean between the two quantities in the two opposite cells that are located in the middle of the outside rows. It is therefore evident that the least quantity in the magic square must occupy a middle cell in one of the four outside rows, and that it *cannot occupy a corner cell.*

Since the middle cell of an outside row must be occupied by the least quantity, and since any of these cells may be made the middle cell of the upper row by rotating the square, we may consider this cell to be so occupied.

Having thus located the least quantity, it is plain that the next higher quantity must be placed in one of the lower corner cells, and since a simple reflection in a mirror would reverse the position of the lower corner cells, it follows that the second smallest quantity may occupy *either* of these corner cells. Next we may write more equations as follows:

$$a + e + n = S \text{ (or summation)}$$
$$d + e + g = S$$
$$h + e + c = S$$

also

$$a + d + h = S$$
$$n + g + c = S$$

therefore

$$3e = S$$

and

$$e = S/3$$

Hence the quantity in the central cell is an arithmetical mean between any two quantities with which it forms a straight row or column.

With these facts in view a magic square may now be constructed as shown in Fig. 225.

Let x, representing the least quantity, be placed in the middle upper cell, and $x + y$ in the lower right-hand corner cell, y being the increment over x.

Since $x + y$ is the mean between x and the quantity in the left-hand central cell, this cell must evidently contain $x + 2y$.

Now writing $x + v$ in the lower left-hand corner cell, (considering v as the increment over x) it follows that the central right-hand cell must contain $x + 2v$.

Next, as the quantity in the central cell in the square is a mean between $x + 2y$ and $x + 2v$, it must be filled with $x + v + y$. It now follows that the lower central cell must contain $x + 2v + 2y$, and the upper left-hand corner cell $x + 2v + y$, and finally the upper right-hand corner cell must contain $x + v + 2y$, thus completing the square which necessarily must be magic with any conceivable values which may be assigned to x, v, and y.

We may assign values to x, v, and y which will produce the numbers 1 to 9 inclusive in arithmetical progression. Evidently x must equal 1, and as there must be a number 2, either v or y must equal 1 also.

Assuming $y = 1$, if $v = 1$ or 2, duplicate numbers would result, therefore *v cannot be less than* 3.

Using these values, viz., $x = 1$, $y = 1$ and $v = 3$, the familiar 3×3 magic square shown in Fig 226 is produced.

Although in Fig. 226 the series of numbers used has an initial number of 1, and also a constant increment of 1, this is only an accidental feature pertaining to this particular square, the real fact being that *a magic square of 3×3 is always composed of three sets each of three numbers.* The difference between the numbers of each trio is uniform, but the difference between the last term of one trio and the first term of the next trio is not necessarily the same as the difference between the numbers of the trios.

For example, if $x = 2$, $y = 5$ and $v = 8$, the resulting square will be as shown in Fig. 227.

x 2s t 2v y	x 2s t	x 2s t v 2y	x 2v y	x	x v 2y	x s 2t 2v y	x s 2t	x s 2t v 2y
x 2s t 2y	x 2s t v y	x 2s t 2v	x 2y	x v y	x 2v	x s 2t 2y	x s 2t v y	x s 2t 2v
x 2s t v	x 2s t 2v 2y	x 2s t y	x v	x 2y 2v	x y	x s 2t v	x s 2t 2v 2y	x s 2t y
x 2t 2v y	x 2t	x 2t v 2y	x s t 2v y	x s t	x s t v 2y	x 2s 2v y	x 2s	x 2s v 2y
x 2t 2y	x 2t v y	x 2t 2v	x s t 2y	x s t v y	x s t 2v	x 2s 2y	x 2s v y	x 2s 2v
x 2t v	x 2t 2v 2y	x 2t y	x s t v	x s t 2v 2y	x s t y	x 2s v	x 2s 2v 2y	x 2s y
x s 2v y	x s	x s v 2y	x 2s 2t 2v y	x 2s 2t	x 2s 2t v 2y	x t 2v y	x t	x t v 2y
x s 2y	x s v y	x s 2v	x 2s 2t 2y	x 2s 2t v y	x 2s 2t 2v	x t 2y	x t v y	x t 2v
x s v	x s 2v 2y	x s y	x 2s 2t v	x 2s 2t 2v 2y	x 2s 2t y	x t v	x t 2v 2y	x t y

Fig. 228.

	x 2s t			x			x s 2t	
	x 2t			x s t			x 2s	
	x s			x 2s 2t			x t	

Fig. 229.

The trios in this square are as follows:

2 — 7 — 12
10 — 15 — 20
18 — 23 — 28

The difference between the numbers of these trios is $y = 5$, and the difference between the homologous numbers is $v = 8$.

A recognition of these different sets of increments is essential to the proper understanding of the magic square. Their existence is masked in the 3×3 square shown in Fig. 226 by the more or less accidental quality that in this particular square the difference between adjacent numbers is always 1. Nevertheless the square given in Fig. 226 is really made up of three trios, as follows:

1st trio 1 — 2 — 3
2d " 4 — 5 — 6
3d " 7 — 8 — 9

in which the difference between the numbers of the trios is $y = 1$, and the difference between the homologous numbers is $v = 3$.

Having thus acquired a clear conception of the structure of a 3×3 magic square, we are in a position to examine a 9×9 compound square intelligently, this square being only an expansion of the 3×3 square, and governed by the same constructive rules.

Referring to Fig. 229 the upper middle cells of the nine subsquares may first be filled, using for this purpose the terms, x, t, and s. Using these as the initial terms of the subsquares the square may then be completed, using y as the increment between the terms of each trio, and v as the increment between the homologous terms of the trios. The completed square is shown in Fig. 228, *in which the assignment of any values to x, y, v, t and s, will yield a perfect, compound 9×9 square.*

Values may be assigned to x, y, v, t and s which will produce the series 1 to 81 inclusive. As stated before in connection with the 3×3 square, x must naturally equal 1, and in order to produce 2, one of the remaining symbols must equal 1. In order to avoid duplicates, the next larger number must at least equal 3, and by

71	64	69	8	1	6	53	46	51
66	68	70	3	5	7	48	50	52
67	72	65	4	9	2	49	54	47
26	19	24	44	37	42	62	55	60
21	23	25	39	41	43	57	59	61
22	27	20	40	45	38	58	63	56
35	28	33	80	73	78	17	10	15
30	32	34	75	77	79	12	14	16
31	36	29	76	81	74	13	18	11

71	8	53	64	1	46	69	6	51
26	44	62	19	37	55	24	42	60
35	80	17	28	73	10	33	78	15
66	3	48	68	5	50	70	7	52
21	39	57	23	41	59	25	43	61
30	75	12	32	77	14	34	79	16
67	4	49	72	9	54	65	2	47
22	40	58	27	45	63	20	38	56
31	76	13	36	81	18	29	74	11

Fig. 230.

77	22	51	56	1	30	71	16	45
24	50	76	3	29	55	18	44	70
49	78	23	28	57	2	43	72	17
62	7	36	68	13	42	74	19	48
9	35	61	15	41	67	21	47	73
34	63	8	40	69	14	46	75	20
65	10	39	80	25	54	59	4	33
12	38	64	27	53	79	6	32	58
37	66	11	52	81	26	31	60	5

77	56	71	22	1	16	51	30	45
62	68	74	7	13	19	36	42	48
65	80	59	10	25	4	39	54	33
24	3	18	50	29	44	76	55	70
9	15	21	35	41	47	61	67	73
12	27	6	38	53	32	64	79	58
49	28	43	78	57	72	23	2	17
34	40	46	63	69	75	8	14	20
37	52	31	66	81	60	11	26	5

Fig. 231.

77	58	69	20	1	12	53	34	45
60	68	76	3	11	19	36	44	52
67	78	59	10	21	2	43	54	35
26	7	18	50	31	42	74	55	66
9	17	25	33	41	49	57	65	73
16	27	8	40	51	32	64	75	56
47	28	39	80	61	72	23	4	15
30	38	46	63	71	79	6	14	22
37	48	29	70	81	62	13	24	5

77	20	53	58	1	34	69	12	45
26	50	74	7	31	55	18	42	66
47	80	23	28	61	4	39	72	15
60	3	36	68	11	44	76	19	52
9	33	57	17	41	65	25	49	73
30	63	6	38	71	14	46	79	22
67	10	43	78	21	54	59	2	35
16	40	64	27	51	75	8	32	56
37	70	13	48	81	24	29	62	5

Fig. 232.

the same reason the next must not be less than 9 and the remaining one not less than 27. Because $1+1+3+9+27=41$, which is the middle number of the series 1—81, therefore just these values must be assigned to the five symbols. The only symbol whose value is fixed, however, is x, the other four symbols may have the values 1—3—9 or 27 assigned to them indiscriminately, thus producing all the possible variations of a 9×9 compound magic square.

If v is first made 1 and $y=2$, and afterwards y is made 1 and $v=2$, the resulting squares will be simply reflections of each other, etc. Six fundamental forms of 9×9 compound magic squares may be constructed as shown in Figs. 230, 231, and 232.

Only six forms may be made, because, excluding x whose value is fixed, only six different couples may be made from the four remaining symbols. Six cells being determined, the rest of the square becomes fixed.

These squares are arranged in three groups of two each, on account of the curious fact that the squares in each pair are mutually convertible into each other by the following process:

If the homologous cells of each 3×3 subsquare be taken in order as they occur in the 9×9 square, a new magic 3×3 square will result. And if this process is followed with all the cells and the resulting nine 3×3 squares are arranged in magic square order a new 9×9 compound square will result.

For example, referring to the upper square in Fig. 230, if the numbers in the central cells of the nine 3×3 subsquares are arranged in magic square order, the resulting square will be the central 3×3 square in the lower 9×9 square in Fig. 230. This law holds good in each of the three groups of two squares (Figs. 230, 231 and 232) and no fundamental forms other than these can be constructed.

The question may be asked: How many variations of 9×9 compound magic squares can be made? Since each subsquare may assume any of eight aspects without disturbing the general order of the complete square, and since there are six radically different, or fundamental forms obtainable, the number of possible variations is 6×8^9!

We will now notice the construction of a 4×4 magic square as represented in Fig. 233. From our knowledge of this magic square we are enabled to write four equations as follows:

$$a+h+p+y=S \text{ (Summation)}$$
$$g+h+n+m=S$$
$$k+o+p+s=S$$
$$t+o+n+d=S$$

By inspection of Fig. 233 it is seen that the sum of the initial terms of these four equations equals S, and likewise that the sum

a	b	c	d
g	h	n	m
k	o	p	s
t	v	x	y

Fig. 233.

a x			a v
			b x
			c x
g x	a y	a t	g v

Fig. 234.

a x	g y	g t	a v
b v	c t	c y	b x
c v	b t	b y	c x
g x	a y	a t	g v

Fig. 235

1	8	12	13
14	11	7	2
15	10	6	3
4	5	9	16

Fig. 236

of their final terms also equals S. Hence $h+n+o+p=S$. It therefore follows:

(1st) That *the sum of the terms contained in the inside* 2×2 *square of a* 4×4 *square is equal to* S.

(2d) Because the middle terms of the two diagonal columns compose this inside 2×2 square, their end terms, or *the terms in the four corner cells of the* 4×4 *square must also equal* S, or:

$$a+d+t+y=S$$

(3d) Because the two middle terms of each of the two inside columns (either horizontal or perpendicular) also compose the central 2×2 square, *their four end terms must likewise equal* S.

We may also note the following equations:

$$b+c+v+x=S$$
$$b+c+a+d=S$$

therefore

$$a + d = v + x,$$

which shows (4th) that *the sum of the terms in any two contiguous corner cells is equal to the sum of the terms in the two middle cells in the opposite outside column.*

Because

$$g + h + n + m = S$$

and

$$o + h + n + p = S$$

it follows that

$$g + m = o + p$$

or, (5th) that *the sum of the two end terms of any inside column,* (either horizontal or perpendicular) *is equal to the sum of the two middle terms in the other parallel column.*

Since

$$t + o + n + d = S$$

and

$$h + o + n + p = S$$

therefore

$$t + d = h + p$$

or (6th) *the sum of the two end terms of a diagonal column is equal to the sum of the two inside terms of the other diagonal column.*

These six laws govern *all* 4×4 magic squares, but the regular or associated squares also possess the additional feature that the sum of the numbers in any two cells that are equally distant from the center and symmetrically opposite to each other in the square equals $S/2$.

Squares of larger dimensions do not seem to be reducible to laws, on account of their complexity.

NOTES ON NUMBER SERIES USED IN THE CONSTRUCTION OF MAGIC SQUARES.

It has long been known that magic squares may be constructed from series of numbers which do not progress in arithmetical order.

Experiment will show, however, that any haphazard series cannot be used for this purpose, but that a definite order of sequence is necessary which will entail certain relationships between different members of the series. It will therefore be our endeavor to determine these relationships and express the same in definite terms.

Let Fig. 237 represent a magic square of 4×4. By our rule No. 4 it is seen that "*the sum of the terms in any two contiguous corner cells is equal to the sum of the terms in the two middle cells in the opposite outside column.*" Therefore in Fig. 237, $a + d = v + s$, and it therefore follows that $a - v = s - d$. In other words, these four quantities form a group with the inter-

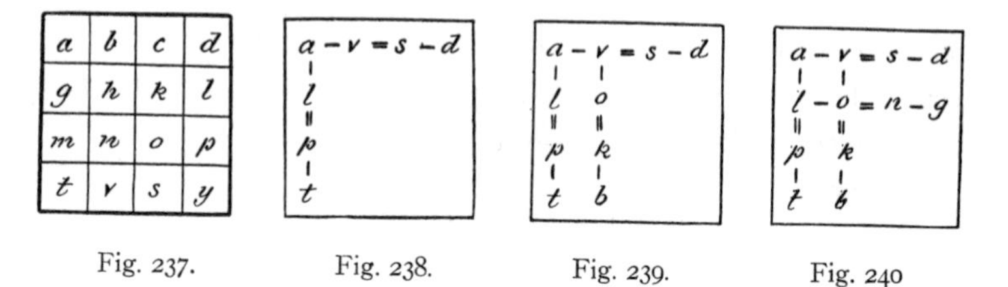

Fig. 237. Fig. 238. Fig. 239. Fig. 240

relationship as shown. By the same rule (No. 4) it is also seen that $a + t = l + p$, and hence also, $a - l = p - t$, giving another group of four numbers having the same form of interrelationship, and since both groups have "a" as an initial number, it is evident that the *increment used in one of these groups must be different from that used in the other, or duplicate numbers would result.* It therefore follows that the numbers composing a magic square are not made up of a single group, but necessarily of more than one group.

Since the term "a" forms a part of two groups, we may write both groups as shown in Fig. 238, one horizontally and the other perpendicularly.

Next, by rule No. 5, it is shown that "*the sum of the two end terms of any inside column (either horizontal or perpendicular) is equal to the sum of the two middle terms in the other parallel column.*" It therefore follows that $v + b = k + o$ or $v - o = k - b$. Using the term v as the initial number, we write this series perpendicularly as shown in Fig. 239. In the same way it is seen that

$l+g=n+o$, or $l-o=n-g$, thus forming the second horizontal column in the square (Fig. 240). Next $p+m=h+k$ or $p-k=h-m$, forming the third horizontal column and in this simple manner the square may be completed as shown in Fig. 241.

It is therefore evident that a 4×4 magic square may be formed of any series of numbers whose interrelations are such as to permit them to be placed as shown in Fig. 241.

The numbers 1 to 16 may be so placed in a great variety of ways, but the fact must not be lost sight of that they only *incidentally* possess the quality of being a single series in straight arithmetical order, being really composed of as many groups as there are cells in a column of the square. Unless this fact is remembered, a clear conception of magical series cannot be formed.

In illustration of the above remarks, three diagrams are given in Figs. 242-244. Figs. 242 and 243 show arrangements of the

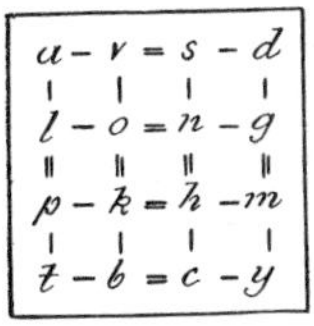

```
a - v = s - d
|   |   |   |
l - o = n - g
‖   ‖   ‖   ‖
p - k = h - m
|   |   |   |
t - b = c - y
```

Fig. 241.

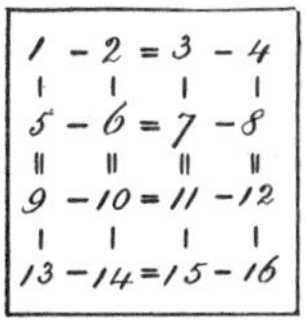

```
 1 -  2 =  3 -  4
 |    |    |    |
 5 -  6 =  7 -  8
 ‖    ‖    ‖    ‖
 9 - 10 = 11 - 12
 |    |    |    |
13 - 14 = 15 - 16
```

Fig. 242.

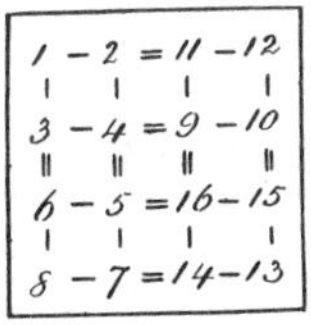

```
 1 -  2 = 11 - 12
 |    |    |    |
 3 -  4 =  9 - 10
 ‖    ‖    ‖    ‖
 6 -  5 = 16 - 15
 |    |    |    |
 8 -  7 = 14 - 13
```

Fig. 243.

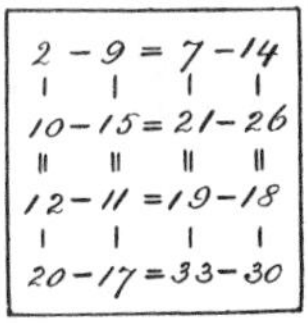

```
 2 -  9 =  7 - 14
 |    |    |    |
10 - 15 = 21 - 26
 ‖    ‖    ‖    ‖
12 - 11 = 19 - 18
 |    |    |    |
20 - 17 = 33 - 30
```

Fig. 244.

numbers 1 to 16 from which the diverse squares Figs. 245 and 246 are formed by the usual method of construction.

Fig. 244 shows an irregular series of sixteen numbers, which, when placed in the order of magnitude run as follows:

2-7-9-10-11-12-14-15-17-18-19-20-21-26-30-33

The magic square formed from this series is given in Fig. 247.

In the study of these number series the natural question presents itself: *Can as many diverse squares be formed from one series as from another?* This question opens up a wide and but little explored region as to the diverse constitution of magic squares. This idea can therefore be merely touched upon in the present article, examples of several different plans of construction being given in illustration and the field left at present to other explorers.

Three examples will be given, Fig. 245 being what is termed

an associated square, or one in which any two numbers that are diametrically opposite and equidistant from the center of the square will be equal in summation to any other pair of numbers so situated. The second, Fig. 246, is a square in which the sum of every diagonal of the four sub-squares of 2×2 is equal, and the fourth, Fig. 248, a square in which the pairs of numbers having similar summations are arranged symmetrically in relation to a perpendicular line through the center of the square.

Returning now to the question, but little reflection is required to show that it must be answered in the negative for the following reasons. Fig. 247 represents a magic square having no special qualities excepting that the columns, horizontal, perpendicular and diagonal, all have the same summation, viz., 66. Hence *any* series

1	14	15	4
8	11	10	5
12	7	6	9
13	2	3	16

Fig. 245.

1	7	14	12
10	16	5	3
15	9	4	6
8	2	11	13

Fig. 246.

2	17	33	14
26	19	11	10
18	21	15	12
20	9	7	30

Fig. 247.

1	11	6	16
14	13	4	3
7	2	15	10
12	8	9	5

Fig. 248.

of numbers that can be arranged as shown in Fig. 241 will yield magic squares as outlined. But that it shall also produce squares that are associated, may or may not be the case accordingly as the series may or may not be capable of still further arrangement.

Referring to Fig. 237, if we amend our definition by now calling it an associated square, we must at once introduce the following continuous equation:

$$a + y = h + o = t + d = n + k = b + s = c + v = g + p = m + l,$$

and if we make our diagram of magic square producing numbers conform to these new requirements, the number of groups will at once be greatly curtailed.

The multiplicity of algebraical signs necessary in our amended diagram is so great that it can only be studied in detail, the complete diagram being a network of minus and equality signs.

The result will therefore only be given here, formulated in the following laws which apply in large measure to all associated squares.

I. Associated magic squares are made of as many series or groups of numbers as there are cells in a column.

II. Each series or group is composed of as many numbers as there are groups.

III. The differences between any two adjoining numbers of a

```
3 - 13 = 18 - 28
|    |    |    |
4 - 14 = 19 - 29
‖    ‖    ‖    ‖
21 - 31 = 36 - 46
|    |    |    |
22 - 32 = 37 - 47
```

Fig. 249.

3	32	37	28
29	36	31	4
46	19	14	21
22	13	18	47

Fig. 250.

series must obtain between the corresponding numbers of all the series.

IV. The initial terms of the series compose another series, as do the second, third, fourth terms and so on.

V. The differences between any adjoining numbers of these secondary series must also obtain between the corresponding terms of all the secondary series.

1	4	7	10	13
8	11	14	17	20
15	18	21	24	27
22	25	28	31	34
29	32	35	38	41

Fig. 251.

25	38	1	14	27
35	13	11	24	22
10	8	21	34	32
20	18	31	29	7
15	28	41	4	17

Fig. 252.

The foregoing rules may be illustrated by the series and associated square shown in Figs. 242 and 245.

Following and consequent upon the foregoing interrelations of these numbers is the remarkable quality possessed by the associated magic square producing series as follows:

If the entire series is written out in the order of magnitude and the differences between the adjacent numbers are written below, the row of differences will be found to be geometrically arranged on each side of the center as will be seen in the following series taken from Fig. 249.

3 – 4 – 13–14–18–19–21–22–28–29–31–32–36–37–46–47
1 9 1 4 1 2 1 (6) 1 2 1 4 1 9 1

In the above example the number 6 occupies the center and the other numbers are arranged in symmetrical order on each side of it. It is the belief of the writer that this rule applies to all associated squares whether odd or even.

The following example will suffice to illustrate the rule as applied to a 5×5 magic square, Fig. 251 showing the series and Fig. 252 the square.

1 . 4 . 7 . 8 . 10 . 11 . 13 . 14 . 15 . 17 . 18 . 20 . 21 . 22 . 24 . 25 . 27 . 28 . 29 . 31 . 32 . 34 . 35 . 38 . 41
3 3 1 2 1 2 1 1 2 1 2 1 | 1 2 1 2 1 1 2 1 2 1 3 3

The diagram shown in Fig. 253 is given to impress upon the reader the idea that a natural series of continuous numbers may be arranged in a great variety of different magic square producing series. A perfect 9×9 square will be produced with any conceivable values that may be assigned to the symbols a, b, c, d and g, used in this diagram. If the square is to be normal we must assign the numbers 1, 1, 3, 9, 27 for these symbols, and a must equal 1. It is then evident that for 2 there is a choice of four cells, as this number may be either $a + b$, $a + c$, $a + d$ or $a + g$. Selecting $a + b$ for 2, makes $b = 1$. There is then a choice of three for 4, and for this number we will choose $a + d$, making $d = 3$. A choice of two, ($a + g$ and $a + c$) now remains for 10. Selecting $a + g$, (and thus making $g = 9$) 28 becomes the fixed value of $a + c$, giving the value of 27 to c. It is thus evident that after locating 1 in any cell (other than the central cell) we may then produce at will ($4 \times 3 \times 2 =$) 24 different 9×9 magic squares. Nevertheless, each of these twenty-four squares will be made on exactly the same plan, and using the same breakmoves; the variations, radical as they may appear to be, are only so because different *series* of the same numbers are employed, of which series, it has been shown, there are at least twenty-four.

If the reader will take Fig. 253 and fill in number values, making "*b*" (successively) = 3, 9, and 27, he will acquire a clear idea of the part taken in magic squares by the series conception.

The work of determining the possible number of 9×9 magic

a c b	a g 2d	a d c 2b	a 2g b	a 2d 2c	a 2g d 2b	a g 2c b	a 2g 2d c	a g d 2c 2b
a g 2d 2c 2b	a d c b	a 2g	a 2d c 2b	a 2g d b	a g 2c	a 2g 2d 2b	a g d 2c b	a c
a d c	a 2g 2c 2b	a 2d c b	a 2g d	a g c 2b	a 2g 2d b	a g d 2c	a 2b	a g 2d 2c b
a 2g 2c b	a 2d c	a 2g 2c d 2b	a g c b	a 2g 2d	a g d c 2b	a b	a g 2d 2c	a d 2b
a 2b 2d	a 2g 2c d b	a g c	a 2g 2d 2c 2b	a b c d g	a	a g 2d c 2b	a d b	a 2g 2c
a 2g 2c d	a g 2b	a 2g 2d 2c b	a g c d	a 2c 2b	a g 2d c b	a d	a 2g c 2b	a 2d b
a g b	a 2g 2d 2c	a g d 2b	a 2c b	a g 2d c	a d 2c 2b	a 2g c b	a 2d	a 2g d c 2b
a 2g 2d c 2b	a g d b	a 2c	a g 2d 2b	a d 2c b	a 2g c	a 2d 2c 2b	a 2g d c b	a g
a g d	a c 2b	a g 2d b	a d 2c	a 2g 2b	a 2d 2c	a 2g d c	a g 2c 2b	a 2g 2d c b

Fig. 253.

squares is now greatly simplified, for all elements are thus determined *saving one,* i. e., *the number of possible modes of progression.*

1 may be located in any of 80 cells and progress may be made in x ways, and 24 variants may be constructed in each case. Therefore, the possible number of different 9×9 squares will be at least

$$80 \times 24 \times x = 1920x.$$

A single example will serve to illustrate the possibilities open to x, the numerical value of which will be left for the present for others to determine. As previously given, let

$$a = 1$$
$$b = 1$$
$$d = 3$$
$$g = 9$$
$$c = 27$$

Then Fig. 254 will represent a 9×9 square based on the arrangement of symbols given in Fig. 253.

29	16	33	20	61	24	65	52	69
72	32	19	36	23	64	27	68	28
31	75	35	22	39	26	67	3	71
74	34	78	38	25	42	2	70	6
9	77	37	81	41	1	45	5	73
76	12	80	40	57	44	4	48	8
11	79	15	56	43	60	47	7	51
54	14	55	18	59	46	63	50	10
13	30	17	58	21	62	49	66	53

Fig. 254.

Considering the numbers 1 to 81 to be arranged in arithmetical order the construction of this square must be governed by the following rule:

Regular spacing: Three successive cells in upward right-hand diagonal.

Breakmoves between

3 and 4
6 " 7
9 " 10
12 " 13 etc.

} Three cells down and one to left. (Extended knight's move.)

and between

27 and 28	}	two cells to the right.
54 " 55		
81 " 1		

In fact, however, the square is built up by the *common rule,* viz.:

Regular spacing: Nine successive cells in upward right-hand diagonal, and *all* breakmoves, two cells to the right, the numbers 1 to 81 being arranged in the following series:

1.2.3	28.29.30	55.56.57
4.5.6	31.32.33	58.59.60
7.8.9	34.35.36	61.62.63 etc., etc.

As shown above, the numbers 1 to 81 may be arranged in at least twenty-four of such magic square producing series, thus giving twenty-four different squares, by the same method of progression, and using the same breakmoves.

L. S. F.

CHAPTER VI.

MAGICS AND PYTHAGOREAN NUMBERS.

> "I have compiled this discourse, which asks for your consideration and pardon not only because the matter itself is by no means easy to be handled, but also because the doctrines herein contained are somewhat contrary to those held by most of the Platonic philosophers." *Plutarch.*

THE mysterious relationships of numbers have attracted the minds of men in all ages. The many-sided Franklin, whose 200th anniversary the philosophical, scientific, and literary worlds have recently celebrated, used to amuse himself with the construction of magic squares and in his memoirs has given an example of his skill in this direction, by showing a very complicated square with the comment that he believes the same to be the most magical magic square yet constructed by any magician.

That magic squares have had in centuries past a deeper meaning for the minds of men than that of simple mathematical curios we may infer from the celebrated picture by Albert Dürer entitled "Melancolia," engraved in 1514. The symbolism of this engraving has interested to a marked degree almost every observer. The figure of the brooding genius sitting listless and dejected amid her uncompleted labors, the scattered tools, the swaying balance, the flowing sands of the glass, and the magic square of 16 beneath the bell, —these and other details reveal an attitude of mind and a connection of thought, which the great artist never expressed in words, but left for every beholder to interpret for himself.

The discovery of the arrangement of numbers in the form of magic diagrams was undoubtedly known to the ancient Egyptians and this may have formed part of the knowledge which Pythagoras brought back from his foreign travels. We have no direct evidence that the Pythagorean philosophers in their studies of the relationship of numbers ever combined them into harmonic figures, yet the

MELANCHOLY.

supposition that they did so is not at all improbable. Such diagrams and their symbolic meanings may well have formed part of the arcana of the esoteric school of Pythagoras, for similar facts were accounted by ancient writers as constituting a part of the aporrheta of the order and the story is told of an unworthy disciple who revealed the secret of the construction of the dodecahedron inscribed within a sphere, this being a symbol of the universe.

Among the best expositions of the Pythagorean philosophy are sections of the "Timæus" and "Republic" of Plato. These dialogues were written after Plato's return from Magna Græcia, where from contact with Archytas of Tarentum and other philosophers, he imbibed so much of the Italian school that his whole system of philosophy became permeated with Pythagorean ideas. It is even suggested that he incorporated into these dialogues parts of the lost writings of Philolaus, whose works he is known to have purchased. No portions of the dialogues named have been more puzzling to commentators than the vague references to different numbers, such as the number 729, which is chosen to express the difference between the kingly man and the tyrant, or the so-called number of the State in the "Republic," or the harmonic number of the soul in the "Timæus" of which Plutarch said that 'it would be an endless toil to recite the contentions and disputes that have from hence arisen among his interpreters." Either our text of these passages is corrupt or Plato is very obscure, throwing out indirect hints which would be intelligible only to those previously informed. Plato states himself in the "Phædrus" that "all writings are to be regarded purely as a means of recollection for him who already knows," and he, therefore, probably wrote more for the benefit of his hearers than for distant posterity.

It is upon the principle of a magic square that I wish to interpret the celebrated passage in the "Republic" referring to the number 729, proceeding from this to a discussion of certain other numbers of peculiar significance in the Pythagorean system. My efforts in this direction are to be regarded as purely fanciful; the same may be said, however, of the majority of other methods of interpretation.

The passage from the "Republic" referred to (Book IX, § 587-8, Jowett's translation) reads as follows:

Socrates. "And if a person tells the measure of the interval which separates the king from the tyrant in truth of pleasure, he will find him, when the multiplication is completed living 729 times more pleasantly, and the tyrant more painfully by this same interval."

Glaucon. "What a wonderful calculation."

Socrates. "Yet a true calculation and a number which closely concerns human life, if human life is concerned with days and nights and months and years."

The number 729 is found to be of great importance all through the Pythagorean system. Plutarch states that this was the number belonging to the sun, just as 243 was ascribed to Venus, 81 to Mercury, 27 to the moon, 9 to the earth, and 3 to Antichthon (the earth opposite to ours). These and many similar numbers were derived from one of the progressions of the Tetractys,—1:2::4:8 and 1:3::9:27. The figures of the above proportions were combined by Plato into one series, 1, 2, 3, 4, 9, 8, 27. (Timæus, § 35). Plutarch in his "Procreation of the Soul," which is simply a commentary upon Plato's "Timæus," has represented the numbers in the form of a triangle; the interior numbers, 5, 13, and 35, representing the sums of the opposite pairs, were also of great importance.

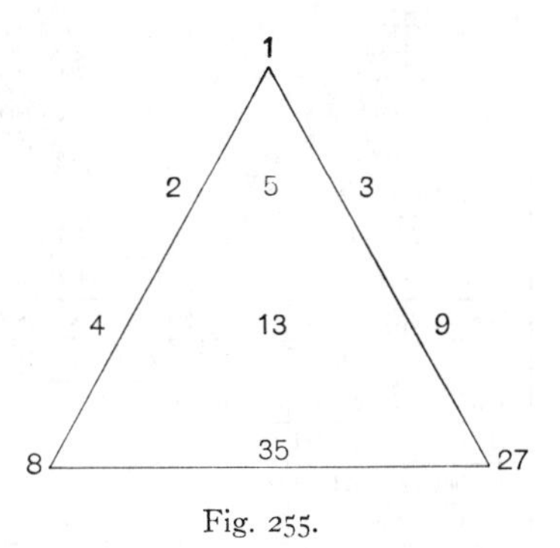

Fig. 255.

The deep significance of the Tetractys in the system of Pythagoras may be inferred from a fragment of an oath contained in the "Golden Verses."

Ναὶ μὰ τὸν ἁμέτερον ψυχᾷ παραδόντα τετρακτὸν
Παγὰν, ἀενάου φύσεως ῥιζώματ' ἔχουσαν.

"Yea, by our Tetractys which giveth the soul the fount and source of ever flowing nature!"

Odd numbers were especially favored by the Pythagoreans and of these certain ones such as 3 and its higher powers were considered to have a higher significance than others and in this way, perhaps, arose the distinction between expressible and inexpressible or ineffable numbers (*ἀριθμοὶ ῥητοὶ καὶ ἄῤῥητοι*). Numbers which expressed some astronomical fact also held high places of honor, as may be seen from a statement by Plutarch (*loc. cit.*) in reference to the Tetractys. "Now the final member of the series, which is 27, has this peculiarity, that it is equal to the sum of the preceding numbers (1+2+3+4+9+8); it also represents the periodical number of days in which the moon completes her monthly course; the Pythagoreans have made it the tone of all their harmonic intervals."

352	381	326	439	468	413	274	303	248	613	642	587	700	729	674	535	564	509	118	147	92	205	234	179	40	69	14
327	353	379	414	440	466	249	275	301	588	614	640	675	701	727	510	536	562	93	119	145	180	206	232	15	41	67
380	325	354	467	412	441	302	247	276	641	586	615	728	673	702	563	508	537	146	91	120	233	178	207	68	13	42
277	306	251	355	384	329	433	462	407	538	567	512	616	645	590	694	723	668	43	72	17	121	150	95	199	228	173
252	278	304	330	356	382	408	434	460	513	539	565	591	617	643	669	695	721	18	44	70	96	122	148	174	200	226
305	250	279	383	328	357	461	406	435	566	511	540	644	589	618	722	667	696	71	16	45	149	94	123	227	172	201
436	465	410	271	300	245	358	387	332	697	726	671	532	561	506	619	648	593	202	231	176	37	66	11	124	153	98
411	437	463	246	272	298	333	359	385	672	698	724	507	533	559	594	620	646	177	203	229	12	38	64	99	125	151
464	409	438	299	244	273	386	331	360	725	670	699	560	505	534	647	592	621	230	175	204	65	10	39	152	97	126
127	156	101	214	243	188	49	78	23	361	390	335	448	477	422	283	312	257	595	624	569	682	711	656	517	546	491
102	128	154	189	215	241	24	50	76	336	362	388	423	449	475	258	284	310	570	596	622	657	683	709	492	518	544
155	100	129	242	187	216	77	22	51	389	334	363	476	421	450	311	256	285	623	568	597	710	655	684	545	490	519
52	81	26	130	159	104	208	237	182	286	315	260	364	393	338	442	471	416	520	549	494	598	627	572	676	705	650
27	53	79	105	131	157	183	209	235	261	287	313	339	365	391	417	443	469	495	521	547	573	599	625	651	677	703
80	25	54	158	103	132	236	181	210	314	259	288	392	337	366	470	415	444	548	493	522	626	571	600	704	649	678
211	240	185	46	75	20	133	162	107	445	474	419	280	309	254	367	396	341	679	708	653	514	543	488	601	630	575
186	212	238	21	47	73	108	134	160	420	446	472	255	281	307	342	368	394	654	680	706	489	515	541	576	602	628
239	184	213	74	19	48	161	106	135	473	418	447	308	253	282	395	340	369	707	652	681	542	487	516	629	574	603
604	633	578	691	720	665	526	555	500	109	138	83	196	225	170	31	60	5	370	399	344	457	486	431	292	321	266
579	605	631	666	692	718	501	527	553	84	110	136	171	197	223	6	32	58	345	371	397	432	458	484	267	293	319
632	577	606	719	664	693	554	499	528	137	82	111	224	169	198	59	4	33	398	343	372	485	430	459	320	265	294
529	558	503	607	636	581	685	714	659	34	63	8	112	141	86	190	219	164	295	324	269	373	402	347	451	480	425
504	530	556	582	608	634	660	686	712	9	35	61	87	113	139	165	191	217	270	296	322	348	374	400	426	452	478
557	502	531	635	580	609	713	658	687	62	7	36	140	85	114	218	163	192	323	268	297	401	346	375	479	424	453
688	717	662	523	552	497	610	639	584	193	222	167	28	57	2	115	144	89	454	483	428	289	318	263	376	405	350
663	689	715	498	524	550	585	611	637	168	194	220	3	29	55	90	116	142	429	455	481	264	290	316	351	377	403
716	661	690	551	496	525	638	583	612	221	166	195	56	1	30	143	88	117	482	427	456	317	262	291	404	349	378

Fig. 256.

This passage indicates sufficiently the supreme importance of the number 27.

If we construct a magic square 27×27 upon the plan of a checker-board—arranging the numbers 1 to 729 first in numerical order, then shifting the 9 largest squares (9×9) into the positions indicated in the familiar 3×3 square, repeating the process with the subdivisions of the 9×9 squares and so on down—we will arrive at the following combination.[1]

It will be noted that we have 365 white squares or days and 364 dark squares or nights—a veritable "checkerboard of nights and days." The number 365, the days of the solar year, very appropriately occupies the centre of the system. The columns, horizontals, and diagonals of the central square 3×3 foot up 1095, or the days of a 3 year period, those of the larger center square 9×9 foot up 3285 the days of a 9 year period, while those of the entire combination 27×27 foot up 9855,[2] the days of a 27 year period,—in other words, periods of years corresponding to the Tetractys 1, 3, 9, 27. We may with safety borrow the language of Plato and say that the above arrangement of numbers "is concerned with days and nights and months and years."

The interpretation of the other passage referred to in the "Republic"—the finding of the number of the State—(Book VIII, § 546) has been a subject of the greatest speculation and by consulting the various editions of Plato it will be found that scarcely any two critics agree upon a solution.[3] As Jowett remarks, it is a puzzle almost as great as that of the Beast in the Book of Revelation. Unfortunately we have no starting-point from which to begin our calculations; this and the very uncertain meanings of many of the Greek terms have caused many commentators to give up the solution of the problem in sheer despair. Aristotle, who was a hearer of Plato's, writes as if having a full knowledge of the mystery; Cicero, however, was unable to solve the riddle and his sentiment became voiced in the proverb *numeris Platonicis nihil obscurius*.

By taking a hint from our magic square and starting with the

---

[1] This method of constructing composite magic squares is, so far as I know, original with the writer. It bears some resemblance to the method of Schubert (see "Compound Magic Squares," p. 44); the numbers of each square, however, increase in periods of threes instead of by sequence.

[2] Not only the perpendiculars, horizontals, and diagonals of this large square foot up 9855, but there are an almost indefinite number of zig-zag lines, which give the same footing.

[3] Schleiermacher, Donaldson, and Schneider suggest 216, and much may be said in favor of this number. Jowett gives 8000 as the possible solution. Others suggest 951, 5040, 17,500, 1728, 10,000, etc.

number 27, I believe we may arrive at as good a solution of the problem as any that I have seen suggested. The following interpretation of the Greek terms is offered.

αὐξήσεις δυνάμεναί τε καὶ δυναστευόμεναι	the square of the number times its root,	$27^2 \times \sqrt[3]{27} = 2187$
ρεῖς ἀποστάσεις	increased by thrice the first terms (of the Tetractys)	$(1+2+3+4+9) \times 3 = 57$
τέτταρας δὲ ὅρους λαβοῦσαι	and four times the whole series	$(1+2+3+4+9+8+27) \times 4 = 216$
ὁμοιούντων τε καὶ ἀνομοιουντων καὶ ἀυξόντων καὶ φθινότνων	of numbers unlike yet bearing the same ratio whether increasing or decreasing	
	(i. e. 1:2::4:8 or 8:4::2:1 It may also refer to the ascending and descending figures of the triangle. 8, 4, 2, 1, 3, 9, 27)	
πάντα προσήγορα καὶ ῥητὰ πρὸς ἄλληλα ἀπέφηναν	makes the sum commensurable and expressible in all its parts.	sum= 2460
	(i e. 2460 is easily divisible by 1, 2, 3, 4, 5, 6, 10, 12 etc.)	
ὧν ἐπίτριτος πυθμὴν,	this sum increased by ⅓	$2460 \times 1\frac{1}{3} = 3280$
πεμπάδι συζυγεὶς	and adding 5	$3280 + 5 = 3285$
τρὶς αὐξηθείς	is multiplied by 3	$3285 \times 3 = 9855$

This solution of the problem, 9855, it will be noted, brings us again but by a different route to the magic number of our large square. The second part of the passage contains a description of the number by which the above calculation may be verified.

δύο ἁρμονίας παρέχεται	(the number) yields two harmonic parts,		
τὴν μὲν ἴσην ἰσάκις,	one of which is a square	$3 \times 3 = 9$	
ἑκατὸν τοσαυτάκις,	multiplied by 100:		$9 \times 100 = 900$
τὴν δὲ ἰσομήκη μὲν,	the other has one side equal to the square		3
τῇ προμήκει δὲ,	and the other oblong		$3 \times 2985 = 8955$
			sum= 9855

The remainder of the passage describes the length of the oblong which we have shown above to be 2985:

ἑκατὸν μὲν ἀριθμῶν ἀπὸ διαμέτρων πεμπάδος,	(the oblong) is 100 times the side of a rectangle having diagonals of 5. (i. e. having sides of 3 and 4.)	$100 \times 3 = 300$
οητῶν δεομένων ἑνὸς ἑκάστων,	less of one each of the expressible parts, i. e. 4 and 5	
ἀῤῥήτων δὲ δυεῖν,	and 2 of the inexpressible	$300-(5+4+3+3) = 285$
ἑκατὸν δὲ κύβων τριάδος	plus 100 times the cube of 3	$(3)^3 \times 100 = 2700$
		sum $= 2985$

Plato states that the number of the State "represents a geometrical figure which has control over the good and evil of births. For when your guardians are ignorant of the right seasons and unite bride and bridegroom out of due time, the children will not be goodly and happy." The number 9855, expressing a period of 27 years, might thus represent the dividing line between the ages when men and women should begin to bear children to the State,—20-27 years for women, 27-34 years for men. (See also "Republic," Book V, § 460). Aristotle in his "Politics" (V, 12. 8) says in reference to the number of the State that when the progression of number is increased by $^1/_3$ and 5 is added, 2 harmonies are produced giving a solid diagram. This, as may be seen from our analysis of the first part of the passage, may have reference to the number 3285, which, being represented by $3^2 \times 365$, may be said to have the dimensions of a solid.

In his "Reflections on Magic Squares" Dr. Carus gives some very striking examples of the relationship between magic squares and the musical figures of Chladni. I would like to touch before concluding upon a closely related subject and show certain connections which exist between the magic square, which we have constructed, and the numbers of the Pythagorean harmonic scale. This scale had, however, more than a musical significance among the

Greek philosophers; it was extended to comprehend the harmony of planetary movements and above all else to represent the manner in which the "soul of the universe" was composed. It is especially in the latter sense that Plato employs the scale in his "Timæus."

In a treatise by Timæus the Locrian upon the "Soul of the World and Nature," we find the following passage: "Now all these proportions are combined harmonically according to numbers, which proportions the demiurge has divided according to a scale scientifically, so that a person is not ignorant of what things and by what means the soul is combined; which the deity has not ranked after the substance of the body. . . ., but he made it older by taking the *first of unities* which is 384. Now of these the first being assumed it is easy to reckon the double and triple; and all the terms, with their complements and eights must amount to 114,695." (Translation by Burge.)

Plato's account of the combination of the soul is very similar to the above, though he seems to have selected 192, (384/2) for the first number. Plutarch in his commentary makes no mention of Timæus, but states that Crantor[4] was the first to select 384, for the reason that it represented the product of $8^2 \times 6$, and is the lowest number which can be taken for the increase by eighths without leaving fractions. Another very possible reason, which I have not seen mentioned, is that 384 is the harmonic ratio of $27^2/2$ or 364.5, a number which expresses very closely the days of the year.

$$243:256::364.5:384.$$

The proportion $243:256(3^5:4^4)$ was employed by the Pythagoreans to mark the ratio[5] which two unequal semitones of the harmonic scale bear to one another.

Batteux has calculated the 36 terms of the Pythagorean scale starting with 384 and his series must be considered correct, for it fulfils the conditions specified by Timæus,—the numbers all footing

---

[4] Crantor lived nearly 100 years after Timæus the Locrian. The treatise upon the "Soul of the World and Nature," which bears the latter's name probably belongs to a much later period.

[5] For further references to this ratio see Plato's "Timæus," § 36, and Plutarch's "Procreation of the Soul," § 18.

up 114,695: A few of the numbers of this harmonic scale marking the "first unity" and several of the semitones will be given.

1st octave	E	384
	C	486
	F	729
2nd octave	C	972
	F	1458
3rd octave	C	1944
	B flat	2187
4th octave	B flat	4374

(For Batteux's full series and method of calculation the reader is referred to Burge's translation of Plato Vol. VI. p. 171).

By referring to our magic square it will be noted that the *first* of *unities*," 384, constitutes the magic number of the small 3×3 square beginning with the number 100. If we arrange the magic numbers of the 81 squares (3×3) in the order of their magnitudes we find that they fall into 9 series of 9 numbers, each series beginning as follows:

I	II	III	IV	V	VI	VII	VIII	IX
87	330	573	816	1059	1302	1545	1788	2031

The intervals between these series are worthy of note.

INTERVALS.

Between	I and II	243	the first member of the ratio 243:256.
"	I " III	486	C of the 1st octave
"	I " IV	729	F " " 1st "
"	I " V	972	C " " 2nd "
"	I " VII	1458	F " " 2nd "
"	I " IX	1944	C " " 3rd "

If we arrange the magic numbers of the large squares (9×9) in the same way, it will be found that they fall into 3 series of 3 numbers, each series beginning

I	II	III
1017	3204	5391

Interval between I and II = 2187 B-flat of the 3rd octave.
" " I " III = 4374 B-flat " " 4th "

Numerous other instances might be given of the very intimate connection between magic squares and various Pythagorean numbers, but these must be left for the curious-minded to develop for themselves. Such connections as we have noted are no doubt in some respects purely accidental, being due to the *intrinsic harmony of numbers* and therefore not implying a knowledge by the ancients of magic squares as we now know them. The harmonic arrangement by the Greeks of numbers in geometrical forms both plane and solid may, however, be accepted, and Plato's descriptions of various numbers obscure and meaningless as they were to succeeding generations, may have been easily comprehended by his hearers when illustrated by a mathematical diagram or model.[6]

Differences between the methods of notation in ancient and modern times have necessarily produced differences in the conception of numerical relations. The expression of numbers among the Greeks by letters of the alphabet was what led to the idea that every name must have a numerical attribute, but the connection of the letters of the name was in many cases lost, the number being regarded as a pure attribute of the object itself. A similar confusion of symbols arose in the representation of various concepts by geometrical forms, such as the five letters of ΥΓΕΙΑ and the symbolization of health by the Pythagoreans under the form of the pentalpha or five-pointed star.

It was the great defect of the Greek schools that in their search for truth, methods of experimental research were not cultivated. Plato in his "Republic" (Book VII, § 530-531) ridicules the empiricists, who sought knowledge by studying the stars or by comparing the sounds of musical strings, and insists that no value is to be placed upon the testimony of the senses. "Let the heavens alone and train the intellect" is his constant advice.

If the examples set by Pythagoras in acoustics and by Archimedes in statics had been generally followed by the Greek philosophers, our knowledge of natural phenomena might have been advanced a thousand years. But as it happened there came to prevail but one idea intensified by both Plato and Aristotle, and handed down through the scholastics even to the present time, that knowl-

---

[6] The description of the number of the State in the "Republic" and that of the Soul in the "Timæus" render such a mode of representation almost necessary. Plutarch ("Procreation of Soul," § 12) gives an illustration of an harmonic diagram 5×7 containing 35 small squares "which comprehends in its subdivisions all the proportions of the first concords of music."

edge was to be sought for only from within. Hence came the flood of idle speculations which characterized the later Pythagorean and Platonic schools and which eventually undermined the structure of ancient philosophy. But beneath the abstractions of these schools one can discover a strong undercurrent of truth. Many Pythagoreans understood by number that which is now termed natural law. Such undoubtedly was the meaning of Philolaus when he wrote "Number is the bond of the eternal continuance of things," a sentiment which the modern physicist could not express more fittingly.

As the first study of importance for the youth of his "Republic" Plato selected the science of numbers; he chose as the second geometry and as the third astronomy, but the point which he emphasized above all was that these and all other sciences should be studied in their "mutual relationships that we may learn the nature of the bond which unites them." "For only then," he states, "will a pursuit of them have a value for our object, and the labor, which might otherwise prove fruitless, be well bestowed." Noble utterance! and how much greater need of this at the present day with our complexity of sciences and tendency towards narrow specialization.

In the spirit of the great master whom we have just quoted we may compare the physical universe to an immense magic square. Isolated investigators in different areas have discovered here and there a few seemingly restricted laws, and paying no regard to the territory beyond their confines, are as yet oblivious of the great pervading and unifying Bond which connects the scattered parts and binds them into one harmonious system. Omar, the astronomer-poet, may have had such a thought in mind, when he wrote:

> "Yes; and a single Alif were the clue—
> Could you but find it—to the treasure-house
> And peradventure to the Master too;
>
> . . . .
>
> Whose secret presence, through creation's veins
> Running quicksilverlike eludes your pains;" etc.

When Plato's advice is followed and the "mutual relationships between our sciences" are understood we may perchance find this clue, and having found it be surprised to discover as great a simplicity underlying the whole fabric of natural phenomena as exists in the construction of a magic square.

C. A. B.

MR. BROWNE'S SQUARE AND LUSUS NUMERORUM.

The 27×27 square of Mr. C. A. Browne, Jr. is interesting because, in additon to its arithmetical qualities commonly possessed by magic squares, it represents some ulterior significance of our calendar system referring to the days of the month as well as the days of the year and cycles of years. It is wonderful, and at first sight mystifying, to observe how the course of nature reflects even to intricate details the intrinsic harmony of mathematical relations; and yet when we consider that nature and pure thought are simply the result of conditions first laid down and then consistently carried out in definite functions of a distinct and stable character, we will no longer be puzzled but understand why science is possible, why man's reason contains the clue to many problems of nature and, generally speaking, why reason with all its wealth of *a priori* thoughts can develop at all in a world that at first sight seems to be a mere chaos of particular facts. The purely formal relations of mathematics, materially considered mere nonentities, constitute the bond of union which encompasses the universe, stars as well as motes, the motions of the Milky Way not less than the minute combinations of chemical atoms, and also the construction of pure thought in man's mind.

Mr. Browne's square is of great interest to Greek scholars because it throws light on an obscure passage in Plato's Republic, referring to a magic square the center of which is 365, the number of days in a year.

The construction of Mr. Browne's square is based upon the simplest square of odd numbers which is 3×3. But it becomes somewhat complicated by being extended to three in the third power which is 27. Odd magic squares, as we have seen, are built up by a progression in staircase fashion, but since those numbers that fall outside the square have to be transferred to their corresponding places inside, the first and last staircases are changed into the knight's move of the chessboard, and only the middle one retains its original staircase form. We must construct the square so that the central figure, which in a 3×3 square is 5, must always fall in the central cell. Accordingly, we must start the square beginning with figure 1 outside of the square in any middle cell immediately bordering upon it, which gives four starting-points from which we may either proceed from the right or the left, either upwards or downwards which yields eight possibilities of the 3×3 square. For the construction of his 27×27 square, Mr. Browne might have taken any of these eight possibilities as his pattern.

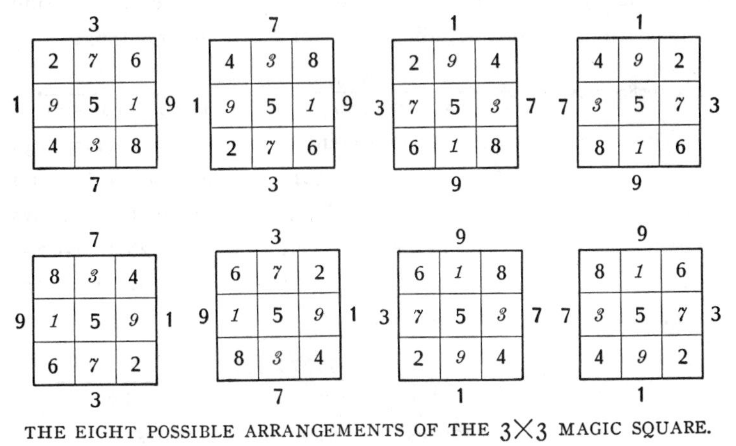

THE EIGHT POSSIBLE ARRANGEMENTS OF THE 3×3 MAGIC SQUARE.
Fig. 257.

He selected the one starting on the top of the square and moving toward the right, and thus he always follows the peculiar arrangement of this particular square. It is the fourth of the eight arrangements shown in Fig. 274. Any one who will take the trouble to trace the regular succession of Mr. Browne's square will find that it is a constant repetition of the knight's move, the staircase move and again a knight's move on a small scale of 3×3 which is repeated on a larger scale 9×9, thus leading to the wonderful regularity which, according to Mr. Browne's interpretation of Plato, astonished the sages of ancient Greece.

Any one who discovers at random some magic square with its

immanent harmony of numbers, is naturally impressed by its apparent occult power, and so it happens that they were deemed supernatural and have been called "magic." They seem to be the product of some secret intelligence and to contain a message of ulterior meaning. But if we have the key to their regularity we know that the harmony that pervades them is necessary and intrinsic.

Nor is the regularity limited to magic squares. There are other number combinations which exhibit surprising qualities, and I will here select a few striking cases.

If we write down all the nine figures in ascending and descending order we have a number which is equal to the square of a number consisting of the figure 9 repeated 9 times, divided by the sum of an ascending and descending series of all the figures thus:

$$12345678987654321 = \frac{999999999 \times 999999999}{1+2+3+4+5+6+7+8+9+8+7+6+5+4+3+2+1}.$$

The secret of this mysterious coincidence is that $11 \times 11 = 121$; $111 \times 111 = 12321$; $1111 \times 1111 = 1234321$, etc., and a sum of an ascending and descending series which starts with 1 is always equal to the square of its highest number. $1+2+1=2\times 2$; $1+2+3+4+3+2+1=4\times 4$, etc., which we will illustrate by one more instance of the same kind, as follows:

$$1234567654321 = \frac{7777777 \times 7777777}{1+2+3+4+5+6+7+6+5+4+3+2+1}.$$

There are more instances of numerical regularities.

All numbers consisting of six equal figures are divisible by 7, and also, as a matter of course, by 3 and 11, as indicated in the following list:

111111 : 7 = 15873
222222 : 7 = 31746
333333 : 7 = 47619
444444 : 7 = 63492
555555 : 7 = 79365
666666 : 7 = 95238
777777 : 7 = 111111
888888 : 7 = 126984
999999 : 7 = 142857

Finally we will offer two more strange coincidences of a *lusus numerorum.*

$$0 \times 9 + 1 = 1$$
$$1 \times 9 + 2 = 11$$
$$12 \times 9 + 3 = 111$$
$$123 \times 9 + 4 = 1111$$
$$1234 \times 9 + 5 = 11111$$
$$12345 \times 9 + 6 = 111111$$
$$123456 \times 9 + 7 = 1111111$$
$$1234567 \times 9 + 8 = 11111111$$
$$12345678 \times 9 + 9 = 111111111$$
$$123456789 \times 9 + 10 = 1111111111.$$

$$1 \times 8 + 1 = 9$$
$$12 \times 8 + 2 = 98$$
$$123 \times 8 + 3 = 987$$
$$1234 \times 8 + 4 = 9876$$
$$12345 \times 8 + 5 = 98765$$
$$123456 \times 8 + 6 = 987654$$
$$1234567 \times 8 + 7 = 9876543$$
$$12345678 \times 8 + 8 = 98765432$$
$$123456789 \times 8 + 9 = 987654321.$$

No wonder that such strange regularities impress the human mind. A man who knows only the externality of these results will naturally be inclined toward occultism. The world of numbers as much as the actual universe is full of regularities which can be reduced to definite rules and laws giving us a key that will unlock their mysteries and enable us to predict certain results under definite conditions. Here is the key to the significance of the *a priori.*

Mathematics is a purely mental construction, but its composition is not arbitrary. On the contrary it is tracing the results of our own doings and taking the consequences of the conditions we have created. Though the scope of our imagination with all its possibilities be infinite, the results of our construction are definitely determined as soon as we have laid their foundation, and the actual world is simply one realization of the infinite potentialities of being. Its regularities can be unraveled as surely as the harmonic relations of a magic square.

Facts are just as much determined as our thoughts, and if we can but gain a clue to their formation we can solve the problem of their nature, and are enabled to predict their occurrence and sometimes even to adapt them to our own needs and purposes.

A study of magic squares may have no practical application, but an acquaintance with them will certainly prove useful, if it were merely to gain an insight into the fabric of regularities of any kind.

P. C.

CHAPTER VII.

SOME CURIOUS MAGIC SQUARES AND COMBINATIONS.

MANY curious and interesting magic squares and combinations have been devised by the ingenious, a selection of which will be given in the following pages, some of the examples being here presented for the first time in print.

The curious irregularities of the 6×6 magic squares were referred to in the first chapter, and many unsuccessful attempts have been made to construct regular squares of this order. An interesting

16	14	33	34	8	6
13	15	36	35	5	7
12	10	17	18	28	26
9	11	20	19	25	27
32	30	1	2	24	22
29	31	4	3	21	23

Fig. 258.

32	31	1	3	21	23
29	30	4	2	24	22
9	11	20	19	25	27
12	10	17	18	28	26
16	15	33	35	5	7
13	14	36	34	8	6

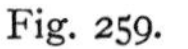

Fig. 259.

6×6 square is illustrated in a work entitled *Games, Ancient and Oriental* by Edward Falkener,* and is here reproduced in Fig. 258. It will be seen however that the two corner diagonals of this square do not sum 111, but by a transposition of the figures this imperfection is corrected in Fig. 259. Other transpositions are also possible which will effect the same result. The peculiarity of this

* Published by Longmans Green & Co., London and New York, 1892.

square consists in its being divided into nine 2×2 squares in each of the four subdivisions of which the numbers follow in arithmetical sequence, and the 2×2 squares are arranged in the order of a 3×3 magic square, according to the progressive value of the numbers 1 to 36. The construction of this 6×6 square is regular only in relation to the totals of the 2×2 squares, as shown in Fig. 260.

Fig. 261 is a remarkable 8×8 square which is given on page 300 of the above mentioned book, and which is presented by Mr. Falkener as "the most perfect magic square of 8×8 that can be constructed." Some of its properties are as follows:

1. The whole is a magic square of 8×8.
2. Each quarter is an associated 4×4 square.
3. The sixteen 2×2 subsquares have a constant summation of 130.

122	10	90
42	74	106
58	138	26

Fig. 260.

4. Each quarter contains four 3×3 squares the corner numbers of which sum 130.
5. Any 5×5 square which is contained within the 8×8 square has its corner numbers in arithmetical sequence.

A very interesting class of squares is referred to in the same work on pages 337-338 and 339 as follows:

"The Rev. A. H. Frost, while a missionary for many years in India, of the Church Missionary Society, interested himself in his leisure hours in the study of these squares and cubes, and in the articles which he published on the subject gave them the name of 'Nasik' from the town in which he resided. He has also deposited 'Nasik' cubes in the South Kensington Museum (London) and he has a vast mass of unpublished materials of an exhaustive nature most carefully worked out.

"Mr. Kesson has also treated the same subject in a different way and more popular form in the *Queen.** He gives them the very appropriate name of Caïssan Squares, a name given to these squares, he says, by Sir William Jones.

"The proper name, however, for such squares should rather be 'Indian,' for not only have the Brahmins been known to be great adepts in the formation of such squares from time immemorial, not only does Mr. Frost give his an Indian name, but one of these squares is represented over the gate of Gwalior, while the natives of

1	59	56	14	2	60	53	15
46	24	27	33	47	21	28	34
32	38	41	19	31	37	44	18
51	9	6	64	50	12	5	63
3	57	54	16	4	58	55	13
48	22	25	35	45	23	26	36
30	40	43	17	29	39	42	20
49	11	8	62	52	10	7	61

Fig. 261.

1	58	3	60	8	63	6	61
16	55	14	53	9	50	11	52
17	42	19	44	24	47	22	45
32	39	30	37	25	34	27	36
57	2	59	4	64	7	62	5
56	15	54	13	49	10	51	12
41	18	43	20	48	23	46	21
40	31	38	29	33	26	35	28

Fig. 262.

India wear them as amulets, and La Loubère, who wrote in 1693, expressly calls them 'Indian Squares.'

"In these Indian squares it is necessary not merely that the summation of the rows, columns and diagonals should be alike, but that *the numbers of such squares should be so harmoniously balanced that the summation of any eight numbers in one direction as in the moves of a bishop or a knight should also be alike.*"

An example of one of these squares is given in Fig. 262 and examination will show it to be of the same order as the "Jaina" square described by Dr. Carus in a previous chapter (pp. 125 ff.), but having enlarged characteristics consequent on its increase in size. It will be seen that the extraordinary properties as quoted

* Published in London, England.

above in italics exist in this square, so that starting from any cell in the square, with a few exceptions, any eight numbers that are covered by eight consecutive similar moves will sum 260. In addition to this the numbers in every 2×2 square, whether taken within the square or constructively, sum 130; thus, $1 + 58 + 16 + 55 = 130$ and $1 + 16 + 61 + 52 = 130$, also $1 + 58 + 40 + 31 = 130$ etc. Furthermore, (as in the Jaina square) the properties of this square will necessarily remain unchanged if columns are taken from one side and put on the other, or if they are removed from the top to the bottom, or *vice versa,* it being a perfectly continuous square in every direction.

The wonderful symmetry of this square naturally invites attention to the method of its construction, which is very simple, as may

1		3		8		6	
16		14		9		11	
17		19		24		22	
32		30		25		27	
	2		4		7		5
	15		13		10		12
	18		20		23		21
	31		29		26		28

Fig. 263.

1	14	4	15
8	11	5	10
13	2	16	3
12	7	9	6

Fig. 264.

be seen by following the natural sequence of the numbers 1 to 32 in Fig. 263 which shows the disposition of the numbers of the first half of the series. The second half is simply a complementary repetition of the first half. The numbers of this square are arranged symmetrically in relation to similarly located cells in diagonally opposite quarters, thus, (referring to Fig. 262) $1 + 64 = 65$ and $4 + 61 = 65$ etc. This feature permits the completion of Fig. 263 by filling in the vacant cells at random with their respective differences between 65 and the various numbers already entered.

Fig. 264 shows a 4×4 square constructed by the same method and having similar properties, with natural limitations due to its small size. This square strikingly resembles the Jaina square as modified by Dr. Carus (see Fig. 222, p. 127) the numbers and arrangement of same in the two corner diagonal columns being identical in both squares, while the other numbers are differently located.

Fig. 265 is an original 8 × 8 square contributed by Mr. L. S. Frierson, which combines to a limited extent some of the curious characteristics of the Franklin and the Jaina or Indian squares. It possesses the following properties:

1. Considered as a whole it is an 8 × 8 magic square.
2. Each quarter is in itself a magic square.
3. The four central horizontal columns make two 4 × 4 magic squares.

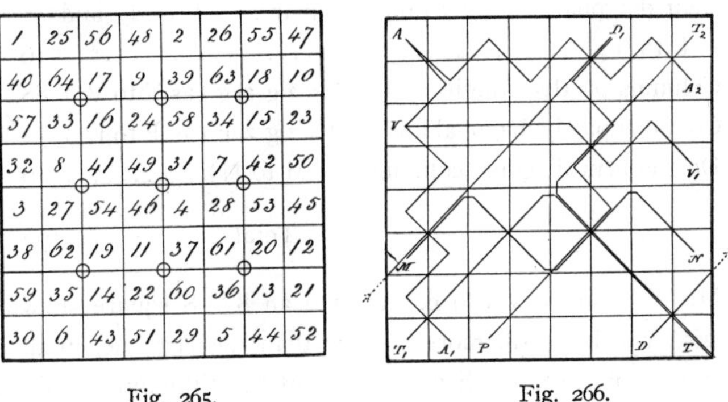

1	25	56	48	2	26	55	47
40	64	17	9	39	63	18	10
57	33	16	24	58	34	15	23
32	8	41	49	31	7	42	50
3	27	54	46	4	28	53	45
38	62	19	11	37	61	20	12
59	35	14	22	60	36	13	21
30	6	43	51	29	5	44	52

Fig. 265.

Fig. 266.

4. It contains twenty-five 2 × 2 squares, each having a constant summation of 130.
5. It also contains twenty-four 3 × 3 squares, the four corner cells of which have a constant summation of 130.
6. *Any* 4 × 4 square has a constant summation of 520.
7. In any 5 × 5 square the four corner cells contain numbers in arithmetical sequence.
8. Any rectangular parallelogram which is concentric with any of the nine subcenters contains numbers in its corner cells that will sum 130, excepting when the diagonals of any of the four subsquares of 4 × 4 form one side of the parallelogram.

9. Any octagon of two cells on a side, that is concentric with any of the nine subcenters will have a constant summation of 260.
10. No less than 192 columns of eight consecutive numbers may be found having the constant summation of 260 as follows (see Fig. 266):

Horizontal columns	8
Perpendicular columns	8
Perpendicular zig-zags (A to A_1)	8
Horizontal zig-zags (A to A_2)	8
Corner diagonals	2
Constructive diagonals (D to D_1)	6
Bent diagonals (as in Franklin squares) (T to T_1 and T to T_2)	16
Columns partly straight and partly zig-zag (as V to V_1)	88
Columns partly diagonal and partly zig-zag (as P to D_1)	32
Double bent diagonal columns (as M to N)	16
Total	192

Mr. Frierson has also constructed an 8×8 square shown in Fig. 267, which is still more curious than the last one, in that it perfectly combines the salient features of the Franklin and the Indian squares, viz., the *bent* and the *continuous diagonals,* besides exhibiting many other interesting properties, some of which may be mentioned as follows:

1. *Any* 2×2 square has a constant summation of 130, with four exceptions.
2. The corner cells of any 3×3 square which lies wholly to the right or left of the axis AB sum 130.
3. The corner cells of any 2×4, 2×6 or 2×8 rectangle perpendicular to AB and symmetrical therewith sum 130.
4. The corner cells of any 2×7 or 3×6 rectangle diagonal to AB sum 130, as $12 + 50 + 45 + 23 = 130$, $49 + 16 + 19 + 46 = 130$ etc., etc..

5. The corner cells of any 5×5 square contain numbers in arithmetical progression.
6. Any constructive diagonal column sums 260.
7. Any *bent* diagonal sums 260.
8. Any reflected diagonal sums 260.

(NOTE: Reflected diagonals are shown in dotted lines on Fig. 267.)

By dividing this square into quarters, and subdividing each quarter into four 2×2 squares, the numbers will be found symmetrically arranged in relation to cells that are similarly located in diagonally opposite 2×2 squares in each quarter, thus: $64 + 1 = 65$, $57 + 8 = 65$ etc.

A

64	57	4	5	56	49	12	13
3	6	63	58	11	14	55	50
61	60	1	8	53	52	9	16
2	7	62	59	10	15	54	51
48	41	20	21	40	33	28	29
19	22	47	42	27	30	39	34
45	44	17	24	37	36	25	32
18	23	46	43	26	31	38	35

B

Fig. 267.

1	61	60	8	9	53	52	16
2	62	59	7	10	54	51	15
63	3	6	58	55	11	14	50
64	4	5	57	56	12	13	49
24	44	45	17	32	36	37	25
23	43	46	18	31	35	38	26
42	22	19	47	34	30	27	39
41	21	20	48	33	29	28	40

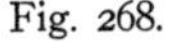

Fig. 268.

Another 8×8 square by Mr. Frierson is given in Fig. 268 which is alike remarkable for its constructive simplicity and for its curious properties. Like Fig. 267 this square combines the principal features of the Indian and the Franklin squares in its *bent* and *continuous* diagonal columns.

To render its structure graphically plain, the numbers 1 to 32 are written within circles. The numbers in the complete square are arranged symmetrically in relation to the two heavy horizontal lines so that when the numbers in *the first half of the series are entered,* the remaining numbers may be filled in at random as explained in connection with the 8×8 Indian square (Fig. 263).

Two other examples of the Frierson squares showing interesting constructive features are given in Figs. 269 and 270. The scheme followed in these squares may also be employed in making magic rectangles, two examples of which are given in Figs. 271 and 272. In Fig. 272 the numbers are arranged in the following series before they are entered in the rectangle:

1 . 5 . 9 . 13 . 17 . 21 . 25 . 29
2 . 6 . 10 . 14 . 18 . 22 . 26 . 30
3 . 7 . 11 . 15 . 19 . 23 . 27 . 31
4 . 8 . 12 . 16 . 20 . 24 . 28 . 32

1	2	44	43	21	22	64	63
3	4	42	41	23	24	62	61
56	55	52	51	13	14	9	10
54	53	50	49	15	16	11	12
25	26	5	6	60	59	40	39
27	28	7	8	58	57	38	37
48	47	29	30	36	35	17	18
46	45	31	32	34	33	19	20

Fig. 269.

1	32	40	57	56	41	17	16
2	31	39	58	55	42	18	15
3	30	38	59	54	43	19	14
4	29	37	60	53	44	20	13
61	36	28	5	12	21	45	52
62	35	27	6	11	22	46	51
63	34	26	7	10	23	47	50
64	33	25	8	9	24	48	49

Fig. 270.

1	4	31	30
5	8	27	26
9	12	23	22
13	16	19	18
20	17	14	15
24	21	10	11
28	25	6	7
32	29	2	3

Fig. 271.

1	19	18	12
2	20	17	11
3	21	16	10
22	4	9	15
23	5	8	14
24	6	7	13

Fig. 272.

Figs. 273 and 274 are ingenious combinations of 4×4 squares also devised by Mr. Frierson. Fig. 273 is a magic cross which possesses many unique features. It is said to contain the almost incredible number of 160,144 different columns of twenty-one numbers which sum 1471.

Some of the properties found in the magic pentagram Fig. 274 may be stated as follows:

Each 4 × 4 rhombus is perfectly magic, with summations of 162. It therefore follows that from any point to the next the num-

Fig. 273.

bers sum 324, and also that every bent row of eight numbers which is parallel with the rows from point to point sums 324.

In each 4 × 4 rhombus there are five others of 2 × 2 whose numbers sum 162, also four others of 3 × 3, the corner numbers of which sum 162.

In each 4×4 rhombus, every number ends with one of two numbers, viz., 0 and 1, 2 and 9, 3 and 8, 4 and 7, 5 and 6.

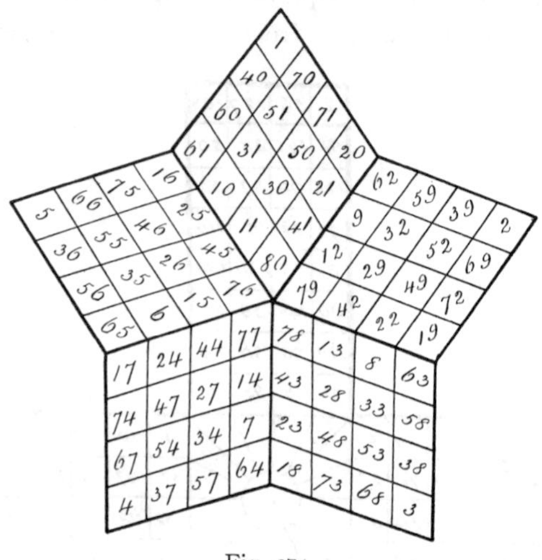

Fig. 274.

Modifications of the concentric magic squares (described in the first chapter) have been devised by Mr. Frierson, two examples of which are shown in Figs. 275 and 276.

11	24	25	14	34	3
18	21	20	15	5	32
22	17	16	19	28	9
23	12	13	26	6	31
1	35	27	33	8	7
36	2	10	4	30	29

Fig. 275.

71	1	51	32	50	2	80	3	79
21	41	61	56	26	13	69	25	57
31	81	11	20	62	65	17	63	19
34	40	60	43	28	64	18	55	27
48	42	22	54	39	75	7	10	72
33	53	15	68	16	44	58	77	5
49	29	67	14	66	24	38	59	23
76	4	70	73	8	37	36	30	35
6	78	12	9	74	45	46	47	52

Fig. 276.

A 5×5 magic square, curiously quartered with four 2×3 magic rectangles, devised by Dr. Planck, is shown in Fig. 277.

The interesting 9×9 magic, Fig. 278, was made by Mr. Frierson. It possesses the following properties:

1. All odd and even numbers are segregated.
2. Any pair of numbers located equally above and below the horizontal axis end in the same integer.
3. The sum of any pair of numbers located equally right and left of the perpendicular axis ends with 2.
4. The twenty-five odd numbers within the circles make a balanced 5×5 square.

S of square = 65

S of rectangles = $\begin{cases} 39 \\ 26 \end{cases}$

25	1	23	6	10
12	14	3	20	16
2	24	13	8	18
11	7	21	9	17
15	19	5	22	4

Fig. 277.

S of 9^2 = 360
S of 5^2 = 205
S of 4^2 = 165

42	58	68	64	1	8	44	34	50
2	66	54	45	11	77	78	26	10
12	6	79	53	21	69	63	46	20
52	7	35	23	31	39	67	55	60
73	65	57	49	41	33	25	17	9
22	27	15	43	51	59	47	75	30
62	36	19	13	61	29	3	76	70
72	56	4	5	71	37	28	16	80
32	48	38	74	81	18	14	24	40

Fig. 278.

S = 2126

539	525	526	536
528	534	533	531
532	530	529	535
527	537	538	524

Fig. 279.

S = 6200

1328	1342	1351	1335	1344
1350	1334	1343	1332	1341
1347	1331	1340	1349	1333
1339	1348	1337	1346	1330
1336	1345	1329	1338	1352

Fig. 280.

5. The sixteen odd numbers between the circles make a balanced 4×4 square.
6. The great square is associated.

It is purposed to treat of magic squares composed exclusively of *prime numbers* in another book. Mr. Chas. D. Shuldham has contributed original 4×4 and 5×5 magics, having the lowest possible summations when *made exclusively of consecutive composite numbers,* as shown in Figs. 279 and 280.

There is nothing curious in the *construction* of these squares, as in this particular they follow the same rules that are applied to all squares that are made from any consecutive arithmetical series. Thus in the square of order 4 given in Fig. 279, 524 takes the place of 1 in an ordinary square, 525 of 2, and so on. They are here submitted to the reader simply as examples of common squares, having the *lowest possible summations that can be made from a series containing no prime numbers.* There are many longer sequences of consecutive composite numbers, from which larger squares might be made, but they run into such high values that the construction of magics therewith becomes laborious.

Dr. C. Planck has kindly contributed the following list of consecutive composite numbers that can be used for squares of order 6 to order 12 under the condition of lowest possible summations.

For Order	6.	15,684	—	15,719	=	36 numbers
" "	7.	19,610	—	19,758	=	49 "
" "	8.	31,398	—	31,461	=	64 "
" "	9.	155,922	—	156,002	=	81 "
" "	10.	370,262	—	370,361	=	100 "
" "	11.	1,357,202	—	1,357,322	=	121 "
" "	12.	2,010,734	—	2,010,877	=	144 "

Many attempts have been made to construct magic squares from a natural series of numbers by locating each succeeding number a knight's move from the last one, until every cell in the square is included in one continuous knight's tour. This difficult problem however has never been solved, and the square in question probably does not exist. Many squares have been made that sum correctly in their lines and columns, but they all fail in their two diagonals and therefore are not strictly magic.

In *Games Ancient and Oriental* (p. 325) one of the most interesting squares of the above description is presented, and it is reproduced here in Fig. 281, the knight's tour being shown in Fig. 282.

This square, like all others of its kind, fails in its two diagonals, but it is remarkable in being quartered, i. e., all of its four corner 4×4's are magic in their lines and columns, which sum 130. Furthermore, if each corner 4×4 is subdivided into 2×2's, each of the latter contains numbers that sum 130. It is stated that this square was made by Mr. Beverly and published in the *Philosophical Magazine* in 1848.

If the use of consecutive numbers is disregarded, a continuous

1	48	31	50	33	16	63	18
30	51	46	3	62	19	14	35
47	2	49	32	15	34	17	64
52	29	4	45	20	61	36	13
5	44	25	56	9	40	21	60
28	53	8	41	24	57	12	37
43	6	55	26	39	10	59	22
54	27	42	7	58	23	38	11

Fig. 281.

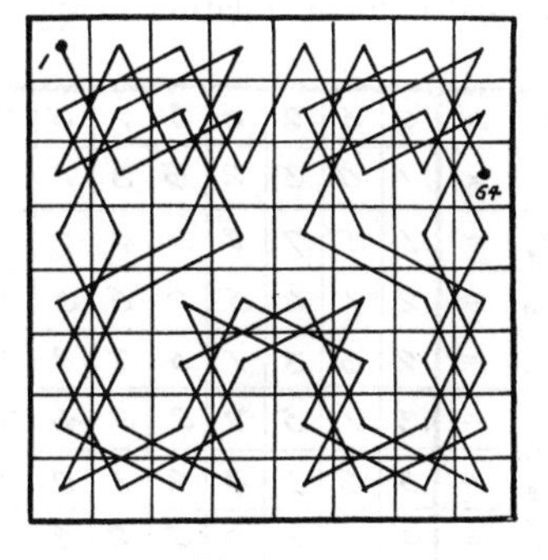

Fig. 282.

knight's tour may be traced through many different magic squares, in which every period of n numbers throughout the tour will sum S. A square having this quality is shown in Fig. 261. The knight's

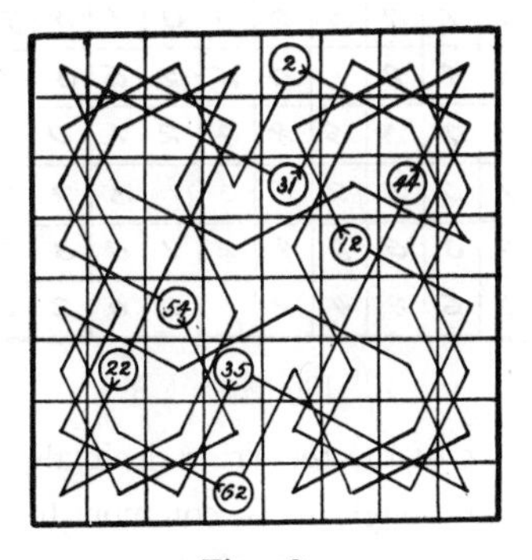

Fig. 283.

tour through this square is given in Fig. 283 in which the starting numbers of each period of eight are marked by circles with arrow heads indicating the direction of progression.

Oddities and curios in magics might be illustrated almost without end, but one more will suffice as a final example.

Fig. 284 shows an 18 × 18 magic made by Mr. Harry A. Sayles, the most interesting feature of which is the method of its production from the values of $^n/_{19}$. The lines of recurring decimals for $^1/_{19}$, $^2/_{19}$, $^3/_{19}$ . . . . . . $^{18}/_{19}$ are arranged one below the other so as to form a magic square. S = 81. It will be seen that the sequence of the digits in all lines is the same, the position of the decimal point in relation to the series being the only difference.

1/19	=	.0	5	2	6	3	1	5	7	8	9	4	7	3	6	8	4	2	1
2/19	=	.1	0	5	2	6	3	1	5	7	8	9	4	7	3	6	8	4	2
3/19	=	.1	5	7	8	9	4	7	3	6	8	4	2	1	0	5	2	6	3
4/19	=	.2	1	0	5	2	6	3	1	5	7	8	9	4	7	3	6	8	4
5/19	=	.2	6	3	1	5	7	8	9	4	7	3	6	8	4	2	1	0	5
6/19	=	.3	1	5	7	8	9	4	7	3	6	8	4	2	1	0	5	2	6
7/19	=	.3	6	8	4	2	1	0	5	2	6	3	1	5	7	8	9	4	7
8/19	=	.4	2	1	0	5	2	6	3	1	5	7	8	9	4	7	3	6	8
9/19	=	.4	7	3	6	8	4	2	1	0	5	2	6	3	1	5	7	8	9
10/19	=	.5	2	6	3	1	5	7	8	9	4	7	3	6	8	4	2	1	0
11/19	=	.5	7	8	9	4	7	3	6	8	4	2	1	0	5	2	6	3	1
12/19	=	.6	3	1	5	7	8	9	4	7	3	6	8	4	2	1	0	5	2
13/19	=	.6	8	4	2	1	0	5	2	6	3	1	5	7	8	9	4	7	3
14/19	=	.7	3	6	8	4	2	1	0	5	2	6	3	1	5	7	8	9	4
15/19	=	.7	8	9	4	7	3	6	8	4	2	1	0	5	2	6	3	1	5
16/19	=	.8	4	2	1	0	5	2	6	3	1	5	7	8	9	4	7	3	6
17/19	=	.8	9	4	7	3	6	8	4	2	1	0	5	2	6	3	1	5	7
18/19	=	.9	4	7	3	6	8	4	2	1	0	5	2	6	3	1	5	7	8

Fig. 284.

A peculiar feature of the recurring decimals used in this square may be mentioned, although it is common to many other such series, with variations. $^1/_{19}$ = .052631578947368421 . . . . . decimal repeats. Starting with the first 5 and dividing by 2 each integer determines the next integer following, thus:

2) 52631578.......... = 2631578..........

The same procession follows for $^n/_{19}$ and also for $1/(19 \times 2^n)$ though the operation will not apply in all cases to the first few numbers of each series.

If the decimal .05263.....1, consisting of 18 figures, is divided into two even sections of 9 figures each, and one section superposed on the other, the sum will be a series of 9's thus:

$$\begin{array}{r} .052631578 \\ 947368421 \\ \hline 999999999 \end{array}$$

The series is thus shown to consist of nine 9's = 81, so that each line of the square, Fig. 284, must sum 81. Also, as any two numbers symmetrically located above and below the horizontal axis of the square sum 9, each column also consists of nine 9's = 81.

It is not easy to understand why each of the two diagonals of this square should sum 81, but if they are written one over the other, each pair of numbers will sum 9.

Considering its constructive origin, and the above mentioned interesting features, this square, notwithstanding its simplicity, may be fairly said to present one of the most remarkable illustrations of the intrinsic harmony of numbers. W. S. A.

CHAPTER VIII.

NOTES ON VARIOUS CONSTRUCTIVE PLANS BY WHICH MAGIC SQUARES MAY BE CLASSIFIED.

AN odd magic square must necessarily have a central cell, and if the square is to be associated, this cell must be occupied by the middle number of the series, $[(n^2 + 1)/2]$ around which the other numbers must be arranged and balanced in pairs, the sum of each pair being $n^2 + 1$. Although in 5×5 and larger odd squares the pairs of numbers are capable of arrangement in a multitude of different ways relative to each other *as pairs,* yet when one number of a pair is located, the position of the other number becomes fixed in order to satisfy the rule that the sum of any two numbers that are diametrically equidistant from the center number must equal twice that number, or $n^2 + 1$.

In an even magic square, however, there is no central cell and no middle number in the series, so the method of construction is not thus limited, butt he pairs of numbers which sum $n^2 + 1$ may be harmoniously balanced either around the center of the square, as in odd squares, or in a variety of other ways.

Mr. L. S. Frierson has cleverly utilized this feature as the basis for a series of constructive plans, according to which the various types of even squares may be classified. He has shown eleven different plans and Mr. Henry E. Dudeney has contributed the twelfth, all of which may be used in connection with 4×4 squares. These twelve constructive plans clearly differentiate the various types of 4×4 squares,—there being for example one plan for the associated or regular squares, another plan for the Franklin squares, another for the pandiagonal or *continuous* squares and so forth, so that a knowledge of these plans makes it easy to classify all 4×4 squares. Six of the eleven plans given by Mr. Frierson cover distinct methods of arrangement, the remaining five plans being made up of various combinations.

PLAN NO. 1.

In this plan, which is the simplest of all, the pairs of numbers that sum $n^2 + 1$ are arranged symmetrically in adjacent cells, form-

16	1	13	4
7	10	6	11
2	15	3	14
9	8	12	5

Fig. 285.

Fig. 286.

ing two vertical columns, as shown in Fig. 285, and diagrammatically in Fig. 286.

PLAN NO. 2.

This plan differs from No. 1 only in the fact that the pairs of

Fig. 287.

Fig. 288.

numbers are placed in alternate instead of in adjacent columns, as seen in Figs. 287 and 288.

PLAN NO. 3.

1	13	4	16
8	12	5	9
14	2	15	3
11	7	10	6

Fig. 289.

Fig. 290.

According to this plan the pairs of numbers are arranged symmetrically on each side of the central axis, one-half of the elements being adjacent to each other, and the other half constructively adjacent as shown in Figs. 289 and 290. This arrangement furnishes the Franklin squares when expanded to 8 × 8, providing that the numbers in *all* 2 × 2 subsquares are arranged to sum 130 (See Figs. 291 and 292). If this condition is not fulfilled, only half of

52	61	4	13	20	29	36	45
14	3	62	51	46	35	30	19
53	60	5	12	21	28	37	44
11	6	59	54	43	38	27	22
55	58	7	10	23	26	39	42
9	8	57	56	41	40	25	24
50	63	2	15	18	31	34	47
16	1	64	49	48	33	32	17

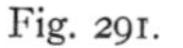

Fig. 291.

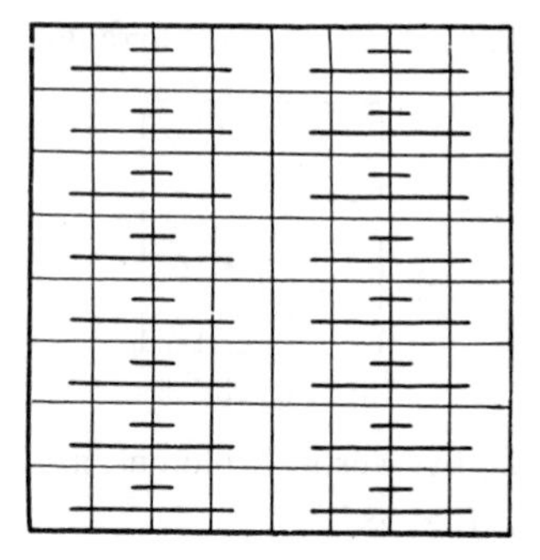

Fig. 292.

the bent diagonals will have proper summations. An imperfect Franklin square of this type may be seen in Fig. 268.

PLAN NO. 4.

In this plan the pairs of numbers are arranged adjacent to each other diagonally, producing four centers of equilibrium (See Figs. 293 and 294).

1	7	14	12
10	16	5	3
15	9	4	6
8	2	11	13

Fig. 293.

Fig. 294.

Magic squares constructed on this plan exhibit in part the features of the Franklin and the pandiagonal squares.

PLAN NO. 5.

The pairs of numbers in this plan are arranged in alternate cells in the diagonal columns, and it produces the *continuous* squares which have been termed Jaina, Nasik and pandiagonal squares. Fig. 295 is the Jaina square as modified by Dr. Carus (Fig. 222, p. 127), and Fig. 296 shows the arrangement of the pairs of numbers.

1	8	10	15
14	11	5	4
7	2	16	9
12	13	3	6

Fig. 295. Fig. 296.

The diagram of the Nasik square (Fig. 262) is a simple expansion of Fig. 296, and the diagram of the Frierson square (Fig. 267) shows a design like Fig. 296 repeated in each of its four quarters.

PLAN NO. 6.

Under this plan the pairs of numbers are balanced symmetrically around the center of the square, and this arrangement is common to all associated squares, whether odd or even. Fig. 297

1	15	14	4
12	6	7	9
8	10	11	5
13	3	2	16

Fig. 297. Fig. 298.

shows a common form of 4×4 square, the diagrammatic plan being given in Fig. 298.

PLAN NO. 7.

Magic squares on this plan are formed by combining plans

16	1	12	5
2	11	6	15
7	14	3	10
9	8	13	4

Fig. 299.

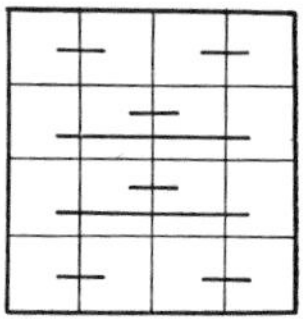

Fig. 300.

Nos. 1 to 3, a square and its diagram being shown in Figs. 299 and 300.

PLAN NO. 8.

This plan covers another combination of plans 1 and 3, and Figs. 301 and 302 show square and diagram.

11	14	3	6
8	9	16	1
10	7	2	15
5	4	13	12

Fig. 301.

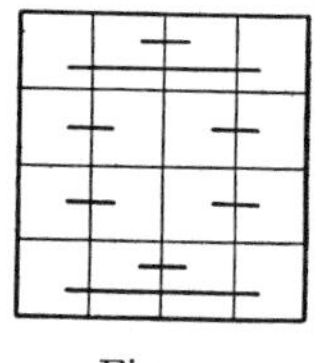

Fig. 302.

PLAN NO. 9.

This is a combination of plans 2 and 3, a square and its diagram being given in Figs. 303 and 304.

5	1	12	16
10	14	3	7
15	11	6	2
4	8	13	9

Fig. 303.

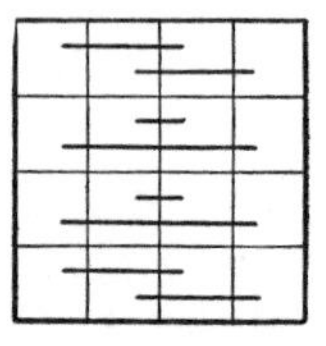

Fig. 304.

PLAN NO. 10.

This is also a combination of plans 2 and 3 and is illustrated in Figs. 305 and 306.

12	4	13	5
1	9	16	8
15	7	2	10
6	14	3	11

Fig. 305.

Fig. 306.

PLAN NO. 11.

One-half of this square is made in accordance with plan No. 2, but in the other half the pairs of numbers are located apart by knight's moves, which is different from any plan hitherto considered. It is impossible to arrange the entire square on the plan of the

1	2	16	15
13	14	4	3
12	7	9	6
8	11	5	10

Fig. 307.

Fig. 308.

knight's move. Figs. 307 and 308 show this square and its constructive plan.

PLAN NO. 12.

We are indebted to Mr. Henry E. Dudeney for the combination shown in Figs. 309-310, thus filling a complete dozen plans which probably cover all types of 4×4 magic squares.

2	15	1	16
11	10	8	5
14	3	13	4
7	6	12	9

Fig. 309.

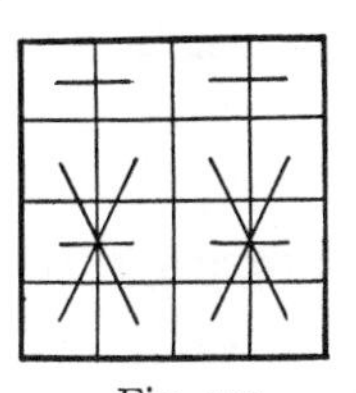

Fig. 310.

In even squares larger than 4×4 these plans naturally exhibit great diversity of design. The following 6×6 squares with their respective plans are given as examples in Figs. 311, 312 to 321, 322.

1	28	27	10	9	36
35	26	25	12	11	2
3	22	21	16	15	34
33	24	23	14	13	4
20	6	8	29	31	17
19	5	7	30	32	18

Fig. 311.

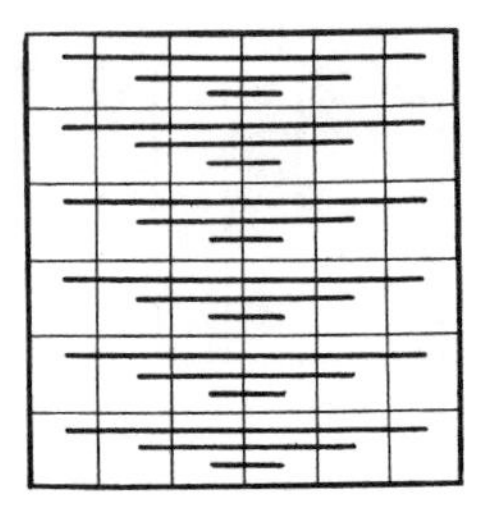

Fig. 312.

1	26	27	12	9	36
35	25	28	11	10	2
3	23	21	14	16	34
33	22	24	15	13	4
20	8	5	30	31	17
19	7	6	29	32	18

Fig. 313.

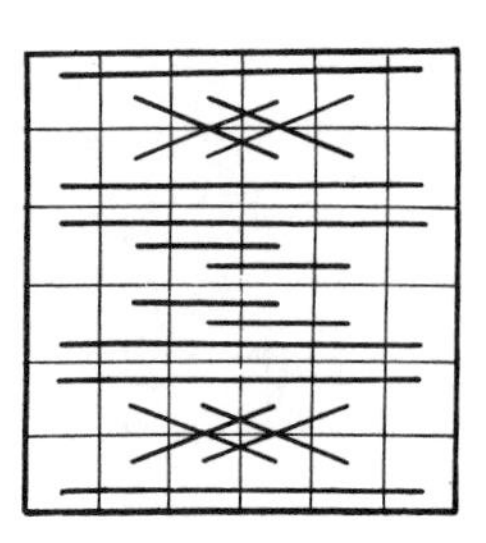

Fig. 314.

1	35	34	3	32	6
30	8	28	27	11	7
24	23	15	16	14	19
13	17	21	22	20	18
12	26	9	10	29	25
31	2	4	33	5	36

Fig. 315.

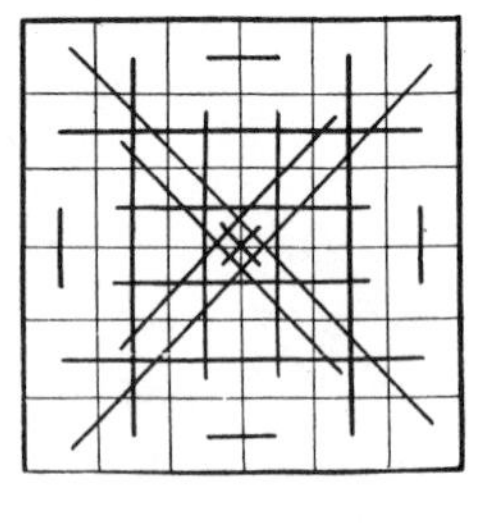

Fig. 316.

Figs. 315 and 317 are identical with 6×6 squares shown on pages 19 and 24. All squares of this class have the same characteristic plans.

The peculiar structure of the squares shown in Figs. 319 and 321 is visualized in their plans (Figs. 320 and 322). Fig. 314 is worthy of notice in having eight pairs of numbers located apart

1	5	33	34	32	6
30	8	28	9	11	25
18	23	15	16	20	19
24	14	21	22	17	13
7	26	10	27	29	12
31	35	4	3	2	36

Fig. 317.

Fig. 318.

1	36	26	23	13	12
35	2	25	24	14	11
3	34	27	28	9	10
33	4	21	22	15	16
20	17	7	6	29	32
19	18	5	8	31	30

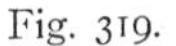

Fig. 319.

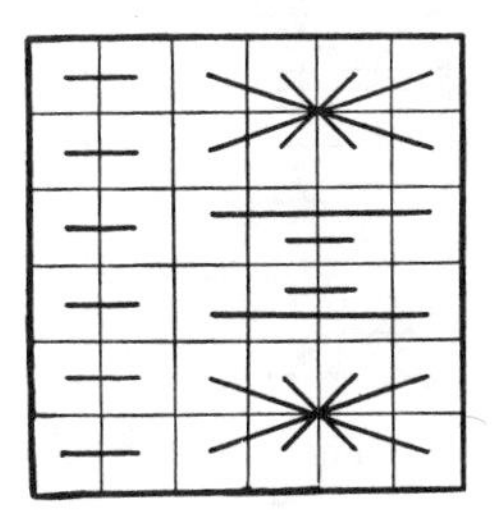

Fig. 320.

25	24	13	12	1	36
26	23	14	11	35	2
21	22	15	16	3	34
27	28	9	10	33	4
5	8	29	32	20	17
7	6	31	30	19	18

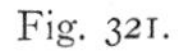

Fig. 321.

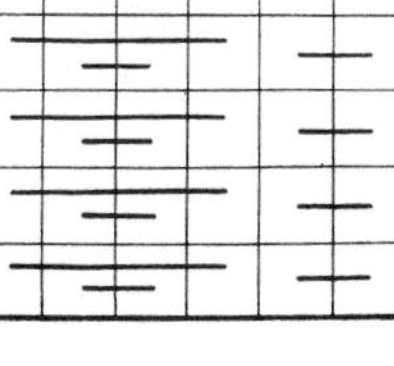

Fig. 322.

by knight's moves. Figs. 323, 324 and 325 illustrate another 6×6 square with its plan and numerical diagram. It will be seen that the latter is symmetrically balanced on each side, differing in this

1	26	28	11	9	36
35	25	27	12	10	2
3	23	21	14	16	34
33	22	24	15	13	4
20	8	6	29	31	17
19	7	5	30	32	18

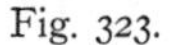
Fig. 323.

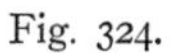
Fig. 324.

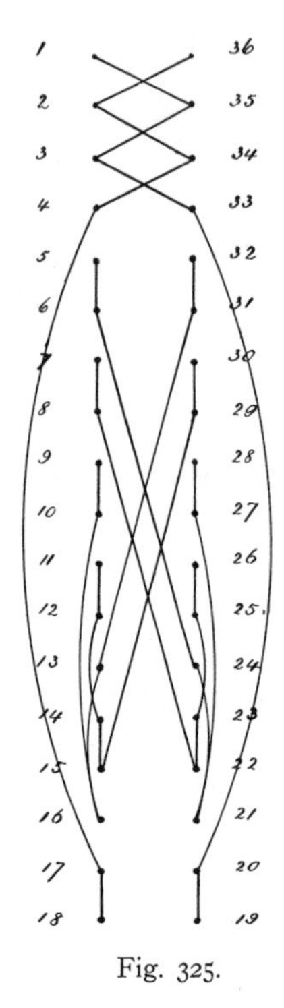

Fig. 325.

respect from the numerical diagrams of the 6×6 squares as described in Chapter I.

Figs. 326-333 are four 6×6 magic squares contributed by Mr. E. Black which show an interesting symmetry in their constructive plans.

35	2	28	9	4	33
14	23	12	25	15	22
17	20	6	31	10	27
5	32	13	24	36	1
29	8	34	3	16	21
11	26	18	19	30	7

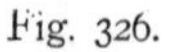
Fig. 326.

Fig. 327.

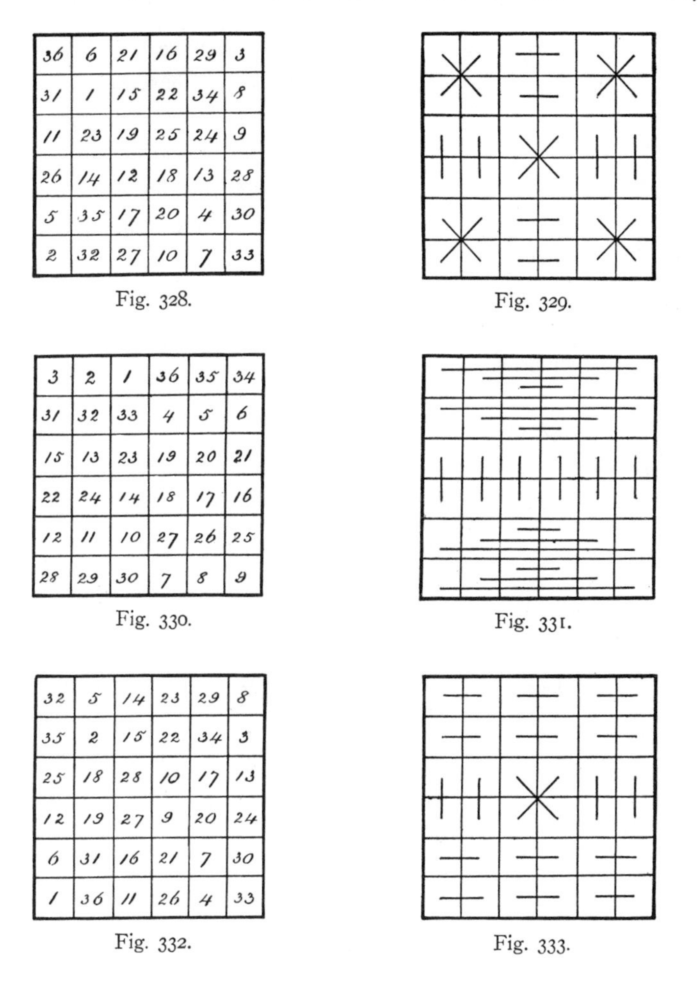

36	6	21	16	29	3
31	1	15	22	34	8
11	23	19	25	24	9
26	14	12	18	13	28
5	35	17	20	4	30
2	32	27	10	7	33

Fig. 328.

Fig. 329.

3	2	1	36	35	34
31	32	33	4	5	6
15	13	23	19	20	21
22	24	14	18	17	16
12	11	10	27	26	25
28	29	30	7	8	9

Fig. 330.

Fig. 331.

32	5	14	23	29	8
35	2	15	22	34	3
25	18	28	10	17	13
12	19	27	9	20	24
6	31	16	21	7	30
1	36	11	26	4	33

Fig. 332.

Fig. 333.

THE MATHEMATICAL VALUE OF MAGIC SQUARES.

The following quotations bearing on the above subject are copied from a paper entitled "Magic Squares and Other Problems on a Chessboard" by Major P. A. MacMahon, R.A., D.Sc., F.R.S., published in *Proceedings of the Royal Institution of Great Britain,* Vol. XVII, No. 96, pp. 50-61, Feb. 4, 1892.

"The construction of magic squares is an amusement of great antiquity; we hear of their being constructed in India and China before the Christian era, while they appear to have been introduced into Europe by Moscopulus who flourished at Constantinople early in the fifteenth century.

"However, what was at first merely a practice of magicians and talisman makers has now for a long time become a serious study for mathematicians. Not that they have imagined that it would lead them to anything of solid advantage, but because the theory was seen to be fraught with difficulty, and it was considered possible that some new properties of numbers might be discovered which mathematicians could turn to account. This has in fact proved to be the case, for from a certain point of view the subject has been found to be algebraical rather than arithmetical and to be intimately connected with great departments of science such as the 'infinitesimal calculus,' the 'calculus of operations,' and the 'theory of groups.'

"No person living knows in how many ways it is possible to form a magic square of any order exceeding 4×4. The fact is that before we can attempt to enumerate magic squares we must see our way to solve problems of a far more simple character.

"To say and to establish that problems of the general nature of the magic square are intimately connected with the infinitesimal calculus and the calculus of finite differences is to sum the matter up."

* * *

It is therefore evident that this field of study is by no means limited, and if this may be said in connection with magic squares the statement will naturally apply with a larger meaning to the consideration of magic cubes.

CHAPTER IX.

MAGIC CUBES OF THE SIXTH ORDER.

IT is stated by Dr. C. Planck in his article on "The Theory of Reversions," Chapter XII, pp. 298 and 304, that the first magic cube of this order was made by the late W. Firth, Scholar of Emanuel, Cambridge, England, in 1889. The pseudo-skeleton of Firth's construction is shown in Fig. 585, on p. 304 and its development into a magic 6^3 is given by Dr. Planck in Fig. 587. He also presents in Fig. 597 in the same chapter another magic 6^3 which he made in 1894 by the artifice of "index-cubes," and gives a full explanation of his method.

Although the cube presented in this chapter by Prof. H. M. Kingery is imperfect in its great diagonals, and therefore not strictly magic, it possesses many novel and interesting features, being an ingenious example of the general principle of the "Franklin" squares carried into the third dimension, and showing, as it does, perfect "bent diagonals." The same method will construct cubes of 10, 14, and other cubes of the $4p + 2$ orders.

The second article in this chapter by Mr. Harry A. Sayles gives a clear and concise solution of the problem by the La Hireian method. Mr. Sayles's cube is strictly magic.

The cube offered in the third article by the late John Worthington, besides being strictly magic, shows the unique feature of having perfect diagonals on the six outside squares. W. S. A.

A "FRANKLIN" CUBE OF SIX.

For a long time after cubes had been constructed and analyzed consisting of odd numbers and those evenly even (divisible by 4), the peculiar properties of the oddly even numbers baffled all attempts to treat them in like manner. While the following construction does not comply with all the criteria laid down for "magic" cubes it has some remarkable features which appear to the writer to deserve attention. It will at least serve to arouse some criticism and discussion, and may contain hints for a complete solution of the problem.

In the first place six magic squares were constructed, exactly similar in plan except that three of them began (at the upper left-hand corner) with odd numbers, each of which was 1 or 1 plus a multiple of 36, and the other three with even numbers, each a multiple of 18. In the first three squares the numbers were arranged in ascending order, in the other three descending. The initial numbers were so chosen that their sum was 651, or $(n/2)(n^3+1)$, which is the proper summation for each dimension of the projected magic cube. In the construction of these original squares, by the way, the diagrams presented in the first chapter of this book proved a great convenience and saved much time.

Each of the six squares so made is "magic" in that it has the same sum (651) for each column, horizontal row and corner diagonal. As the initial numbers have the same sum the similarity of the squares, with ascending arrangement in one half and descending in the other half, insures the same totals throughout for numbers occupying corresponding cells in the several squares; e. g., taking the third number in the upper row of each square and adding the six together we reach the sum 651, and so for any other position of the thirty-six.

In constructing our cube we may let the original six squares serve as the horizontal layers or strata. We have seen that the vertical columns in the cube must by construction have the correct summation. Furthermore, as the successive right-and-left rows in the horizontal squares constitute the rows of the vertical squares facing the front or back of the cube, and as the columns in the horizontal squares constitute the rows of the vertical squares facing right or left, it is easily seen that each of these twelve vertical squares has the correct summation for all its columns and rows.

Here appears the first imperfection of our cube. Neither the

I

1	215	214	3	212	6
210	8	208	207	11	7
204	203	15	16	14	199
13	17	201	202	200	18
12	206	9	10	209	205
211	2	4	213	5	216

II

198	20	21	196	23	193
25	191	27	28	188	192
31	32	184	183	185	36
186	182	34	33	35	181
187	29	190	189	26	30
24	197	195	22	194	19

III

37	179	178	39	176	42
174	44	172	171	47	43
168	167	51	52	50	163
49	53	165	166	164	54
48	170	45	46	173	169
175	38	40	177	41	180

145	71	70	147	68	150
66	152	64	63	155	151
60	59	159	160	158	55
157	161	57	58	56	162
156	62	153	154	65	61
67	146	148	69	149	72

IV

144	74	75	142	77	139
79	137	81	82	134	138
85	86	130	129	131	90
132	128	88	87	89	127
133	83	136	135	80	84
78	143	141	76	140	73

V

126	92	93	124	95	121
97	119	99	100	116	120
103	104	112	111	113	108
114	110	106	105	107	109
115	101	118	117	98	102
96	125	123	94	122	91

VI

Fig. 334.

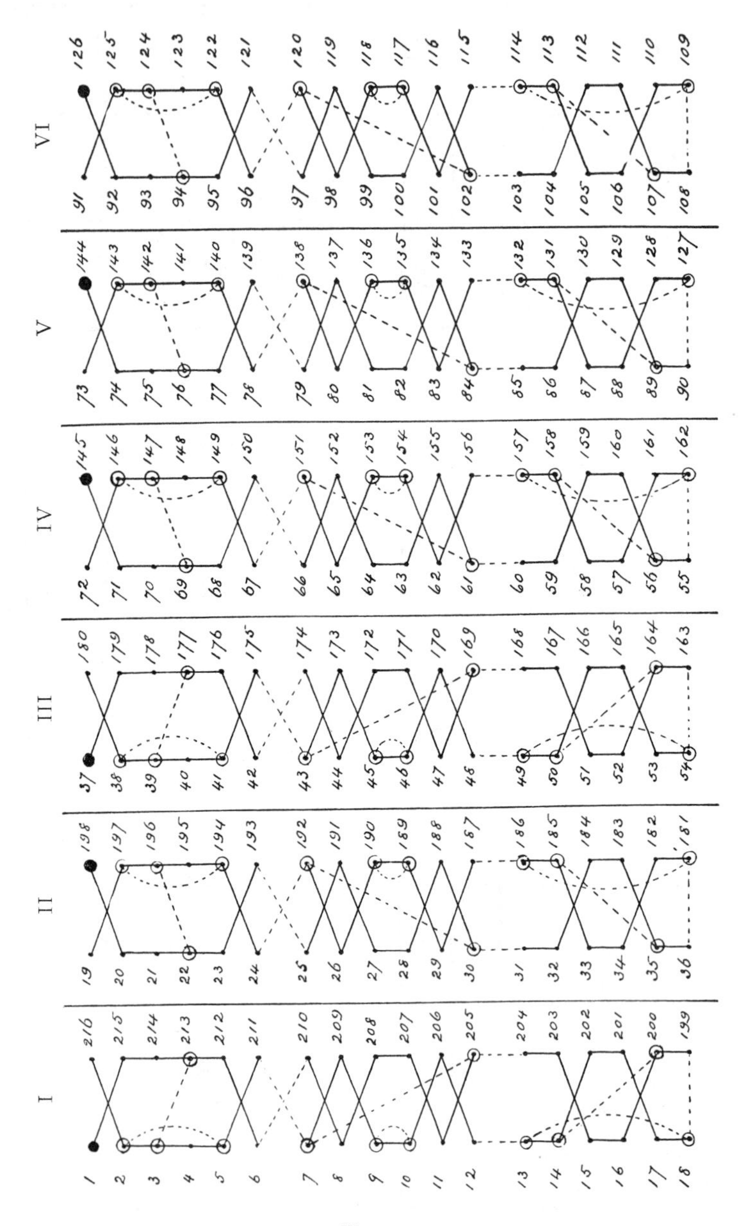
I
216 215 214 213 212 211 210 209 208 207 206 205 204 203 202 201 200 199
1 2 3 4 5 6 7 8 9 10 11 12 13 14 15 16 17 18
II
198 197 196 195 194 193 192 191 190 189 188 187 186 185 184 183 182 181
19 20 21 22 23 24 25 26 27 28 29 30 31 32 33 34 35 36
III
180 179 178 177 176 175 174 173 172 171 170 169 168 167 166 165 164 163
37 38 39 40 41 42 43 44 45 46 47 48 49 50 51 52 53 54
IV
145 146 147 148 149 150 151 152 153 154 155 156 157 158 159 160 161 162
72 71 70 69 68 67 66 65 64 63 62 61 60 59 58 57 56 55
V
144 143 142 141 140 139 138 137 136 135 134 133 132 131 130 129 128 127
73 74 75 76 77 78 79 80 81 82 83 84 85 86 87 88 89 90
VI
126 125 124 123 122 121 120 119 118 117 116 115 114 113 112 111 110 109
91 92 93 94 95 96 97 98 99 100 101 102 103 104 105 106 107 108

Fig. 335.

diagonals of the vertical squares nor those of the cube itself have the desired totals, though their *average* footing is correct. It is true further that the footings of the two cubic diagonals originating at opposite extremities of the same plane diagonal average 651, though neither alone is right.

At this point, however, we come upon an interesting fact. While the cubic diagonals vary, the two half-diagonals originating at opposite extremities of either plane diagonal in either the upper or the lower face, and meeting at the center of the cube, together have the sum 651. These correspond in the cube to the "bent diagonals" of Franklin's "square of squares." Of course a moment's reflection will show that this feature is inevitable. The original squares were so constructed that in their diagonals the numbers equidistant from the middle were "complementary," that is, taken together they equaled 217, or $n^3 + 1$, n representing the number of cells in a side of the square. In taking one complementary pair from each of three successive squares to make our "bent diagonal" we must of necessity have $3 \times 217 = 651$.

As in the Franklin squares, so in this cube do the "bent diagonals" parallel to those already described have the same totals. A plane square may be thought of as being bent around a cylinder so as to bring its upper edge into contact with the lower, and when this is done with a Franklin square it will be seen that there is one of these "bent diagonals" for each row. In like manner, if it were possible by some fourth-dimension process analogous to this to set our cube upon itself, we should see that there were six (or in general n) "bent diagonals" for each diagonal in each of the horizontal faces, or 24 in all, and all having the same sum, 651.

The occurrences of S may be tabulated as follows:

In the vertical columns	36	or	n^2
In the rows from front to back	36	or	n^2
In the rows from right to left	36	or	n^2
In the diagonals of the original square	12	or	$2n$
In the cubic "bent diagonals"	24	or	$4n$
	144	or	$3n^2+6n$

The column of n values at the right represents the "general" numbers, found in cubes of 10, 14, etc., as well as in that of 6.

All these characteristics are present no matter in what order the original squares are piled, which gives us 720 permutations. Furthermore, only one form of magic square was employed, and Mr. Andrews has given diagrams to illustrate at least 128 forms, any one of which might have been used in the construction of our cube.* Still further, numerous transpositions within the squares are possible—always provided the vertical totals are guarded by making the same transpositions in two squares, one ascending and the other descending. From this it is easy to see that the numbers 1–216 may be arranged in a very great number of different ways to produce such a cube.

So much for the general arrangement. If we so pile our original squares as to bring together the three which begin with odd numbers and follow them with the others (or *vice versa*) we find some new features of interest. In the arrangement already discussed none of the vertical squares has the correct sum for any form of diagonal. The arrangement now suggested shows "bent diagonals" for the vertical squares facing right and left as follows: Each of the outside squares—at the extreme right or left—has four "bent diagonals" facing the upper and four facing the lower edge. These have their origin in the first, second, fourth and fifth rows moving upward or downward, i. e., in the first two rows of each group—those yielded by original squares starting with odd and those with even numbers. Each of the four inside vertical squares has but two "bent diagonals" facing its upper and two facing its lower edge, and these start in the first and fourth rows—the first of each group of three. This will be true no matter in what order the original squares are piled, provided the odd ones are kept together and the evens together. This will add 32 (8 for each of the two outer and 4 for each of the four inner squares) to the 144 appearances of the sum 651 tabulated above, making 176; but this will apply, of course, only to the cube in which the odd squares are successive and the even squares successive. As the possible permutations of three objects

* See pp. 22 and 23.

number 6, and as each of these permutations of squares beginning with odd numbers can be combined with any one of the equal number of permutations of the even squares, a total of 36 arrangements is possible.

While the straight diagonals of these squares do not give the required footing the two in each square facing right or left average that sum: thus the diagonals of the left-hand square have totals of 506 and 796, of the second square 708 and 594, third 982 and 320, fourth 596 and 706, fifth 798 and 504, and the right-hand square 986 and 316, each pair averaging 651. I have not yet found any arrangement which yields the desired total for the diagonals, either straight or bent, of the vertical squares facing back or front; nor do their diagonals, like those just discussed, average 651 for any single square, though that is the exact average of the whole twelve.

By precisely similar methods we can construct cubes of 10, 14, 18, and any other oddly-even numbers, and find them possessed of the same features. I have written out the squares for the magic cube of 10, but time would fail to carry actual construction into higher numbers. Each column and row in the 10-cube foots up 5005, in the 14-cube 19,215, in the 30-cube 405,015, and in a cube of 42 no less than 1,555,869! Life is too short for the construction and testing of squares and cubes involving such sums.

That it is possible to build an absolutely "perfect" cube of 6 is difficult to affirm and dangerous to deny. The present construction fails in that the ordinary diagonals of the vertical squares and of the cube itself are unequal, and the difficulty is made to appear insuperable from the fact that while the proper summation is 651, an odd number, all the refractory diagonals are even in their summation.

The diagrams in Figure 335 are especially valuable because they show how the numbers of the natural series 1–216 are arranged in the squares which constitute the cube. This is a device of Mr. Andrews's own invention, and certainly is ingenious and beautiful. The diagrams here given for squares of six can be expanded on well-defined principles to apply to those of any oddly-even number, and several of them are printed in Chapter I.

It will be noticed that the numbers 1–108 are placed at the left of the diagrams, and those from 109 to 216 inclusive at the right in inverse order. Consequently the sum of those opposite each other is everywhere 217. In each diagram are two pairs of numbers connected by dotted lines and marked ○. These in every case are to be interchanged. Starting then at the heavy dot at the top we follow the black line across to 215, down to 212 (substituting 3 for 213) and back to 6; then across on the dotted line to 210 and along the zigzag black line to 8, 208, 207, 11 and 7 (interchanged with 205); down the dotted line to 204, then to 203, 15, 16, 14 (in place of 200), 199; then across the diagram and upward, observing the same methods, back to 216. This gives us the numbers which constitute our square No. I, written from left to right in successive rows. In like manner the diagrams in column II give us square No. II, and so on to the end. It is worthy of notice that in the fourth column of diagrams the numbers are written in the reverse of their natural order. This is because it was necessary in writing the fourth square to begin with the number 145 (which naturally would be at the bottom of the diagram) in order to give the initial numbers the desired sum of 651.

H. M. K.

A MAGIC CUBE OF SIX.

The two very interesting articles on Oddly-Even Magic Squares by Messrs. D. F. Savage and W. S. Andrews, which appear in Chapter X, might suggest the possibilities of extending those methods of construction into magic cubes. It is an interesting proposition and might lead to many surprising results.

Although the cube to be described here is not exactly of the nature mentioned above, it follows similar principles of construction and involves features quite unusual to cubes of this class.

The six respective layers of this cube are shown in Fig. 336. All of its 108 columns, and its four great diagonals give the constant summation of 651. If we divide this into 27 smaller cubes, which we will call cubelets, of eight cells each, the six faces, and also two diagonal planes of any cubelet give constant summations. For example, we will note the central cubelet of the first and

4	139	161	26	174	147
85	166	107	188	93	12
98	152	138	3	103	157
179	17	84	165	184	22
183	21	13	175	89	170
102	156	148	94	8	143

1

193	58	80	215	39	66
112	31	134	53	120	201
125	71	57	192	130	76
44	206	111	30	49	211
48	210	202	40	116	35
129	75	67	121	197	62

2

18	153	136	163	23	158
99	180	1	82	104	185
181	19	95	176	171	9
100	154	149	14	90	144
167	5	108	189	172	10
86	140	162	27	91	145

3

207	72	55	28	212	77
126	45	190	109	131	50
46	208	122	41	36	198
127	73	68	203	117	63
32	194	135	54	37	199
113	59	81	216	118	64

4

155	20	150	15	169	142
101	182	96	177	88	7
6	87	106	187	92	173
141	168	160	25	11	146
151	16	137	83	105	159
97	178	2	164	186	24

5

74	209	69	204	34	61
128	47	123	42	115	196
195	114	133	52	119	38
60	33	79	214	200	65
70	205	56	110	132	78
124	43	191	29	51	213

6

Fig. 336.

second layer, which is shown diagrammatically in Fig. 337. Its summations are as follows.

The six faces:

57	138	138	84	57	192
192	3	3	165	111	30
30	165	192	30	84	165
111	84	57	111	138	3
390	390	390	390	390	390

The two diagonal planes:

57	192
30	111
165	84
138	3
390	390

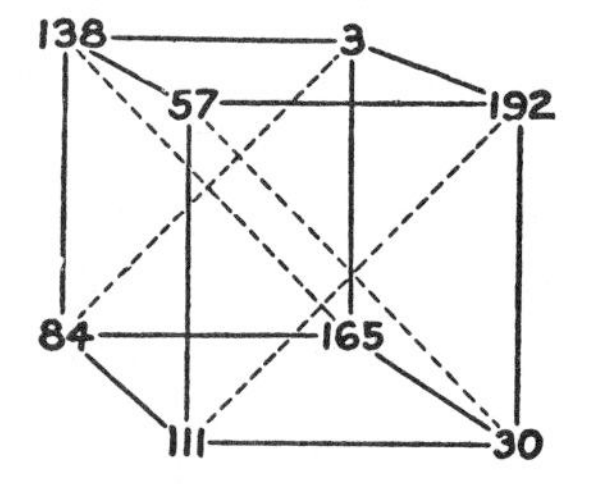

Fig. 337.

Also, if the sum of the eight cells in each of the cubelets be taken as a whole, we have a $3 \times 3 \times 3$ cube with 37 summations, each amounting to 2604.

The construction of this cube is by La Hireian method, using two primary cubes shown in Figs. 338 and 339. Fig. 338 contains 27 cubelets, each containing eight cells with eight equal numbers; the numbers in the respective cubelets ranking in order as the series, 1, 2, 3,....27. These 27 cubelets are arranged according to the methods of any $3 \times 3 \times 3$ cube. This gives us a primary cube with all the features of the final cube.

Fig. 339 is also divided ino 27 cubelets, each of which must contain the series 0, 27, 54, 81, 108, 135, 162, 189. The arrangement of the numbers in these 27 cubelets must be such as will give the primary cube all the required features of the final cube. The eight numbers of the cubelet series are, for convenience, divided by 27, and give the series 0, 1, 2, 3, 4, 5, 6, 7, which can easily be brought back to the former series after the primary cube is constructed.

To construct the cubelet, we divide the above series into two sets of four numbers each, so that the sums of the two sets are equal, and the complementaries of one set are found in the other. This division is 0, 5, 6, 3 and 7, 2, 1, 4, which separates the complemen-

4	4	26	26	12	12
4	4	26	26	12	12
17	17	3	3	22	22
17	17	3	3	22	22
21	21	13	13	8	8
21	21	13	13	8	8

1

4	4	26	26	12	12
4	4	26	26	12	12
17	17	3	3	22	22
17	17	3	3	22	22
21	21	13	13	8	8
21	21	13	13	8	8

2

18	18	1	1	23	23
18	18	1	1	23	23
19	19	14	14	9	9
19	19	14	14	9	9
5	5	27	27	10	10
5	5	27	27	10	10

3

18	18	1	1	23	23
18	18	1	1	23	23
19	19	14	14	9	9
19	19	14	14	9	9
5	5	27	27	10	10
5	5	27	27	10	10

4

20	20	15	15	7	7
20	20	15	15	7	7
6	6	25	25	11	11
6	6	25	25	11	11
16	16	2	2	24	24
16	16	2	2	24	24

5

20	20	15	15	7	7
20	20	15	15	7	7
6	6	25	25	11	11
6	6	25	25	11	11
16	16	2	2	24	24
16	16	2	2	24	24

6

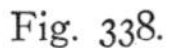

Fig. 338.

taries and gives two sets, each amounting to 14. We can place one set in any desired order on one face, and it only remains to place the four complementaries in the opposite face, so that the four lines connecting complementary pairs are parallel.

These cubelets are arranged in the primary cube with the 0, 5, 6, 3 faces placed in the 1st, 3d, and 5th layers, and the 7, 2, 1, 4 faces placed in the 2d, 4th, and 6th layers, which arrangement satisfies the summations perpendicular to the layers.

0	5	5	0	6	5
3	6	3	6	3	0
3	5	5	0	3	5
6	0	3	6	6	0
6	0	0	6	3	6
3	5	5	3	0	5

1

7	2	2	7	1	2
4	1	4	1	4	7
4	2	2	7	4	2
1	7	4	1	1	7
1	7	7	1	4	1
4	2	2	4	7	2

2

0	5	5	6	0	5
3	6	0	3	3	6
6	0	3	6	6	0
3	5	5	0	3	5
6	0	3	6	6	0
3	5	5	0	3	5

3

7	2	2	1	7	2
4	1	7	4	4	1
1	7	4	1	1	7
4	2	2	7	4	2
1	7	4	1	1	7
4	2	2	7	4	2

4

5	0	5	0	6	5
3	6	3	6	3	0
0	3	3	6	3	6
5	6	5	0	0	5
5	0	5	3	3	5
3	6	0	6	6	0

5

2	7	2	7	1	2
4	1	4	1	4	7
7	4	4	1	4	1
2	1	2	7	7	2
2	7	2	4	4	2
4	1	7	1	1	7

6

Fig. 339.

It now remains to adjust the pairs in the cubelets to suit the summations in the layers and the four diagonals. We first arrange the pairs that will give the diagonal summations, and by doing so, we set the position of four numbers in each of the layers 3 and 4, and eight numbers in each of the layers 1, 2, 5 and 6. We then arrange the remaining numbers in the layers 1, 3 and 5 to suit the twelve summations of each layer, which consequently locates the numbers for layers 2, 4 and 6, since complementary pairs must lie perpendicularly to the cubes layers. This gives us a primary cube such as that shown in Fig. 339.

The numbers in each cell of Fig. 339 must then be multiplied by 27, and added to the respective cells in Fig. 338, which combination gives us the final cube shown in Fig. 336. H. A. S.

MAGIC CUBE OF SIX.

In the cube, whose horizontal squares are shown in Fig. 340, the sum of each of the normal rows (those perpendicular to the faces of the cube) is 651, and the sum of each of the sixteen diagonals connecting the corners of the cube is the same.

These diagonals include the entire diagonals of the surfaces of the cube and the four diagonals of the solid running from corner to corner through the center of the cube.

DIAGONALS.

Top Square.	106	116	115	103	104	107
	109	12	11	202	205	112
Bottom Square.	111	117	118	98	97	110
	108	13	14	207	204	105
Front Square.	112	131	132	82	84	110
	107	31	29	190	189	105
Rear Square.	106	130	136	83	88	108
	109	30	25	191	185	111
Left Square.	106	37	40	182	181	105
	112	126	121	89	92	111
Right Square.	109	34	38	183	177	110
	107	127	125	90	94	108

Diagonals of the Solid.					
106	152	147	70	66	110
109	143	139	77	78	105
107	153	156	63	61	111
112	46	42	172	171	108

FIRST OR TOP SQUARE.

106	8	7	212	209	109
199	116	113	16	12	195
196	114	115	11	15	200
21	203	202	103	100	22
17	205	208	99	104	18
112	5	6	210	211	107

SECOND SQUARE.

166	130	129	32	30	164
37	152	148	137	143	34
33	151	150	142	140	35
128	41	47	157	154	124
126	46	44	155	153	127
161	131	133	28	31	167

THIRD SQUARE.

163	135	136	25	27	165
36	145	149	144	138	39
40	146	147	139	141	38
121	48	42	156	159	125
123	43	45	158	160	122
168	134	132	29	26	162

FOURTH SQUARE.

55	192	191	83	81	49
93	60	57	176	174	91
89	62	63	172	175	90
182	74	77	70	65	183
180	75	73	68	71	184
52	188	190	82	85	54

FIFTH SQUARE.

50	185	186	86	88	56
92	61	64	169	171	94
96	59	58	173	170	95
179	79	76	67	72	178
181	78	80	69	66	177
53	189	187	87	84	51

SIXTH OR BOTTOM SQUARE.

111	1	2	213	216	108
194	117	120	9	13	198
197	119	118	14	10	193
20	206	207	98	101	19
24	204	201	102	97	23
105	4	3	215	214	110

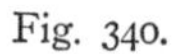

Fig. 340.

The foregoing cube was constructed in the following manner.

The foundation of this construction is the cube of 3 which is shown in Fig. 341.

FIRST OR TOP SQUARE

19	5	18
17	21	4
6	16	20

SECOND OR MIDDLE SQUARE.

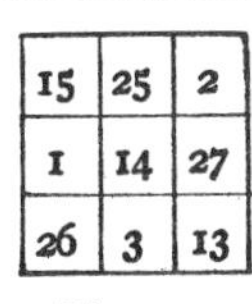

15	25	2
1	14	27
26	3	13

THIRD OR BOTTOM SQUARE

8	12	22
24	7	11
10	23	9

Fig. 341.

FIRST, OR TOP, AND SECOND SQUARES.

144	144	32	32	136	136
144	144	32	32	136	136
128	128	160	160	24	24
128	128	160	160	24	24
40	40	120	120	152	152
40	40	120	120	152	152

THIRD AND FOURTH SQUARES.

112	112	192	192	8	8
112	112	192	192	8	8
0	0	104	104	208	208
0	0	104	104	208	208
200	200	16	16	96	96
200	200	16	16	96	96

FIFTH AND SIXTH SQUARES.

56	56	88	88	168	168
56	56	88	88	168	168
184	184	48	48	80	80
184	184	48	48	80	80
72	72	176	176	64	64
72	72	176	176	64	64

Fig. 342. THE BASIC CUBE.

The sum of each normal row in the above cube, whether running from left to right, from rear to front or from top to bottom, is 42; and the sum of each diagonal of which the central term 14 is a member, as 19 14 9, 5 14 23, 15 14 13, etc., is also 42.

Deduct 1 from each term of the above cube and multiply the remainder by 8. With each of these multiples construct a cubic group consisting of eight repetitions of the multiple. Substitute

FIRST OR TOP SQUARE.

3	2	8	6	5	3
5	1	4	7	2	8
8	7	3	5	3	1
4	6	8	2	2	5
5	3	3	2	8	6
2	8	1	5	7	4

SECOND SQUARE.

6	7	1	3	4	6
4	8	5	2	7	1
1	2	6	4	6	8
5	3	1	7	7	4
4	6	6	7	1	3
7	1	8	4	2	5

THIRD SQUARE.

3	2	4	8	7	3
1	4	7	3	4	8
7	8	2	5	1	4
6	5	8	3	3	2
8	5	1	2	8	3
2	3	5	6	4	7

FOURTH SQUARE.

6	7	5	1	2	6
8	5	2	6	5	1
2	1	7	4	8	5
3	4	1	6	6	7
1	4	8	7	1	6
7	6	4	3	5	2

FIFTH SQUARE.

2	3	8	7	2	5
8	5	4	6	3	1
2	1	2	8	8	6
3	5	5	3	4	7
8	6	5	1	2	5
4	7	3	2	8	3

SIXTH OR BOTTOM SQUARE.

7	6	1	2	7	4
1	4	5	3	6	8
7	8	7	1	1	3
6	4	4	6	5	2
1	3	4	8	7	4
5	2	6	7	1	6

Fig. 343. THE GROUP CUBE.

FIRST OR TOP SQUARE.

147	146	40	38	141	139
149	145	36	39	138	144
136	135	163	165	27	25
132	134	168	162	26	29
45	43	123	122	160	158
42	48	121	125	159	156

SECOND SQUARE.

150	151	33	35	140	142
148	152	37	34	143	137
129	130	166	164	30	32
133	131	161	167	31	28
44	46	126	127	153	155
47	41	128	124	154	157

THIRD SQUARE.

115	114	196	200	15	11
113	116	199	195	12	16
7	8	106	109	209	212
6	5	112	107	211	210
208	205	17	18	104	99
202	203	21	22	100	103

FOURTH SQUARE.

118	119	197	193	10	14
120	117	194	198	13	9
2	1	111	108	216	213
3	4	105	110	214	215
201	204	24	23	97	102
207	206	20	19	101	98

FIFTH SQUARE.

58	59	96	95	170	173
64	61	92	94	171	169
186	185	50	56	88	86
187	189	53	51	84	87
80	78	181	177	66	69
76	79	179	178	72	67

SIXTH OR BOTTOM SQUARE.

63	62	89	90	175	172
57	60	93	91	174	176
191	192	55	49	81	83
190	188	52	54	85	82
73	75	180	184	71	68
77	74	182	183	65	70

Fig. 344. The Complete Cube.

each of these groups for that term of the cube from which it was derived, and the result will be a cube with six terms in each row. The horizontal squares of this cube are shown in Fig. 342, the second square being the same as the first, the fourth as the third, and the sixth as the fifth.

The sum of the terms in each normal row of the preceding cube is 624, and the sum of each diagonal which includes two terms from the central group of the cube is also 624. It follows that the middle two squares in each normal direction are magical and that each diagonal of the solid has the same sum as the normal rows. This cube is called the *basic* cube.

Another magic cube with six terms in each row was next constructed. This cube is called the *group* cube. Each position which in the basic cube is occupied by a cubic group of eight equal numbers is occupied in the group cube by a cubic group consisting of the numbers 1, 2, 3, 4, 5, 6, 7, 8. All of the rows and diagonals which have equal sums in the basic cube will have equal sums in the group cube.

Adding together the terms which occupy corresponding positions in the basic cube and the group cube the result is the complete cube shown in Fig. 344, containing the numbers from 1 to $6^3 = 216$.

In the complete cube the middle two squares in each direction are magical while the outer squares are not.

To bring these magical squares to the surface the squares of each set of parallel squares may be permuted as follows:

Original order 1, 2, 3, 4, 5, 6,
Permuted order ....3, 2, 1, 6, 5, 4.

The result is the final cube shown in Fig. 340.

The above permutation is subject to two conditions. The several sets of parallel squares must all be permuted in the same manner. Any two parallel squares which in the original cube are located on opposite sides of the middle plane of the cube and at an equal distance from it, in the permuted cube must be located on opposite sides of the middle plane of the cube and at an equal distance from it. These conditions are for the protection of the diagonals. J. W.

CHAPTER X.

VARIOUS KINDS OF MAGIC SQUARES.

OVERLAPPING MAGIC SQUARES.

A PECULIAR species of compound squares may be called overlapping magic squares. In these the division is not made as usual by some factor of the root into four, nine, sixteen or more subsquares of equal area, but into several subsquares or panels not all of the same size, some lying contiguous, while others overlap. The simplest specimens have two minor squares of equal measure apart in opposite corners, and in the other corners two major squares which overlap at the center, having as common territory a middle square 2×2, 3×3, or larger, or only a single cell. Such division can be made whether the root of the square is a composite or a prime number, as 4-5-9; 4-6-10; 5-6-11; 6-9-15; 8-12-20 etc. The natural series 1 to n^2 may be entered in such manner that each subsquare shall be magic by itself, and the whole square also magic to a higher or lower degree. For example the 9-square admits of division into two minor squares 4×4, and two major squares 5×5 which overlap in the center having one cell in common. For convenience, the process of construction may begin with an orderly arrangement of materials.

The series 1 to 81 is given in Fig. 345, which may be termed a *primitive square.* The nine natural grades of nine terms each, appear in direct order on horizontal lines. It is evident that any natural series 1 to n^2 when thus arranged will exhibit n distinct grades of n terms each, the common difference being unity in the horizontal direction, n vertically, $n+1$ on direct diagonals, and $n-1$ on transverse diagonals. This primitive square is therefore something more than a mere assemblage of numbers, for, on dividing it as proposed, there is seen in each section a set of terms which may be handled as regular grades, and with a little manipulation may become magical. The whole square with all its component parts may be tilted over to right or left 45°, so that all grades will be turned into a diagonal direction, and all diagonals will become rectangular rows, and *presto,* the magic square appears in short order. The principle has been admirably presented and employed in various connections on pp. 17 and 113. It is a well-known fact that the primitive square gives in its middle rows an average and equal summation; it is also a fact not so generally recognized, or so distinctly stated, that *all*

1	2	3	4	5	6	7	8	9
10	11	12	13	14	15	16	17	18
19	20	21	22	23	24	25	26	27
28	29	30	31	32	33	34	35	36
37	38	39	40	41	42	43	44	45
46	47	48	49	50	51	52	53	54
55	56	57	58	59	60	61	62	63
64	65	66	67	68	69	70	71	72
73	74	75	76	77	78	79	80	81

Fig. 345.

the diagonal rows are already correct for a magic square. Thus in this 9-square the direct diagonal, 1, 11, 21, 31 etc. to 81 is a mathematical series, $4\frac{1}{2}$ normal couplets = 369. Also the parallel partial diagonal 2, 12, 22, 32, etc. to 72, eight terms, and 73 to complete it, = 369. So of all the broken diagonals of that system; so also of all the nine transverse diagonals; each contains $4\frac{1}{2}$ normal couplets or the value thereof = 369. The greater includes the less, and these features are prominent in the subsquares. By the expeditious plan indicated above we might obtain in each section some squares of fair magical quality, quite regular and symmetrical, but when paired they would not be equivalent, and it is obvious that the coupled squares must have an equal summation of rows, whatever may be their difference of complexion and constitution. The major squares are like those once famous Siamese twins, Eng and Chang, united by a vinculum, an organic part of each, through which vital currents must flow; the central cell containing the middle term 41, must be their bond of union, while it separates the other pair. The materials being parceled out and ready to hand, antecedents above and consequents below, an equitable allotment may be made of normal couplets to each square. Thus from N. W. section two grades may be taken as they stand horizontally, or vertically, or diagonally or any way symmetrically. The consequents belonging to those, found in S. E. section will furnish two grades more and complete the square. The other eight terms from above and their consequents from below will empty those compartments and supply the twin 4-square with an exact equivalent. Some elaborate and elegant specimens, magic to a high degree, may be obtained from the following distribution:

1st grade 1, 3, 11, 13 (all odd), 2, 4, 10, 12 (all even);
2d grade 19, 21, 29, 31 and 20, 22, 28, 30.

Then from N. E. section two grades may be taken for one of the major squares; thus 5, 6, 7, 8, 9 and 23, 24, 25, 26, 27 leaving for the twin square, 14, 15, 16, 17, 18 and 32, 33, 34, 35, 36. To each we join the respective consequents of all those terms forming 4th and 5th grades, and they have an equal assignment. But each requires a middle grade, and the only material remaining is that whole middle grade of the 9-square. Evidently the middle portion, 39, 40, 41, 42, 43 must serve for both, and the 37, 38, and their partners 44, 45 must be left out as undesirable citizens. Each having received its quota may organize by any plan that will produce a magic and bring the middle grade near the corner, and especially the number 41 into a corner cell.

In the 5-square Fig. 346 we may begin anywhere, say the cell below the center and write the 1st grade, 14, 15, 16, 17, 18, by a uniform oblique step moving to the left and downward. From the end of this grade a new departure is found by counting two cells down or three cells up if more convenient, and the 2d grade, 32,

33, 34, 35, 36 goes in by the same step of the 1st grade. All the grades follow the same rule. The leading terms 14, 32, 39, 46, 64 may be placed in advance, as they go by a uniform step of their own, analogous to that of the grades; then there will be no need of any "break-move," but each grade can form on its own leader wherever that may stand, making its proper circuit and returning to its starting point. The steps are so chosen and adjusted that every number finds its appointed cell unoccupied, each series often crossing the path of others but always avoiding collision. The resulting square is magic to a high degree. It has its twelve normal couplets arranged geometrically radiating around that unmatched middle term 41 in the central cell. In all rectangular rows and in all diagonals, entire and broken, the five numbers give by addition

50	(39)	33	16	67
34	17	68	(46)	40
(64)	47	41	35	18
42	36	(14)	65	48
15	66	49	43	(32)

Fig. 346.

(23)	45	58	73	6
55	70	(5)	31	44
13	30	41	52	(69)
38	(51)	77	12	27
76	9	24	(37)	59

Fig. 347.

the constant S = 205. There are twenty such rows. Other remarkable features might be mentioned.

For the twin square Fig. 347 as the repetition of some terms and omission of others may be thought a blemish, we will try that discarded middle grade, 37, 38, 41, 44, 45. The other grades must be reconstructed by borrowing a few numbers from N. W. section so as to conform to this in their sequence of differences, as Mr. Frierson has ably shown (Fig. 249, p. 141). Thus the new series in line 5-6-9-12-13, 23-24-27-30-31, 37-38-(41)-44-45 etc. has the differences 1 3 3 1 repeated throughout, and the larger grades will necessarily have the same, and the differences between the grades will be reciprocal, and thus the series of differences will be balanced geometrically on each side of the center, as well as the normal couplets. Therefore we proceed with confidence to construct the 5-square Fig. 347 by the same rule as used in Fig. 346, only applied in contrary directions, counting two cells to right and one upward. When completed it will be the reciprocal of Fig. 346 in pattern, equivalent in summation, having only the term 41 in common and possessing similar magical properties. It remains to be seen how those disorganized grades in the N. W. section can be made available for the two minor squares. Fortunately, the fragments allow this distribution:

Regular grades 1, 2, 3, 4,—irregular grades 7, 8, 10, 11
19, 20, 21, 22 25, 26, 28, 29

These we proceed to enter in the twin squares Figs. 348 and 349. The familiar two-step is the only one available, and the last half of each grade must be reversed, or another appropriate permutation employed in order to secure the best results. Also the 4th grade comes in before the 3d. But these being consequents, may

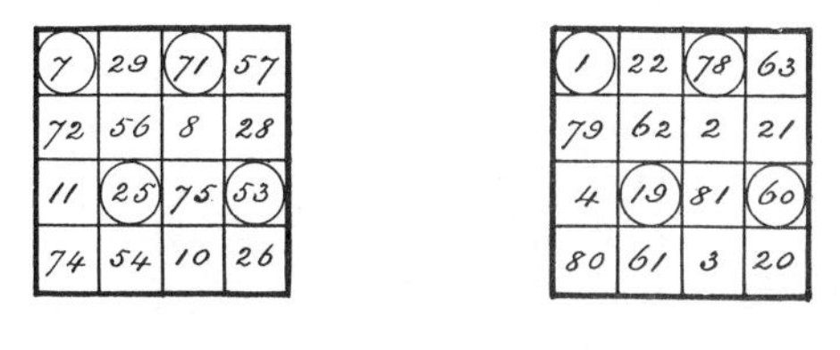

7	29	71	57
72	56	8	28
11	25	75	53
74	54	10	26

Fig. 348.

1	22	78	63
79	62	2	21
4	19	81	60
80	61	3	20

Fig. 349.

go in naturally, each diagonally opposite its antecedent. The squares thus made are magical to a very high degree. All rectangular and all diagonal rows to the number of sixteen have the constant $S = 164$. Each quadrate group of four numbers $= 164$. There are nine of these overlapping 2-squares. The corner numbers or two numbers taken on one side together with the two directly opposite $= 164$. The corner numbers of any 3-squares $= 164$. There are four of these overlapping combinations arising from the peculiar distribution of the eight normal couplets.

These squares may pass through many changes by shifting whole rows from side to side, that is to say that we may choose any cell as starting point. In fact both of them have been thus changed when taking a position in the main square. The major squares shown in Figs. 346 and 347 pass through similar changes in order to bring the number 41 to a corner. With these four subsquares all in place we have the 9-square, shown in Fig. 350, containing the whole series 1 to 81. The twenty continuous rows have the constant $S = 164 + 205 = 369$. Besides the 4-squares in N. W. and S. E. there is a 4-square in each of the other corners overlapping the 5-square, not wholly magic but having eight normal couplets placed geometrically opposite, so that taken by fours symmetrically they $= 164$. The four corner numbers $31 + 36 + 22 + 73 = 164$.

This combination may be taken as typical of the odd squares which have a pair of subsquares overlapping by a single cell. Whatever peculiarities each individual may exhibit they must all conform

75	53	11	25	14	65	48	42	36
10	26	74	54	49	43	32	15	66
71	57	7	29	33	16	67	50	39
8	28	72	56	68	46	40	34	17
52	69	13	30	41	35	18	64	47
12	27	38	51	77	80	20	3	61
37	59	76	9	24	4	60	81	19
73	6	23	45	58	79	21	2	62
31	44	55	70	5	1	63	78	22

Fig. 350.

to the requirement of equal summation in coupled subsquares; and for the distribution of values the plan of taking as a unit of measure the normal couplet of the general series is so efficacious and of such universal application that no other plan need be suggested. These principles apply also to the even squares which have no central cell but a block of four cells at the intersection of the axes. For example, the 14-square, Fig. 351, has two minor subsquares 6×6, and two major squares 8×8, with a middle square 2×2. This indicates a convenient subdivision of the whole area into 2-squares. Thus in N. W. section we have sixteen blocks; it is a quasi-4-square, and the compartments may be numbered from 1 to 16 following some approved pattern of the magic square, taking such point of departure as will bring 16 to the central block. This is called 1 for the S. E. section in which 2, 3, etc. to 16 are located as before. Now as these single numbers give a constant sum in every line, so will any mathematical series that may replace them in the same order as 1st, 2d, 3d terms etc. Thus in 1 the numbers 1, 2, 3, 4, in 2; 5, 6, 7, 8, and so on by current groups, will give correct results. In this case the numbers 1 to 18, and 19 to 36 with their consequents should be reserved for the twin minor squares. So that here in the N. W. section we begin with 37, 38, in 1 instead of 1, 2, leaving

47	149	65	131	56	142	44	154	7	18	193	4	185	184
48	150	66	132	55	141	43	153	186	6	187	194	1	17
57	139	39	157	50	148	62	136	9	15	183	8	181	195
58	140	40	158	49	147	61	135	188	16	13	190	182	2
145	51	133	63	138	60	160	38	12	196	10	3	191	179
146	52	134	64	137	59	159	37	189	180	5	192	11	14
143	53	155	41	152	46	130	68	108	90	103	93	115	81
144	54	156	42	151	45	129	67	107	89	104	94	116	82
25	36	175	22	167	166	99	97	121	75	126	72	114	84
168	24	169	176	19	35	100	98	122	76	125	71	113	83
27	33	165	26	163	177	73	123	85	111	96	102	78	120
170	34	31	172	164	20	74	124	86	112	95	101	77	119
30	178	28	21	173	161	91	105	79	117	70	128	88	110
171	162	23	174	29	32	92	106	80	118	69	127	87	109

Fig. 351.

the 3, 4 spaces to be occupied by the consequents 159, 160. Then in 2 we continue 39, 40 (instead of 5, 6) and so following the path of the primary series, putting two terms into each 2-square, and arriving with 67, 68 at the middle square. Then the coupled terms go on 69, 70—71, 72 etc. by some magic step across the S. E. section reaching the new No. 16 with the terms 97, 98. This exhausts the antecedents. Each 2-square is half full. We may follow a reversed track putting in the consequents 99, 100 etc. returning to the starting point with 159, 160. It is evident that all the 2-squares are equivalent, and that each double row of four of them = 1576, but it does not follow that each single row will = 788. In fact they do so, but that is due to the position of each block as direct or reversed or inverted according to a chart or theorem employed in work of this kind. The sixteen rectangular rows, the two entire diagonals and those which pass through the centers of the 2×2 blocks sum up correctly. There are also many bent diagonals and

225	216	3	222	5	7	73	143	75	141	77	139	79	152	138
10	1	223	4	221	219	153	83	151	85	149	87	147	88	74
6	220	11	18	212	211	89	129	91	127	93	136	126	81	145
218	8	213	210	12	17	137	97	135	99	133	100	90	82	144
2	224	14	15	215	208	101	119	103	124	118	95	131	150	76
217	9	214	209	13	16	125	107	123	108	102	96	130	84	142
77	149	71	155	69	157	112	117	110	105	121	134	92	148	78
52	174	64	162	70	156	111	113	115	106	120	98	128	86	140
181	45	180	46	186	40	116	109	114	122	104	132	94	146	80
53	173	66	160	168	154	37	167	39	29	36	194	193	24	202
178	48	163	63	72	58	189	59	187	195	192	30	35	20	206
55	171	169	158	38	161	44	159	62	32	33	197	190	200	26
176	50	68	57	188	65	182	67	164	196	191	31	34	199	27
184	165	41	172	43	170	47	146	49	21	204	23	25	207	198
61	42	185	54	183	56	179	80	177	205	22	203	201	28	19

Fig. 352.

zigzag rows of eight numbers that = 788. Each quarter of the square = 1576 and any overlapping 4-square made by four of the blocks gives the same total. The minor squares are *inlaid*. Thus in the N. E. square if the twenty numbers around the central block be dropped out and the three at each angle be brought together around the block we shall have a 4-square magical to a high degree. In fact this is only reversing the process of construction.

Fig. 352 is a 15-square which develops the overlapping principle to an unusual extent. There are two minor squares 6×6, and two major squares 9 × 9 with a middle square 3 × 3 in common. The whole area might have been cut up into 3-squares. The present division was an experiment that turned out remarkably well. The general series, 1 to 225 is thus apportioned. For N. W. 6-square the numbers 1 to 18 and 208 to 225; for S. E. 19 to 36 and 190 to 207; that is just eighteen normal couplets to each. For S. W. 9-square the numbers 37 to 72 and 154 to 189; for N. E. 73 to 108 and 118 to 153; for the middle square, 109 to 117. Figs. 353 and 354 show the method of construction. The nine middle terms are first arranged as a 3-square, and around this are placed by a well-known process (Fig. 103, p. 47) eight normal couplets 101 + 125 etc. forming a border and making a 5-square. By a similar process

74	153	83	151	85	149	87	147	88
145	90	137	97	135	99	133	100	81
144	131	102	125	107	123	108	95	82
76	130	121	112	117	110	105	96	150
142	92	120	111	113	115	106	134	84
78	128	104	116	109	114	122	98	148
140	94	118	101	119	103	124	132	86
80	126	89	129	91	127	93	136	146
138	73	143	75	141	77	139	79	152

Fig. 353.

1	223	4	221	219	10
220	11	18	212	211	6
8	213	210	12	17	218
224	14	15	215	208	2
9	214	209	13	16	217
216	3	222	5	7	225

Fig. 354.

this is enlarged to a 7-square, and this again to a 9-square, Fig. 353. Each of these concentric, or bordered, or overlapping squares is magic by itself. The twin square N. E. is made by the same process with the same 3-square as nucleus. In order to bring this nucleus to the corner of each so that they may coalesce with a bond of union, both of the squares are turned inside out. That is, whole rows are carried from bottom to top and from left to right. Such transposition does not affect the value of any rectangular row, but it does affect the diagonals. In this case the corner numbers, 74, 138 and 152 become grouped around the other corner 88, each of the couplets having the same diagonal position as before. Thus we obtain a 7-square with double border or panel on the North and East, still magic. This 7-square may now be moved down and out a little, from the border so as to give room to place its bottom row above, and its left column to the right, and we have a 5-square with panels of four rows. Again we move a little down and out leaving space for the bottom and left rows of the 5-square and thus the 3-square advances to the required position, and the four squares still overlap and retain all of their magical properties. The twin square S. W. passes through analogous transformation. The minor squares were first built up as bordered 4^2's as shown in Fig. 354 and then the single border was changed to double panel on two sides, but they might have gone in without change to fill the corners of the main square. As all this work was done by the aid of movable numbered blocks the various operations were more simple and rapid than any verbal description can be. The 15-square (Fig. 352) as a whole has the constant $S = 1695$ in thirty rectangular rows and two diagonals, and possibly some other rows will give a correct result. If the double border of fifty-two normal couplets be removed the remaining 11-square, 4-7-11 will be found made up of two 4-squares and two overlapping 7-squares with middle 3-square, all magic. Within this is a volunteer 7-square, of which we must not expect too much, but its six middle rows and two diagonals are correct, and the corner 2×2 blocks pertaining to the 4-squares although not composed of actual couplets have the value thereof, $224 + 228$. However, without those blocks we have two overlapping 5-squares all right. By the way, these 4-squares have a very high degree of magic, like those shown in Fig. 350, with their 2-squares and 3-squares so curiously overlapping. Indeed, this recent study had its origin some years ago from observing these special features of the 4-square at its best state. The same traits were recognized in the 8's and other congeners; also some remarkable results found in the oddly-even squares when filled by current groups, as well as in the quartered squares, led gradually to the general scheme of overlapping squares as here presented. D. F. S.

ODDLY-EVEN MAGIC SQUARES.

A convenient classification of magic squares is found by recognition of the root as either a prime number or evenly-even, or oddly-even, or oddly-odd. These four classes have many common traits, but owing to some characteristic differences, a universal rule of construction has hitherto seemed unattainable. The oddly-even squares especially, have proved intractable to methods that are readily applicable to the other classes, and it is commonly believed that they are incapable of attaining the high degree of magical character which appears in those others.

As some extensive explorations, recently made along those lines, have reached a very high latitude, the results will now be presented, showing a plan for giving to this peculiar sort, more than the ordinary magical properties.

Problem: To make oddly-even squares which shall have proper summation in all diagonal and rectangular rows except two, which two shall contain S—1 and S+1 respectively. This problem is solved by the use of auxiliary squares.

If n is an oddly-even root, and the natural series 1, 2, 3 etc. to n^2 is written in current groups of four terms, thus:

1.2.3.4.—5.6.7.8.—9.10.11.12.—13.14.15.16. etc.
0.1.2.3.—0.1.2.3.—0. 1. 2. 3.— 0. 1. 2. 3. etc.

---

1 5 9 13 etc.

then from each current group a series 0.1.2.3 may be subtracted, leaving a series 1.5.9.13 etc. to n^2—3, a regular progression of $n^2/4$ terms available for constructing a square whose side is $n/2$. As there are four such series, four such squares, exactly alike, readily made magic by well-known rules, when fitted together around a center, will constitute an oddly-even square possessing the magical character to a high degree. This will serve as the principal auxiliary. Another square of the same size must now be filled with the series 0.1.2.3 repeated $n^2/4$ times. The summation $3n/2$ being always odd, cannot be secured at once in every line, nor equally divided in the half lines, but all diagonal and all rectangular rows, except two of the latter, can be made to sum up correctly. Hence the completed square will show a minimum of imperfection.

In illustration of these general principles, a few examples will be given, beginning properly with the 2-square, smallest of all and first of the oddly-even. This is but an embryo, yet it exhibits in its nucleated cells some germs of the magical character, capable of indefinite expansion and growth, not only in connection with those of its own sort, but also with all the other sorts. Everything being reduced to lowest terms, a very general, if not a universal principle of construction may be discovered here. Proceeding strictly by rule, the series 1.2.3.4. affords only the term 1. repeated four times, and the series 0.1.2.3. taken once. The main auxiliary (Fig. 355) is a genuine quartered 2-square, equal and identical and regular and continuous every way. S=2.

1	1
1	1

Fig. 355.

0	1
2	3

Fig. 356.

0	2
3	1

Fig. 357.

0	3
2	1

Fig. 358.

1	2
3	4

Fig. 359.

The second auxiliary (Fig. 356) taking the terms in direct order, has eight lines of summation, showing equality, S=3, in all four diagonals, while the four rectangular rows give inequalities 1.5 and 2.4; an exact balance of values. This second auxiliary may pass through eight reversed, inverted or revolved phases, its semi-magic character being unchanged. Other orders may be employed, as shown in Figs. 357 and 358, bringing equality into horizontal or vertical rows, but not in both directions at the same time. Now any one of these variables may combine with the constant shown in Fig. 355, developing as many as twenty-four different arrangements of the 2-square, one example of which is given in Fig. 359. It cannot become magic unless all its terms are equal; a series whose common difference is reduced to zero. As already suggested, this 2-square plays an important part in the present scheme for producing larger squares, pervading them with its kaleidoscopic changes, and forming, we may say, the very warp and woof of their substance and structure.

The 6-square now claims particular attention. The main auxiliary, Fig. 360, consists of four 3-squares, each containing the series 1.5.9.13 etc. to 33. The 3-square is infantile; it has but one plan of construction; it is indeed regular and can not be otherwise, but it is imperfect. However, in this combination each of the four has a different aspect, reversed or inverted so that the inequalities of partial diagonals exactly balance. With this adjustment of subsquares the 6-square as a whole becomes a perfect quartered square, S=102; it is a quasi 2-square analogous to Fig. 355.

The four initial terms, 1.1.1.1 symmetrically placed, are now to be regarded as one group, a 2-square scattered into the four quarters; so also with the other groups 5.5.5.5 etc. Lines connecting like terms in each quarter will form squares or other

13	33	5	5	33	13
9	17	25	25	17	9
29	1	21	21	1	29
29	1	21	21	1	29
9	17	25	25	17	9
13	33	5	5	33	13

Fig. 360.

0	2	2	0	3	2
3	1	1	3	0	1
0	2	2	0	3	2
3	0	1	3	1	1
0	3	2	0	2	2
3	0	1	3	1	1

Fig. 361.

rectangles, a pattern, as shown in Fig. 363, with which the second auxiliary must agree. The series 0.1.2.3 is used nine times to form this second square as in Fig. 361. There are two conditions: to secure in as many lines as possible the proper summation, and also an adjustment to the pattern of Fig. 360. For in order that the square which is to be produced by combination of the two auxiliaries shall contain all the terms of the original series, 1 to 36, a group 0.1.2.3 of the one must correspond with the group 1.1.1.1 of the other, so as to restore by addition the first current group 1.2.3.4. Another set 0.1.2.3 must coincide with the 5.5.5.5; another with the 9.9.9.9 and so on with all the groups. The auxiliary Fig. 361 meets these conditions. It has all diagonals correct, and also all rectangular rows, except the 2d and 5th verticals, which sum up respectively 8 and 10.

Consequently, the finished square Fig. 362 shows inequality in the corresponding rows. However, the original series has been restored, the current groups scattered according to the pattern, and although not strictly magic it has the inevitable inequality reduced to a minimum. The faulty verticals can be easily equalized by transposing the 33 and 34 or some other pair of numbers therein, but the four diagonals that pass through the pair will then become incorrect, and however these inequalities may be shifted about they can never be wholly eliminated. It is obvious that many varieties of the finished square having the same properties may be obtained by reversing or revolving either of the auxiliaries, and many more by some other arrangement of the subsquares. It will be observed

13	35	7	5	36	15
12	18	26	28	17	10
29	3	23	21	4	31
32	1	22	24	2	30
9	20	27	25	19	11
16	33	6	8	34	14

Fig. 362.

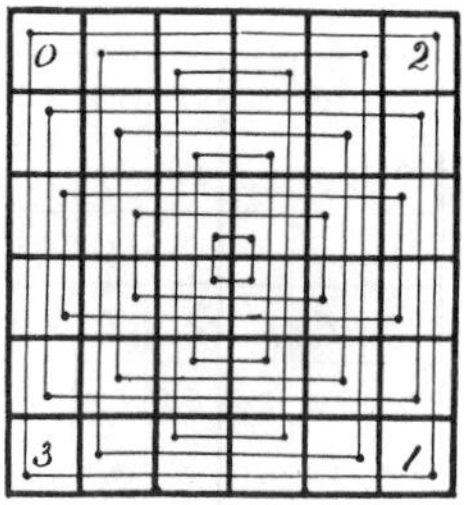

Fig. 363.

that in Fig. 360 the group 21 is at the center, and that each 3-square may revolve on its main diagonal, 1 and 25, 9 and 33, 29 and 5 changing places. Now the subsquares may be placed so as to bring either the 5 or the 13 or the 29 group at the center, with two changes in each case. So that there may be $8\times8\times8=512$ variations of this kind. There are other possible arrangements of the subsquares that will preserve the balance of the partial diagonals, but the pattern will be partly rhomboidal and the concentric figures tilted to right and left. These will require special adaptation of the second auxiliary.

We come now to the 10-square, no longer hampered as in the 6-square, by the imperfection of the subsquares. The main auxiliary Fig. 364 consists of four 5-squares, precisely alike, each containing the series 1.5.9 etc. to 97, S=245, in every respect regular and continuous. All four face the same way, but they might have been written right and left, as was *necessary* for the 3-square. The groups 1.1.1.1, 5.5.5.5 etc. are analogously located, and the pattern consists of equal squares, not concentric but overlapping. The 10-square as a whole is regular and continuous. S=490.

73	29	85	41	17	73	29	85	41	17
45	1	77	33	89	45	1	77	33	89
37	93	49	5	61	37	93	49	5	61
9	65	21	97	53	9	65	21	97	53
81	57	13	69	25	81	57	13	69	25
73	29	85	41	17	73	29	85	41	17
45	1	77	33	89	45	1	77	33	89
37	93	49	5	61	37	93	49	5	61
9	65	21	97	53	9	65	21	97	53
81	57	13	69	25	81	57	13	69	25

Fig. 364.

0	3	1	0	3	2	2	0	2	2
3	0	2	3	0	1	1	3	1	1
0	3	1	0	3	2	2	0	2	2
3	0	2	3	0	1	1	3	1	1
0	3	1	0	3	2	2	0	2	2
3	0	3	3	0	1	1	2	1	1
0	3	0	0	3	2	2	1	2	2
3	0	3	3	0	1	1	2	1	1
0	3	0	0	3	2	2	1	2	2
3	0	3	3	0	1	1	2	1	1

Fig. 365.

The second auxiliary Fig. 365 is supposed to have at first the normal arrangement in the top line 0.3.0.0.3.2.2.1.2.2. which would lead to correct results in the rectangular rows, but an alternation of values in all diagonals, 14 or 16. This has been equalized by exchange of half the middle columns, right and left, making all the diagonals = 15, but as the portions exchanged are unequal those two columns are unbalanced. The exchange of half columns might have taken place in the 1st and 8th, or in the 2d and 6th, either the upper or the lower half, or otherwise symmetrically, the same results following.

The resultant square Fig. 366 contains all the original series, 1 to 100; it has the constant S=505 in thirty-eight out of the total of forty rows. When made magic by transposition of 15 and 16, or some other pair of numbers in those affected columns, the four diagonals that pass through such pair must bear the inequality. Here, as in the previous example, the object is to give the second

73	32	86	41	20	75	31	85	43	19
48	1	79	36	89	46	2	80	34	90
37	96	50	5	64	39	95	49	7	63
12	65	23	100	53	10	66	24	98	54
81	60	14	69	28	83	59	13	71	27
76	29	88	44	17	74	30	87	42	18
45	4	77	33	92	47	3	78	35	91
40	93	52	8	61	38	94	51	6	62
9	68	21	97	56	11	67	22	99	55
84	57	16	72	25	82	58	15	70	26

Fig. 366.

auxiliary equal summation in all diagonals at the expense of two verticals, and then to correct the corresponding error of the finished square by exchange of two numbers that differ by unity.

In all cases the main auxiliary is a quartered square, but the second auxiliary is not; hence the completed square cannot have the half lines equal, since S is always an odd number. However, there are some remarkable combinations and progressions. For instance in Fig. 366 the half lines in the top row are 252 + 253; in the second row 253+252; and so on, alternating all the way down. Also in the top row the alternate numbers 73+86+20+31+43=253 and the 32, 41 etc. of course = 252. The same peculiarity is found in all the rows. Figs. 364 nad 365 have similar combinations. Also

Figs. 360, 361 and 362. This gives rise to some Nasik progressions. Thus in Fig. 364 from upper left corner by an oblique step one cell to the right and five cells down: 73+29+85+41 etc. ten terms, practically the same as the top row = 490. This progression may be taken right or left, up or down, starting from any cell at pleasure. In Fig. 365 the ten terms will always give the constant S = 15 by the knight's move (2, 1) or (1, 2) or by the elongated step (3, 4). Fig. 366 has not so much of the Nasik property. The oblique step one to the right and five down, 73+29+86 44 etc. ten terms = 505. This progression may start from any cell moving up and down, right and left by a sort of zigzag. The second auxiliary is richest in this Nasik property, the main auxiliary less so, as it is made by the knight's move; and the completed square still less so, as the other two neutralize each other to some extent. A vast number of variations may be obtained in the larger squares, as the subsquares admit of so many different constructive plans.

The examples already presented may serve as models for the larger sizes; these are familiar and easily handled, and they clearly show the rationale of the process. If any one wishes to traverse wider areas and to set down more numbers in rank and file, no further computations are required. The terms 0.1.2.3 are always employed: the series 1.5.9 etc. to 97, and after that 101.105.109 and so on. The principal auxiliary may be made magic by any approved process as elegant and elaborate as desired, the four subsquares being facsimiles. The second auxiliary has for all sizes an arrangement analogous to that already given which may be tabulated as follows:

6-square,	0 3 0 — 2 2 2	top row
10-square,	0 3 0 0 3 — 2 2 1 2 2	" "
14-square,	0 3 3 0 0 0 3 — 2 2 2 1 2 2 1	" "
18-square,	0 3 3 3 0 0 0 0 3 — 2 2 2 2 1 2 1 1 2	" "
etc.		

The top row being thus written, under each term is placed its complement, and all succeeding rows follow the same rule, so that the 1st, 3d, 5th etc. are the same, and the 4th, 6th, 8th etc. are repetitions of the 2d. This brings all the 0.3 terms on one side and all the 1.2 terms on the opposite. In columns there is a regular alternation of like terms; in horizontals the like terms are mostly consecutive, thus bringing the diagonals more nearly to an equality so that they may be corrected by wholesale at one operation. This systematic and somewhat mechanical arrangement insures correct summation in rows and columns, facilitates the handling of diagonals, and provides automatically for the required pattern of the 2-squares, in which both the auxiliaries and the completed square must agree. In making a square from the table it should be observed that an exchange of half columns is required, either the upper or the lower half, preferably of the middle columns; but as we have seen in the 10-square, several other points may be found suitable for the exchange.

1	5	13	9
13	9	1	5
1	5	13	9
13	9	1	5

Fig. 367.

0	3	0	3
1	2	1	2
3	0	3	0
2	1	2	1

Fig. 368.

1	8	13	12
14	11	2	7
4	5	16	9
15	10	3	6

Fig. 369.

This plan and process for developing to so high a degree of excellence, the oddly-even squares, starting with the 2-square, and constantly employing its endless combinations, is equally applicable to the evenly-even squares. They do not need it, as there are many well-known, convenient and expeditious methods for their construction. However, in closing we will give a specimen of the 4-square, type of all that class, showing the pervading influence therein of the truly ubiquitous 2-square.

The primaries Figs. 367 and 368 as well as the complete square Fig. 369 singly and together fill the bill with no discount. Each is a quartered square, magic to a high degree. Each contains numerous 2-squares, four being compact in the quarters and five others overlapping. And there are many more variously scattered abroad especially in Fig. 368. While these specimens seem to conform exactly to foregoing rules they were actually made by continuous process using the knight's move (2, 1) and (1, 2). The pattern is rhomboidal.

In all the combinations here presented, and especially in these last specimens, the 2-square is pervasive and organic. "So we have a symmetry," as one of our philosophical writers has said—"which is astonishing, and might be deemed magical, if it were not a matter of intrinsic necessity." D. F. S.

NOTES ON ODDLY-EVEN MAGIC SQUARES.

The foregoing article on oddly-even squares by Mr. D. F. Savage is a valuable contribution to the general literature on magic squares. Mr. Savage has not only clearly described a clever and unique method of constructing oddly-even squares, but he has also lucidly demonstrated the apparent limit of their possible perfection.

The arrangement of concentric quartets of four consecutive numbers in his 6×6 square is strikingly peculiar, and in studying this feature it occurred to the writer that it might be employed in the development of these squares by a direct and continuous process, using the arithmetical series 1 to n^2 taken in groups of four consecutive terms, 1.2.3.4. — 5.6.7.8. etc.

The constructive method used by Mr. Savage is based on the well-known and elegant plan of De la Hire, but the two number series which he has chosen for the first and second auxiliary squares are unusual, if not entirely new. It is difficult to see how these unique squares could have been originally evolved by any other method than that adopted by Mr. Savage, and the different constructive scheme presented herewith must be regarded as only a natural outcome of the study of his original plan. It may also tend to throw a little additional light on the "ubiquitous 2×2 square" and to make somewhat clearer the peculiar features that obtain in these oddly-even squares.

Referring to Fig. 370 (which is a reflected inversion of Fig. 361 and therefore requires no further explanation) it will be seen that this square contains nine quadrate groups of the series 0.1. 2.3., the numbers in each group being scattered in each of the

3×3 quarters, and in concentric relationship to the 6×6 square. The numbers of these quadrate groups are not, however, distributed in any apparent order as viewed numerically, although the diagram

3	0	1	3	1	1
0	3	2	0	2	2
3	0	1	3	1	1
0	2	2	0	3	2
3	1	1	3	0	1
0	2	2	0	3	2

Fig. 370.

of their consecutive forms, which will be referred to later on, reveals the symmetry of their arrangement.

Any middle outside cell of the 3×3 quarters containing a

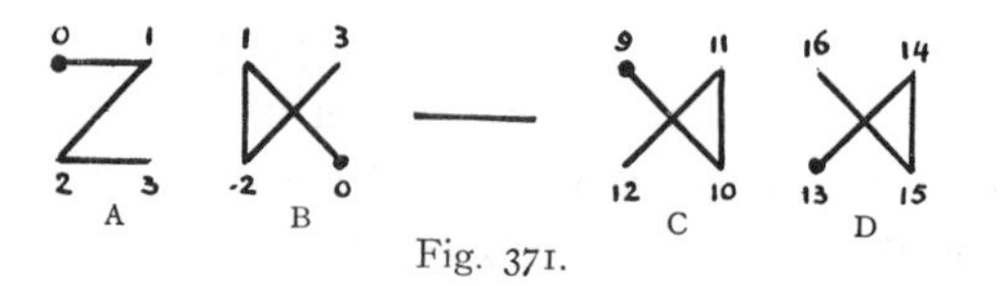

Fig. 371.

cypher can be used as a starting point for a 6×6 square, and inspection will show four such cells in Fig. 370.

Selecting the second cell from the left in the upper line to start

	1			2	
9					11
16		6	8		14
13		7	5		15
12					10
	3			4	

Fig. 372.

32	1	22	24	2	30
9	20	27	25	19	11
16	33	6	8	34	14
13	35	7	5	36	15
12	18	26	28	17	10
29	3	23	21	4	31

Fig. 373.

from, the numbers in the quadrate concentric group of which this cell is a member will be seen to have the formation shown in Fig. 371A, so the first group of four numbers (1.2.3.4) in the series

1 to 36 are similarly placed in Fig. 372, running also in the same *relative* numerical order.

To secure magic results in the completed square, each succeeding entry in the 3×3 quarters must follow the last entry in *magic square order.* For the next entry in Fig. 372 there is consequently a choice of two cells. Selecting the lower right-hand corner cell of the 3×3 quarter of Fig. 370 used at the start, it is seen to be occupied by 1, and the formation of the quadrate concentric group is as shown in Fig. 371B. The terms 5.6.7.8 are therefore entered in Fig. 372 in similarly located cells, and as before, in the

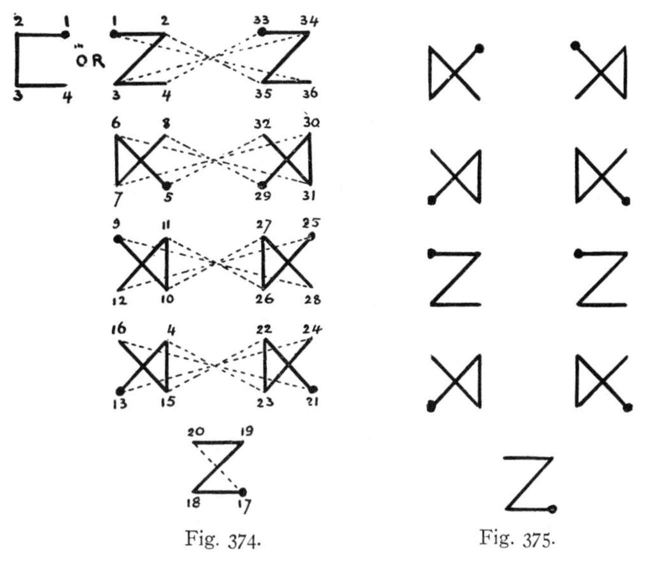

Fig. 374. Fig. 375.

same relative numerical order. The next quadrate group of 9.10. 11.12 have the order shown in Fig. 371C,—13.14.15.16 are arranged as in Fig. 371D, and so on until all of the 36 cells are filled. The resulting finished square is shown in Fig. 373.

Fig. 374 shows the different forms of the nine consecutive quadrate groups contained in Fig. 373, written in regular order, and it discloses the harmonious relationship of the couplets.

There are two alternative forms for the first group, as shown in Fig. 374. If the square is to be pan-diagonal or continuous at the expense of the summation of two vertical columns, the right-hand form must be used, but if the square is to be strictly magic at the expense of making four diagonals incorrect, then the left-hand form is correct.

This graphic presentation of number order is instructive, as it shows at a glance certain structural peculiarities which are not apparent on the face of the square.

8	25	22	24	26	6
33	20	3	1	19	35
16	9	30	32	10	14
13	11	31	29	12	15
36	18	2	4	17	34
5	27	23	21	28	7

Fig. 376.

Another of the many variants of this 6×6 square may be made by starting from the fourth cell of the second line in Fig. 370, this being also a middle outside cell of a 3×3 square.

Under this change the forms of the quadrate groups are shown in Fig. 375, the resulting square being given in Fig. 376.

20	41	86	32	73	75	31	85	43	19
89	36	79	1	48	46	2	80	34	90
63	5	50	93	40	39	95	49	7	64
56	97	23	68	9	10	66	24	98	54
25	72	14	59	84	83	57	13	71	27
26	70	16	58	82	81	60	15	69	28
55	99	21	67	11	12	65	22	100	53
62	6	52	94	37	38	96	51	8	61
91	35	77	3	47	45	4	78	33	92
18	44	88	30	74	76	29	87	42	17

+1 —1

Fig. 377.

When these 6×6 squares are made pan-diagonal, i. e., perfect in all their diagonals, the normal couplets are arranged in harmonic relation throughout the square, the two paired numbers that equal n^2+1 being always located in the same diagonal and equally spaced $n/2$ cells apart. If the square is made strictly magic, however, this harmonic arrangement of the couplets is naturally disturbed in the imperfect diagonals.

The above remarks and rules will of course apply generally to 10×10 and larger squares of this class. A 10×10 square modified from Mr. Savage's example to secure the harmonic arrangement of the couplets, as above referred to, is given in Fig. 377. W. S. A.

NOTES ON PANDIAGONAL AND ASSOCIATED MAGIC SQUARES.

The reader's attention is invited to the plan of a magic square of the thirteenth order shown in Fig. 378 which is original with the

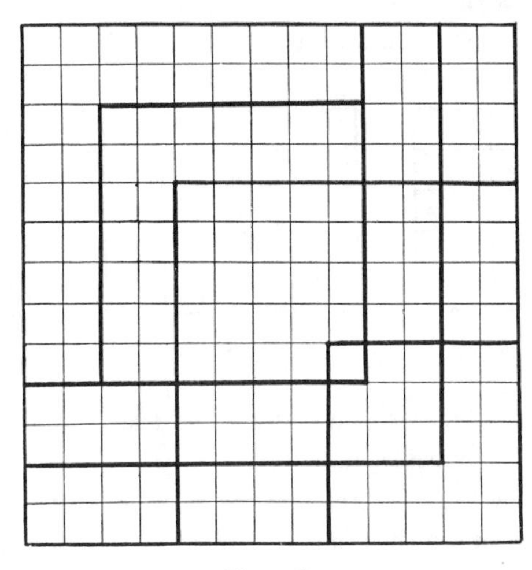

Fig. 378.

writer. It is composed of four magic squares of the fourth order, two of the fifth order, two of the seventh order, two of the ninth order, one of the eleventh order and finally the total square of the thirteenth order, thus making twelve perfect magics in one, several of which have cell numbers in common with each other.

To construct this square it became necessary to take the arithmetical series 1, 2, 3.... 169 and resolve it into different series capable of making the sub-squares. A close study of the constitution of all these squares became a prerequisite, and the following observations are in a large part the fruit of the effort to accomplish the square shown. This article is intended however to cover more particularly the constitution of squares of the fifth order. The results naturally apply in a large degree to all magic squares, but especially to those of uneven orders.

It has of course been long known that magic squares can be built with series other than the natural series 1, 2, 3.... n^2, but the perplexing fact was discovered, that although a magic square might result from one set of numbers when arranged by some rule, yet when put together by another method the construction would fail to give magic results, although the second rule would work all right with another series. It therefore became apparent that these rules were in a way only *accidentally right.* With the view of explaining

a				
		a		
				a
	a			
			a	

Fig. 379.

a	b	c	d	g
d	g	a	b	c
b	c	d	g	a
g	a	b	c	d
c	d	g	a	b

Fig. 380.

x	y	s	t	v
s	t	v	x	y
v	x	y	s	t
y	s	t	v	x
t	v	x	y	s

Fig. 381.

these puzzling facts, we will endeavor to analyze the magic square and discover, if possible, its *raison d'être.*

The simplest, and therefore what may be termed a "primitive" square, is one in which a single number is so disposed that every column contains this number once and only once. Such a square is shown in Fig. 379, which is only one of many other arrangements by which the same result will follow. In this square every column has the same summation (a) and it is therefore, in a limited sense, a magic square.

Our next observation is that the empty cells of this figure may be filled with other quantities, resulting, under proper arrangement, in a square whose every column will still have a constant summation. Such a square is shown in Fig. 380 in which every column sums $a+b+c+d+g$, each quantity appearing once and only once in each row, column, and diagonal. These squares however have the fatal defect of duplicate numbers, which can not be tolerated. This defect can be removed by constructing another primitive square, of five other numbers (Fig. 381), superimposing one square upon the other, and adding together the numbers thus brought together. This idea is De la Hire's theory, and it lies at the very foundation of magical science. If however we add a to x in one cell and in another cell add them together again, duplicate numbers will still result, but this can be obviated by making the geometrical pattern in one square the *reverse* of the same pattern in the other square. This idea is illustrated in Figs. 380 and 381, wherein the positions of a and v are reversed. Hence, in the addition of cell numbers in two such squares a series of diverse numbers must result. These series are necessarily magical because the resulting square is so. We can now lay down the first law regarding the constitution of magical series, viz., *A magic series is made by the addition, term to term, of x quantities to x other quantities.*

As an example, let us take five quantities, a, b, c, d and g, and add them successively to five other quantities x, y, s, t and v, and we have the series:

$$\begin{array}{ccccc} a+x & a+y & a+s & a+t & a+v \\ b+x & b+y & b+s & b+t & b+v \\ c+x & c+y & c+s & c+t & c+v \\ d+x & d+y & d+s & d+t & d+v \\ g+x & g+y & g+s & g+t & g+v \end{array}$$

This series, with *any values* given to the respective symbols, will produce magic squares if properly arranged. It is therefore a *universal series,* being convertible into any other possible series.

We will now study this series, to discover its peculiar properties if we can, so that hereafter it may be possible at a glance to determine whether or not a given set of values can produce magical results. First, there will be found in this series a property which may be laid down as a law, viz.:

There is a constant difference between the homologous numbers of any two rows or columns, whether adjacent to each other or not. For example, between the members of the first row and the corresponding members of the second row there is always the constant difference of $a-b$. Also between the third and fourth rows there is a constant difference $c-d$, and between the second and third columns we find the constant difference $y-s$ etc., etc. Second, it will be seen that any column can occupy any vertical position in the system and that any row could exchange place with any other row. (As any column could therefore occupy any of five positions in the system, in the arrangement of columns we see a total of

$$5\times 4\times 3\times 2\times 1=120 \text{ choices.}$$

Also we see a choice of 120 in the rows, and these two factors indicate a total of 14,400 different arrangements of the 25 numbers and a similar number of variants in the resulting squares, to which point we will revert later on.)

This *uniformity of difference* between homologous numbers of

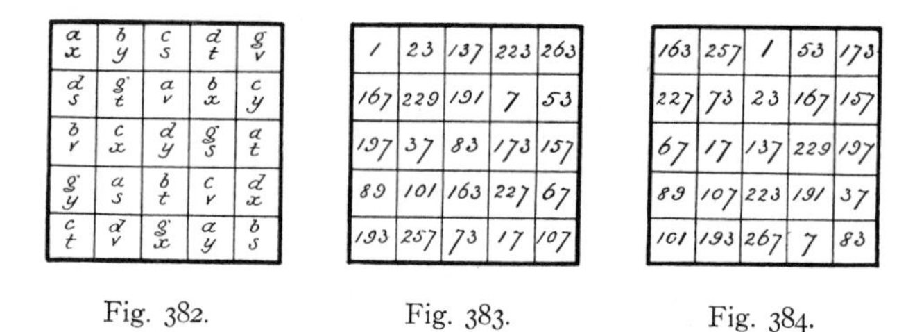

Fig. 382. Fig. 383. Fig. 384.

any two rows, or columns, appears to be the *only essential* quality of a magical series. It will be further seen that this must necessarily be so, because of the process by which the series is made, i. e., the successive addition of the terms of one series to those of the other series.

As the next step we will take two series of five numbers each, and, with these quantities we will construct the square shown in Fig. 382 which combines the two primitives, Figs. 380 and 381.

By observation we see that this is a *pure* square, i. e., in no row, column, or diagonal is any quantity *repeated* or *lacking*. Because any value may be assigned to each of the ten symbols used, it will be seen that this species of square depends for its peculiar properties *upon the geometrical arrangement of its members and not on their arithmetical values*; also that the five numbers represented by the symbols *a, b, c, d, g,* need not bear any special ratio to each other, and the same heterogeneity may obtain between the numbers represented by *x, y, s, t, v.*

There is however another species of magic square which is termed "associated" or "regular," and which has the property that the sum of any two diametrically opposite numbers equals twice the contents of the central cell. If we suppose Fig. 382 to be such a square we at once obtain the following equations:

$$(1) \qquad (d+s)+(d+x)=2d+2y \therefore x+s=2y$$
$$(2) \qquad (d+t)+(d+v)=2d+2y \therefore t+v=2y$$
$$(3) \qquad (c+y)+(g+y)=2d+2y \therefore c+g=2d$$
$$(4) \qquad (a+y)+(b+y)=2d+2y \therefore a+b=2d$$

Hence it is evident that if we are to have an associated square, the element *d* must be an arithmetical mean between the quantities *c* and *g* and also between *a* and *b*. Also, *y* must be a mean between *x* and *s*, and between *t* and *v*. It therefore follows that an associated square can only be made when the proper *arithmetical relations* exist between the numbers used, while the construction of a continuous or pandiagonal square depends upon the *method of arrangement* of the numbers.

The proper relations are embraced in the above outline, i. e., that the *central term of each of the five (or x) quantities shall be a mean between the diametrically opposite pair.* For example, 1, 4, 9, 14, 17, or 1, 2, 3, 4, 5, or 1, 2, 10, 18, 19, or 1, 10, 11, 12, 21 are all series which, when combined with similar series, will yield magical series from which associated magic squares may be constructed.

The failure to appreciate this distinction between pandiagonal and associated squares is responsible for much confusion that exists, and because the natural series $1, 2, 3, 4 \ldots n^2$ *happens,* as it were, accidentally to be such a series as will yield associated squares, *empirical rules have been evolved for the production of squares which are only applicable to such a series,* and which consequently fail when another series is used. For example, the old time Indian rule of regular diagonal progression when applied to a *certain class* of series will yield magic results, but when applied to another class of series it fails utterly!

As an example in point, the following series, which is composed of prime numbers, will yield the continuous or Nasik magic square shown in Fig. 383, but a square made from the same numbers arranged according to the old rule is not magic in its diagonals as shown in Fig. 384.

1	7	37	67	73
17	23	53	83	89
101	107	137	167	173
157	163	193	223	229
191	197	227	257	263

The fundamentally *partial* rules, given by some authors, have elevated the *central row* of the proposed numbers into a sort of axis on which they propose to build. This central row of the series is thrown by their rules into one or the other diagonal of the completed square. The fact that this central row adds to the correct summation is, as before stated, simply an accident accruing to the normal series. The central row does *not* sum correctly in many magical series, and rules which throw this row into a diagonal are therefore incompetent to take care of such series.

Returning to the general square, Fig. 382, it will be seen that because each row, column and diagonal contains every one of the ten quantities composing the series, the sum of these ten quantities equals the summation of the square. Hence it is easy to make a square whose summation shall be any desired amount, and also at the same time to make the square contain certain predetermined numbers.

For example, suppose it is desired to make a square whose summation shall be 666, and which shall likewise contain the numbers 6, 111, 3 and 222. To solve this problem, two sets of five numbers each must be selected, the sum of the two sets being 666, and the sums of some members in pairs being the special numbers wished. The two series of five numbers each in this case may be

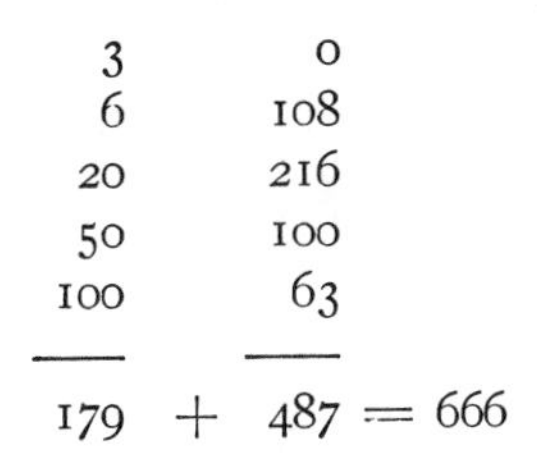

3	0
6	108
20	216
50	100
100	63

$$179 + 487 = 666$$

from which by regular process we derive the magic square series

3	6	20	50	100
111	114	128	158	208
219	222	236	266	316
103	106	120	150	200
66	69	83	113	163

containing the four predetermined numbers. The resulting magic

3	114	236	150	163
266	200	66	6	128
69	20	158	316	103
208	219	106	83	50
120	113	100	111	222

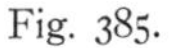

Fig. 385.

1	59	8	15	19
14	12	13	21	42
33	4	48	11	6
45	5	26	16	10
9	22	7	39	25

Fig. 386.

Fig. 387.

square is shown in Fig. 385, the summation of which is 666 and which is continuous or pandiagonal. As many as eight predetermined numbers can be made to appear together with a predetermined summation, in a square of the fifth order, but in this case duplicate numbers can hardly be avoided if the numbers are selected at random. We may go still further and force four predetermined numbers into four certain cells of any chosen column or row as per following example:

A certain person was born on the 1st day of the 8th month, was married at the age of 19, had 15 children and is now 102 years old. Make a pandiagonal square whose $S = 102$ and in which numbers 1, 8, 15, 19 shall occupy the first, third, fourth and fifth cells of the upper row.

Referring to the universal square given in Fig. 382,

Let $a = 0$ $\quad x = 1$
$c = 3$ $\quad s = 5$
$d = 9$ $\quad t = 6$
$g = 6$ $\quad v = 13$

These eight quantities sum 43, so that the other pair (b and y) must sum 59, $(43 + 59 = 102)$. Making therefore $b = 20$ and $y = 39$, and replacing these values in Fig. 382, we get the desired square shown in Fig. 386.

As previously shown, continuous squares are dependent on the geometrical placing of the numbers, while associated squares depend also upon the arithmetical qualities of the numbers used. In this connection it may be of interest to note that *a square of third order can not be made continuous,* but *must be associated*; a square of the fourth order may be made *either continuous or associated, but can not combine these qualities*; in a square of the fifth order *both qualities may belong to the same square.* As I showed in the first article of this chapter, very many continuous or Nasik squares of the fifth order may be constructed, and it will now be proven that associated Nasik squares of this order can only be made in fewer numbers.

In a continuous or "pure" square each number of the sub-series must appear once and only once in each row, column, and diagonal (broken or entire). Drawing a square, Fig. 387, and placing in it an element x as shown, the cells in which this element *cannot* then be placed are marked with circles. In the second row only two cells are found vacant, thus giving only two choices, indicating two forms of the square. Drawing now another square, Fig. 388, and filling its first row with five numbers, represented by the symbols t, v, x, y and s, and choosing one of the two permissible cells for x in the second row, it will be seen that there can be but *two* variants when once the first row is filled, the contents of every cell in the square being forced as soon as the choice between the two cells in the second row is made for x. For the other subsidiary square, Fig. 389, with numbers represented by the symbols, a, b, c, d and g, there is *no choice,* except in the filling of the first row. If this row is filled, for example, as shown in Fig. 389, all the other cells in this square *must* be filled in the manner shown in order that it may fit Fig. 388.

Now, therefore, taking the five symbols x, y, s, t, v, any one of them may be placed in the first cell of the first line of Fig. 388. For the second cell there will remain a choice of four symbols, for the third cell three, for the fourth cell two, for the fifth cell no choice, and finally in the second line there will be a choice of two cells. In the second subsidiary there will be, as before, a choice of five, four, three and finally two, and no choice in the second row. Collecting these choices we have $(5\times4\times3\times2\times2)\times(5\times4\times3\times2) = 28{,}800$, so that exactly 28,800 continuous or Nasik squares of the fifth order may be made from any series derived from ten numbers.

t	v	x	y	s
x	y	s	t	v
s	t	v	x	y
v	x	y	s	t
y	s	t	v	x

Fig. 388.

a	b	c	d	g
d	g	a	b	c
b	c	d	g	a
g	a	b	c	d
c	d	g	a	b

Fig. 389.

1	5	2	3	4
3	4	1	5	2
5	2	3	4	1
4	1	5	2	3
2	3	4	1	5

Fig. 390.

Only one-eighth of these, or 3600, will be really diverse since any square shows eight manifestations by turning and reflection.

The question now arises, how many of these 3600 diverse Nasik squares are also associated? To determine this query, let us take the regular series 1, 2, 3....25 made from the ten numbers

1	2	3	4	5
0	5	10	15	20

Making the first subsidiary square with the numbers 1, 2, 3, 4, 5, (Fig. 390) as the square is to be associated, the central cell must contain the number 3. Selecting the upward left-hand diagonal to work on, we can place either 1, 2, 4 or 5 in the next upward cell of this diagonal (a choice of four). Choosing 4, we *must* then write 2 in its associated cell. For the upper corner cell there remains a choice of two numbers, 1 and 5. Selecting 1, the location of 5 is forced. Next, by inspection it will be seen that the number 1 may be placed in either of the cells marked. □, giving two choices. Selecting the upper cell, every remaining cell in the square becomes *forced.* For this square we have therefore only

$$4 \times 2 \times 2 = 16 \text{ choices.}$$

For the second subsidiary square (Fig. 391) the number 10 must occupy the central cell. In the left-hand upper diagonal adjacent cell we can place either 0, 5, 15 or 20 (four choices). Selecting 0 for this cell, 20 becomes fixed in the cell associated with that containing 0. In the upper left-hand corner cell we can place either 5 or 15 (two choices). Selecting 15, 5 becomes fixed. Now we cannot in this square have any further choices, because all other 15's *must* be located as shown, and so with all the rest of the numbers, as may be easily verified. The total number of choices in this

15	10	5	0	20
5	0	20	15	10
20	15	10	5	0
10	5	0	20	15
0	20	15	10	5

Fig. 391.

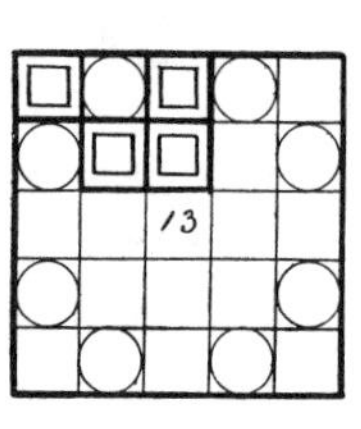

Fig. 392.

1	47	6	43	5	48
35	17	30	21	31	16
36	12	41	8	40	13
7	45	2	49	3	44
29	19	34	15	33	20
42	10	37	14	38	9

Fig. 393.

square are therefore $4 \times 2 = 8$, and for both of the two subsidiaries, $16 \times 8 = 128$. Furthermore, as we have seen that each square has eight manifestations, there are really only $128/8 = 16$ *different plans* of squares of this order which combine the associated and Nasik features.

If a continuous square is expanded indefinitely, any square block of twenty-five figures will be magic. Hence, with any given square, twenty-five squares may be made, only one of which can be associated. There are therefore $16 \times 25 = 400$ variants which can be made according to the above plan. We have however just now shown that there are 3600 different plans of continuous squares of this order. Hence it is seen that only one plan in nine ($3600/400 = 9$) of continuous squares can be made *associated* by shifting the lines and columns. Bearing in mind the fact that eight variants of a square may be made by turning and reflection, it is interesting to note that if we wish a square of the fifth order to be both associated and continuous, we can locate unity in any one of the four cells marked □ in Fig. 392, but by no constructive process can the desired result be effected, if unity is located in any cells marked ○. Then having selected the cell for 1, the cell next to 1 in the same column with the central cell (13) must contain one of the four numbers 7, 9, 17, or 19. The choices thus entailed yield our estimated number of sixteen diverse associated Nasik squares, which may be naturally increased eight times by turning and reflection.

That we must place in the same row with 1 and 13, one of the four numbers 7, 9, 17, or 19 is apparent when it is noted that of the series

1	2	3	4	5
0	5	10	15	20

having placed 3 and 10 in the central cells of the two subsidiaries, and 0 and 1 in two other cells, we are then compelled to use in the same line either 5 or 15 in one subsidiary and either 2 or 4 in the other subsidiary, the combination of which four numbers affords only 7 and 17, or 9 and 19.

With these facts now before us we are better prepared to construct those squares in which only prime numbers are used, etc. Reviewing a list of primes it will be seen that every number excepting 2 and 5 ends in either 1, 3, 7 or 9. Arranging them therefore in regular order according to their terminal figures as

1	11	31	41	
3	13	23	43	
7	17	37	47	etc.

we can make an easier selection of desired numbers.

A little trial develops the fact that it is impossible to make five rows of prime numbers, showing the same differences between every row, or members thereof, and therefore a *set* of differences must be found, such as 6, 30, 30, 6 (or some other suitable *set*). Using the above set of differences, the series of twenty-five primes shown on page 234 may be found. In this series it will be seen that similar differences exist between the homologous numbers of any row, or column, and it is therefore only necessary to arrange the numbers by a regular rule, in order to produce the magic square in Fig. 383.

These facts throw a flood of light upon a problem on which gallons of ink have been wasted, i. e., the production of pandiagonal and regular squares of the sixth order. It is impossible to distribute six marks among the thirty-six cells of this square so that one and only one mark shall appear in every column, row and diagonal. Hence a *primitive* pandiagonal magic square of this

157	13	23	147	109	31	111	138	36	66	102	100	72
145	25	17	153	61	139	59	32	134	104	68	98	70
16	154	144	26	57	56	30	112	136	99	105	60	110
22	148	156	14	113	114	140	58	34	65	71	133	37
97	73	94	76	151	18	21	89	146	135	35	29	141
79	91	78	92	27	82	150	155	11	63	107	33	137
74	96	75	95	143	159	15	20	88	115	55	101	69
90	80	93	77	19	24	81	149	152	54	116	103	67
164	6	3	167	85	142	158	12	28	64	106	108	62
7	163	168	86	1	132	44	39	125	50	48	118	124
162	8	84	2	169	38	126	131	45	120	122	52	46
5	83	161	10	166	129	43	40	128	123	117	49	51
87	165	9	160	4	41	127	130	42	47	53	121	119

Fig. 394.

order is excluded by a geometrical necessity. In this case the natural series of numbers is not adapted to construct pandiagonal squares of this order. That the difficulty is simply an arithmetical one is proven by the fact that 6×6 pandiagonal squares can be made with *other series,* as shown in Fig. 393. We are indebted to Dr. C. Planck for this interesting square which is magic in its six rows, six columns and twelve diagonals, and is also four-ply and nine-ply, i. e., any square group of four or nine cells respectively, sums four or nine times the mean. It is constructed from a series made by arranging the numbers 1 to 49 in a square and eliminating all numbers in the central line and column, thus leaving thirty-six numbers as follows:*

1	2	3	5	6	7
8	9	10	12	13	14
15	16	17	19	20	21
29	30	31	33	34	35
36	37	38	40	41	42
43	44	45	47	48	49

Fig. 394 shows the completed square which is illustrated in skeleton form in Fig. 378. All the subsquares are faultless except the small internal 3×3, in which one diagonal is incorrect.

L. S. F.

SERRATED MAGIC SQUARES.

The curious form of magic squares which is to be described here possesses a striking difference from the general form of magic squares.

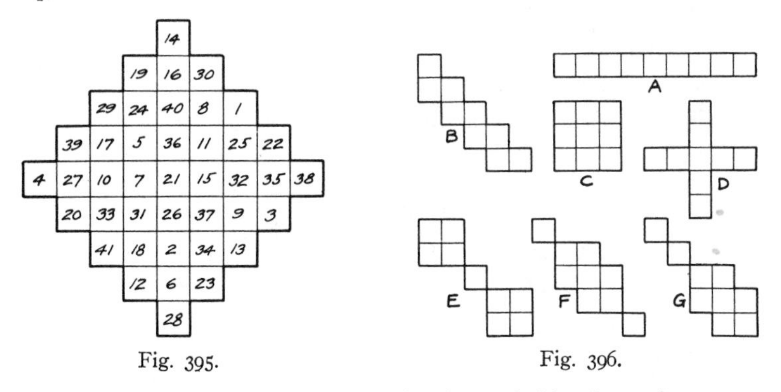

Fig. 395. Fig. 396.

To conform with the saw-tooth edges of this class of squares, I have ventured to call them "serrated" magic squares.

A square containing the series 1, 2, 3, 4,....41 is shown in Fig. 395. Its diagonals are the horizontal and vertical series of nine numbers, as A in Fig. 396. Its rows and columns are zigzag as

* For further information regarding squares of this type wherein n is of the form $4p + \text{B}$. See p. 267.

shown at B, and are sixteen in number, a quantity which is always equal to the number of cells which form the serrations.

All of this class of squares must necessarily contain the two above features.

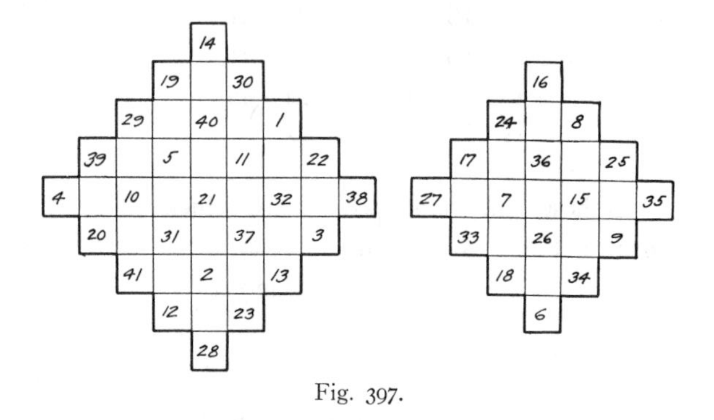

Fig. 397.

But, owing to its Nasik formation, Fig. 395 possesses other features as follows:

There are nine summations each of the square and cruciform, as at C and D in Fig. 396, the centers of which are 40, 11, 32, 5, 21, 37, 10, 31 and 2 respectively. Of E and F there are six summations each, and of the form G there are twelve summations.

This square was formed by the interconcentric position of the

1	2	3	4	5	6	7	8	9
10	11	12	13	14	15	16	17	18
19	20	21	22	23	24	25	26	27
28	29	30	31	32	33	34	35	36
37	38	39	40	41				

Fig. 398.

two Nasik squares shown in Fig. 397, and the method of selecting their numbers is clearly shown in Fig. 398.

There are numerous other selections for the sub-squares and the summations are not necessarily constant. This is shown by the following equations.

Let N and n equal the number of cells on a side of the large and small squares respectively, and let S equal the summations.

Then, when the means of each sub-square are equal

$$S=\frac{(1+N^2+n^2)(N+n)}{2}$$

When the large square has the first of the series and the small square has the last of the series

$$S=\frac{N(1+N^2)}{2}+\frac{n(1+n^2)}{2}+N^2n$$

When the large square has the last of the series and the small square has the first of the series

$$S=\frac{N(1+N^2)}{2}+\frac{n(1+n^2)}{2}+Nn^2$$

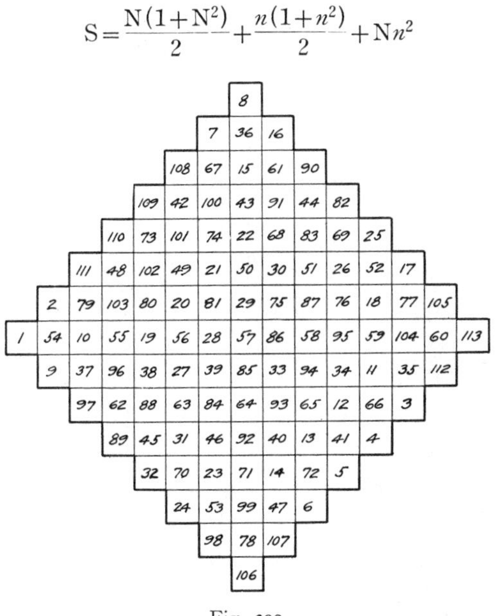

Fig. 399.

Only in such squares that fit the first equation, is it possible to have complementary pairs balanced about the center; in other words known as regular or associated squares.

Fig. 399 is one of this class and has summations of 855. In this case the mean of the series was used in the 7×7 sub-square and the remaining extremes made up the 8×8 square.

Figs. 400, 401, and 402 are the smallest possible examples of serrated squares. Fig. 400 is regular and is formed with the first of the above mentioned equations, and its summations are 91. Fig.

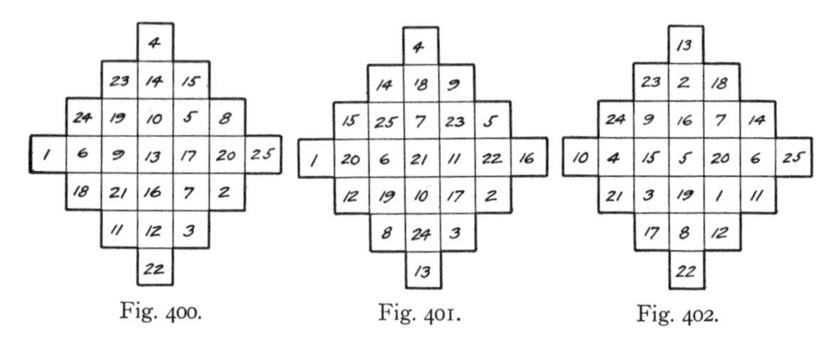

Fig. 400. Fig. 401. Fig. 402.

401 is formed with the second equation and its summations are 97. Fig. 402 is formed with the third equation and its summations are 85.

H. A. S.

LOZENGE MAGIC SQUARES.

Recently the writer has noticed in a weekly periodical a few examples of magic squares in which all of the odd numbers are arranged sequentially in the form of a square, the points of which meet the centers of the sides of the main square and the even numbers filling in the corners as shown in Fig. 405.

These articles merely showed the completed square and did not show or describe any method of construction.

A few simple methods of constructing these squares are described below, which may be found of some interest.

To construct such squares, n must necessarily be odd, as 3, 5, 7, 9, 11 etc.

A La Hireian method is shown in Figs. 403, 404, and 405, in which the first two figures are primary squares used to form the main square, Fig. 405. We begin by filling in the cells of Fig. 403, placing 1 in the top central cell and numbering downward 1, 2, 3 to 7 or n. We now repeat these numbers pan-diagonally down to the left filling the square.

Fig. 404 is filled in the same manner, only that we use the series

0, 1, 2, to 6 or n—1 in our central vertical column, and repeat these pan-diagonally down to the right. The cell numbers in Fig. 404 are then multiplied by 7 or n and added to the same respective cell numbers of Fig. 403, which gives us the final square Fig. 405.

5	6	7	1	2	3	4
6	7	1	2	3	4	5
7	1	2	3	4	5	6
1	2	3	4	5	6	7
2	3	4	5	6	7	1
3	4	5	6	7	1	2
4	5	6	7	1	2	3

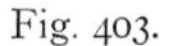

Fig. 403.

3	2	1	0	6	5	4
4	3	2	1	0	6	5
5	4	3	2	1	0	6
6	5	4	3	2	1	0
0	6	5	4	3	2	1
1	0	6	5	4	3	2
2	1	0	6	5	4	3

Fig. 404.

26	20	14	1	44	38	32
34	28	15	9	3	46	40
42	29	23	17	11	5	48
43	37	31	25	19	13	7
2	45	39	33	27	21	8
10	4	47	41	35	22	16
18	12	6	49	36	30	24

Fig. 405.

Another method is shown in Fig. 406 where we have five sub-squares placed in the form of a cross. The central one of these is filled consecutively from 1 to n^2. We then take the even numbers of the upper quarter, in this case 2, 8 and 4, and place them in the same respective cells in the lower sub-square. The lower quarter

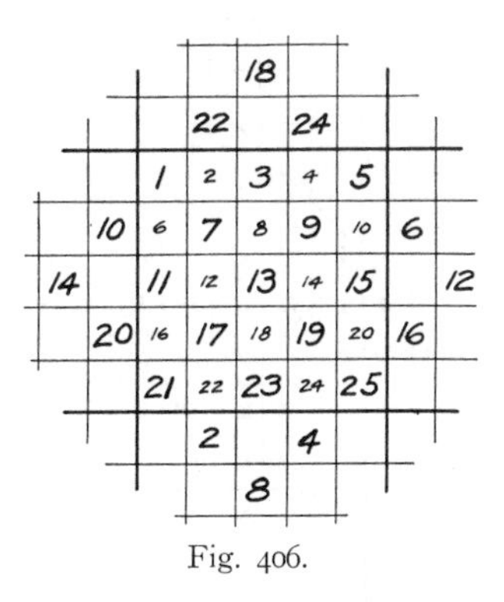

Fig. 406.

or 22, 18 and 24, are placed in the upper square. Likewise the left-hand quarter is placed in the right-hand square, and the right-hand quarter in the left-hand square. This gives us the required square, which is shown in heavy numbers.

A third method is to write the numbers consecutively, in the form of a square, over an area of adjacent squares as in Fig. 407.

The mean of the series must be placed in the center cell of the central or main square and the four next nearest to the center must find their places in the corner cells of the main square, which con-

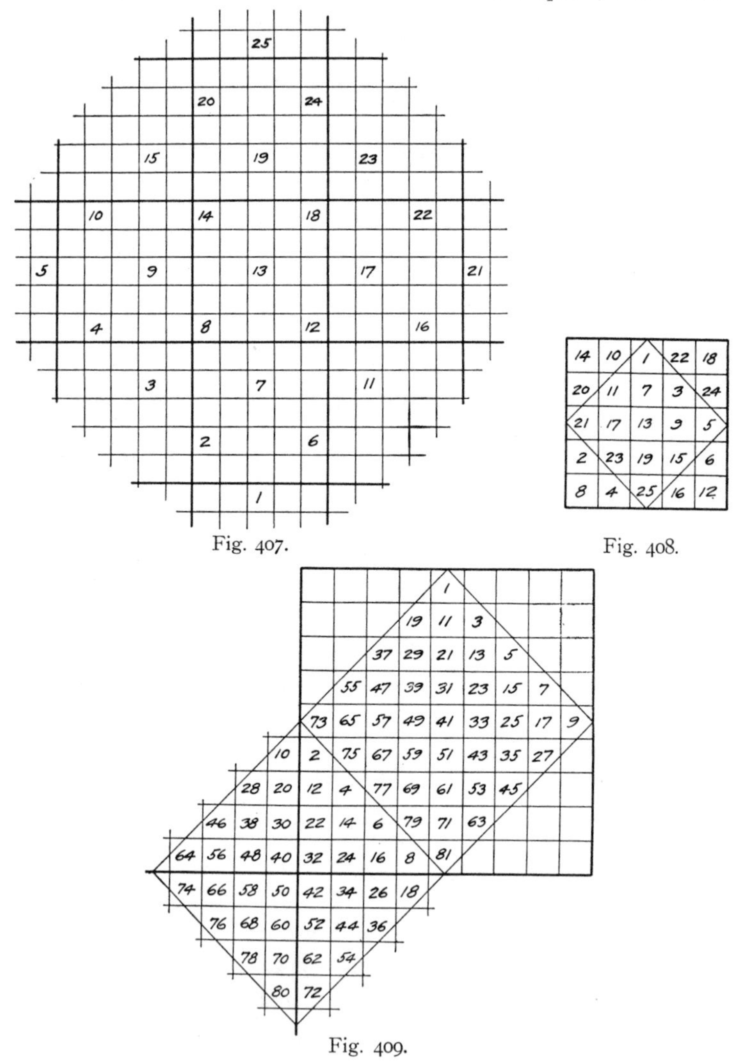

Fig. 407.

14	10	1	22	18
20	11	7	3	24
21	17	13	9	5
2	23	19	15	6
8	4	25	16	12

Fig. 408.

Fig. 409.

sequently governs the spacing in writing the series. We then remove all these numbers to the same respective cells in the main square, and this gives us the square shown in Fig. 408.

This last method is not preferable, owing to the largeness of the primary arrangement, which becomes very large in larger squares. It might however be used in the break-move style where the steps are equal to the distance from the center cell to the corner cell, and the breakmoves are one cell down when 1 is at the top.

What seems to be the most simple method is shown in Fig. 409 where the odd numbers are written consecutively in the main square, and directly following in the same order of progression the even numbers are written.

42	34	26	18	1	74	66	58	50
52	44	36	19	11	3	76	68	60
62	54	37	29	21	13	5	78	70
72	55	47	39	31	23	15	7	80
73	65	57	49	41	33	25	17	9
2	75	67	59	51	43	35	27	10
12	4	77	69	61	53	45	28	20
22	14	6	79	71	63	46	38	30
32	24	16	8	81	64	56	48	40

Fig. 410.

The even numbers necessarily run over into three adjacent sub-squares. These are removed to the same respective cells in the main square, the result of which is shown in Fig. 410.

The summations of Fig. 405 are 175, the summations of Figs. 406 and 408 are 65, and the summations for Fig. 410 are 369. Also, all complementary pairs are balanced about the center.

H. A. S.

CHAPTER XI.

SUNDRY CONSTRUCTIVE METHODS.

A NEW METHOD FOR MAKING MAGIC SQUARES OF ODD ORDERS.

IN an endeavor to discover a general rule whereby all forms of magic squares might be constructed, and thereby to solve the question as to the possible number of squares of the fifth order, a method was devised whereby squares may be made, for whose construction the rules at present known to the writer appear to be inadequate.

A *general rule,* however, seems as yet to be unattainable; nor does the solution of the possible number of squares of an order higher than four seem to be yet in sight, though, because of the discovery, so to speak, of hitherto unknown variants, the goal must, at least, have been brought nearer to realization.

The new method now to be described does not pretend to be other than a partial rule, i. e., a rule by which most, but possibly not *all* kinds of magic squares may be made. It is based on De la Hire's method, i. e., on the implied theory that a normal magic square is made up of two primary squares, the one superimposed on the other and the numbers in similarly placed cells added together. This theory is governed by the fact that a given series of numbers may be produced by the consecutive addition of the terms of two or more diverse series of numbers. For example, the series of natural numbers from one to sixteen may be regarded (*a*) as a single series, as stated, or (*b*) as the result of the addition, successively, of all the terms of a series of eight terms to those of another series of two terms. For example, if series No. 1 is composed of 0-1-2-3-4-5-6 and 7 and series No. 2 is composed of 1 and 9, all the numbers from 1 to 16 may be thus produced. Or (*c*) a series of four numbers, added successively to all the terms of another series of four numbers, will likewise produce the same result, as for example 0-1-2 and 3, and 1-5-9 and 13.

Without undertaking to trace out the steps leading up to the rule to be described, we will at once state the method in connection with a 5×5 square. First, two primary squares must be made, which will hereafter be respectively referred to as the A and B primary squares. If the proposed magic square is to be associated, that is, if its complementary couplets are to be arranged geometrically equidistant from the center, the central cell of each square must naturally

	3			
3				
		3		
				3
			3	

Fig. 411.

			10	
				10
		10		
10				
	10			

Fig. 412.

				3
	3			
		3		
			3	
3				

Fig. 413.

be occupied by the central number of the series of which the square is composed. The two series in this case may be 1-2-3-4-5 and 0-5-10-15-20. The central number of the first series being 3 and of the second series 10, these two numbers must occupy the central cells of their respective squares.

In each of these squares, each of the terms of its series must be represented five times, or as many times as the series has terms. Having placed 3 and 10 in their respective central cells, four other cells in each square must be similarly filled. To locate these cells, any geometrical design may be selected which is *balanced* about the central cell. Having done this in primary square A the *reverse* of the same design must be taken for primary square B, two examples being shown in Figs. 411 and 412 and Figs. 413 and 414.

Having selected a design, the next step will be to fill the *central* row, which may be done by writing in any of the four empty cells in this row, any of the four remaining terms of the series. The

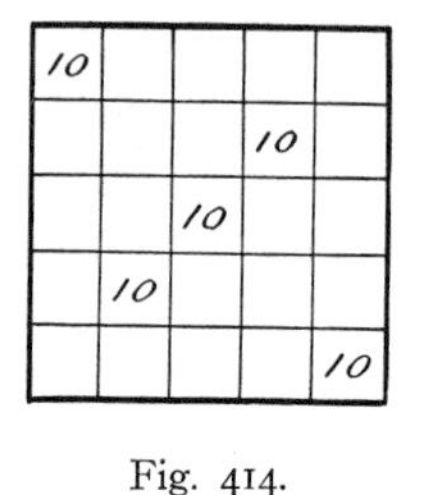

10				
			10	
		10		
	10			
				10

Fig. 414.

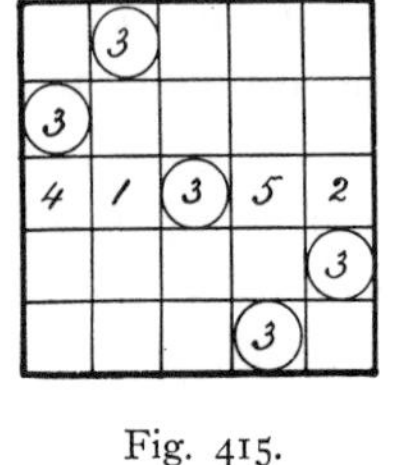

	3			
3				
4	1	3	5	2
				3
			3	

Fig. 415.

1	3	5	2	4
3	5	2	4	1
4	1	3	5	2
5	2	4	1	3
2	4	1	3	5

Fig. 416.

opposite cell to the one so filled must then be filled with the complementary number of the one last entered. Next, in either of the two remaining empty cells, write either of the remaining two terms

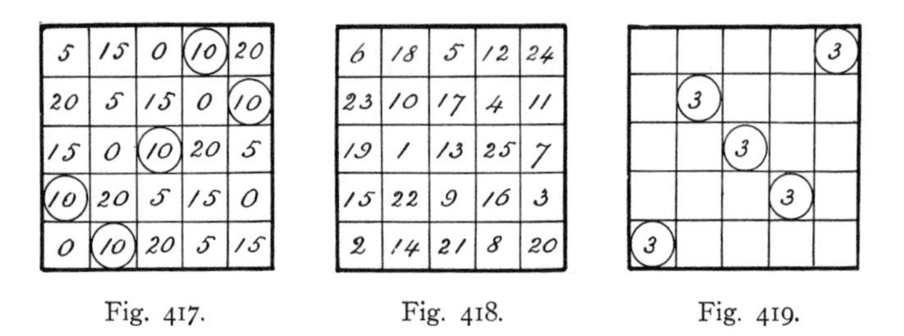

5	15	0	10	20
20	5	15	0	10
15	0	10	20	5
10	20	5	15	0
0	10	20	5	15

Fig. 417.

6	18	5	12	24
23	10	17	4	11
19	1	13	25	7
15	22	9	16	3
2	14	21	8	20

Fig. 418.

				3
	3			
		3		
			3	
3				

Fig. 419.

of the series, and, in the last empty cell the then remaining number, which will complete the central row as shown in Fig. 415. All the other rows in the square must then be filled, using the same *order*

4	5	1	2	3
2	3	4	5	1
1	2	3	4	5
5	1	2	3	4
3	4	5	1	2

Fig. 420.

10	5	0	20	15
0	20	15	10	5
20	15	10	5	0
15	10	5	0	20
5	0	20	15	10

Fig. 421.

14	10	1	22	18
2	23	19	15	6
21	17	13	9	5
20	11	7	3	24
8	4	25	16	12

Fig. 422.

of numbers as in this *basic* row, and the square will be completed as shown in Fig. 416. The second square can then be made up with the numbers of its series in exactly the same way, as shown in Fig. 417.

Adding together the terms of Figs. 416 and 417, will give the associated 5×5 magic square shown in Fig. 418, which can not be made by any previously published rule known to the writer. Another example may be given to impress the method on the student's mind, Fig. 419 showing the plan, Figs. 420 and 421 the A and B primary

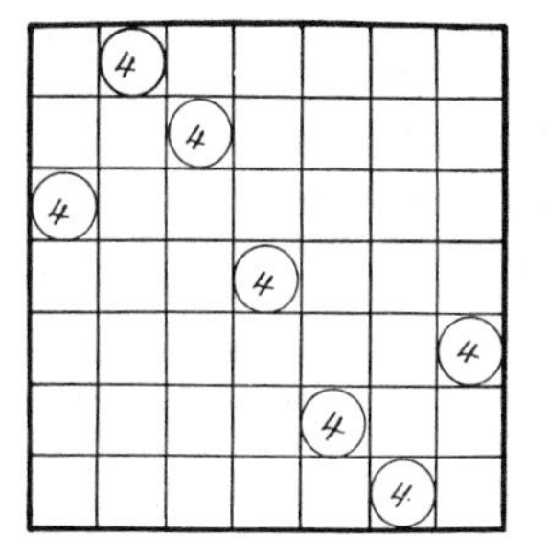

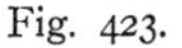
Fig. 423.

7	4	1	2	3	5	6
6	7	4	1	2	3	5
4	1	2	3	5	6	7
5	6	7	4	1	2	3
1	2	3	5	6	7	4
3	5	6	7	4	1	2
2	3	5	6	7	4	1

Fig. 424.

squares, and Fig. 422 the resulting magic square. Any odd square can be readily made by this method, a 7×7 being shown. Fig. 423 shows the plan, Figs. 424 and 425 being the primary squares and 426 the complete example. Returning to the 5×5 square, it will be seen that in filling out the central row of the A primary square

35	14	28	7	42	21	0
14	28	7	42	21	0	35
0	35	14	28	7	42	21
28	7	42	21	0	35	14
21	0	35	14	28	7	42
7	42	21	0	35	14	28
42	21	0	35	14	28	7

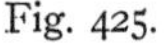
Fig. 425.

42	18	29	9	45	26	6
20	35	11	43	23	3	40
4	36	16	31	12	48	28
33	13	49	25	1	37	17
22	2	38	19	34	14	46
10	47	27	7	39	15	30
44	24	5	41	21	32	8

Fig. 426.

Fig. 415, for the first of the four empty cells, there is a choice of 16, and next a choice of four. Also for the B primary square there are the same choices. Hence we have

$$(16 \times 4)^2 = 4096 \text{ choices.}$$

In addition to this, by *reversing* the *patterns* in the two primary squares, the above number can be doubled.

It is therefore evident that with any chosen geometrical plan, 8192 variants of associated 5×5 squares can be produced, and as at least five distinct plans can be made, 40,960 different 5×5 associated squares can thus be formed. This however is not the limit, for the writer believes it to be a law that all "*figures of equilibrium*"

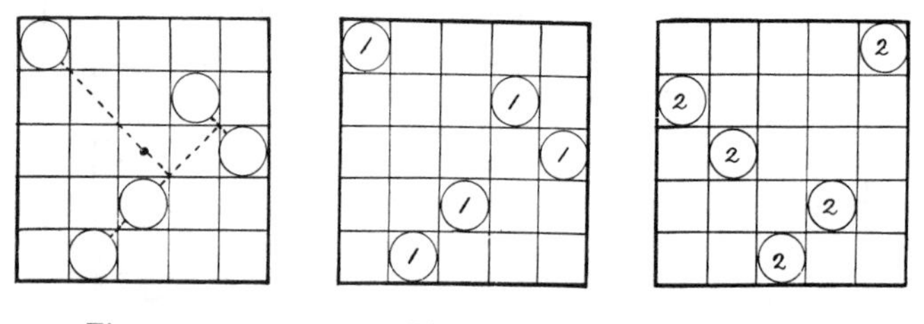

Fig. 427. Fig. 428. Fig. 429.

will produce magic squares as well as *geometrically* balanced diagrams or plans.

Referring to Fig. 427, if the circles represent equal weights connected as by the dotted lines, the system would balance at the center of the square. This therefore is a "figure of equilibrium" and it may be used as a basis for magic squares, as follows: Fill the

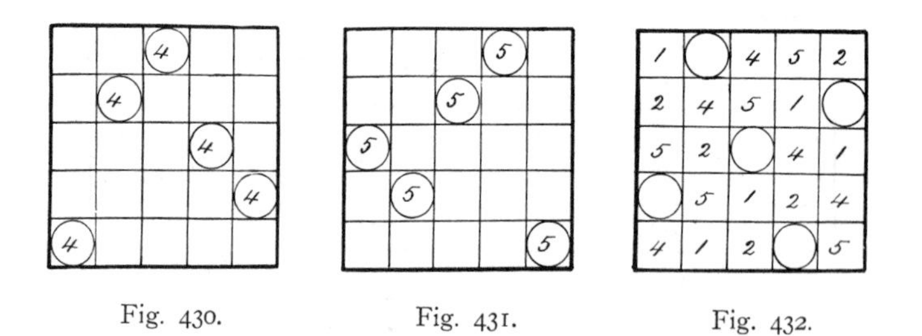

Fig. 430. Fig. 431. Fig. 432.

marked cells with a number, as for example 1 as in Fig. 428; then with the other numbers of the series, (excepting only the central number) make three other similar "figures of equilibrium" as shown separately in Figs. 429, 430 and 431, and collectively in Fig. 432. The five cells remaining empty will be geometrically balanced, and must be filled with the middle terms of the series (in this instance

3) thus completing the A primary square as shown in Fig. 433. Fill the B primary square with the series 0-5-10-15-20 in the same manner as above described and as shown in Fig. 434. The com-

1	3	4	5	2
2	4	5	1	3
5	2	3	4	1
3	5	1	2	4
4	1	2	3	5

Fig. 433.

5	0	15	10	20
10	20	0	15	5
20	15	10	5	0
15	5	20	0	10
0	10	5	20	15

Fig. 434.

6	3	19	15	22
12	24	5	16	8
25	17	13	9	1
18	10	21	2	14
4	11	7	23	20

Fig. 435.

bination of Figs. 433 and 434 produces the associated magic square given in Fig. 435.

There are at least five different "figures of equilibrium" that

0	5	15	20	10
5	15	20	10	0
15	20	10	0	5
20	10	0	5	15
10	0	5	15	20

Fig. 436.

3	1	2	4	5
5	3	1	2	4
4	5	3	1	2
2	4	5	3	1
1	2	4	5	3

Fig. 437.

2	4	1	3	5
3	5	2	4	1
4	1	3	5	2
5	2	4	1	3
1	3	5	2	4

Fig. 438.

can be drawn in a 5×5 square, and these can be readily shown to give as many variants as the geometrical class, which as before noted yielded 40,960 different squares. The number may therefore

3	6	17	24	15
10	18	21	12	4
19	25	13	1	7
22	14	5	8	16
11	2	9	20	23

Fig. 439.

2	9	16	23	15
8	20	22	14	1
19	21	13	5	7
25	12	4	6	18
11	3	10	17	24

Fig. 440.

now be doubled, raising the total to 81,920 associated 5×5 magic squares that are capable of being produced by the rules thus far considered.

The student must not however imagine that the possibilities of this method are now exhausted, for a further study of the subject will show that a geometrical pattern or design may often be used not only with its own reverse as shown, but also with another *entirely different design*, thus rendering our search for the universal rule still more difficult.

4	2	5	3	1
3	1	4	2	5
2	5	3	1	4
1	4	2	5	3
5	3	1	4	2

Fig. 441.

2	3	4	5	1
4	5	1	2	3
1	2	3	4	5
3	4	5	1	2
5	1	2	3	4

Fig. 442.

3	1	4	2	5
5	3	1	4	2
2	5	3	1	4
4	2	5	3	1
1	4	2	5	3

Fig. 443.

For example the pattern shown in Fig. 436 may be combined in turn with its reverse shown in Fig. 437 and also with Fig. 438, making the two associated magic squares shown in Figs. 439 and 440.

In consideration of this as yet unexplored territory, therefore,

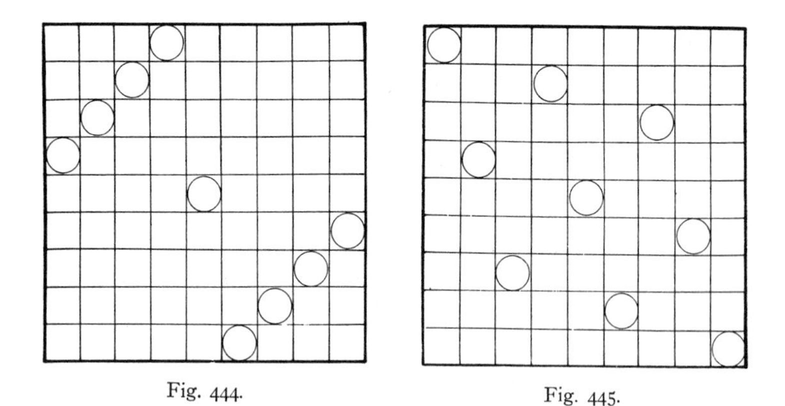

Fig. 444. Fig. 445.

the rules herein briefly outlined can only be considered as partial, and fall short of the "universal" rule for which the writer has been seeking. Their comprehensiveness however is evidenced by the fact that *any square* made by any other rule heretofore known to the writer, may be made by these rules, and also a great variety of other squares which may only be made with great difficulty, if at all, by the older methods.

To show the application of these rules to the older methods, a few squares given in Chapter I may be analyzed.

Figs. 441, 442 and 443 show the plans of 5 × 5 squares given in Figs. 22, 23 and 41 in the above mentioned chapter.

Their comprehensiveness is still further emphasized in squares of larger size, as for example in the 7 × 7 square shown in Fig. 426. Two final examples are shown in Figs. 444 and 445 which give plans of two 9 × 9 squares which if worked out will be found to be unique and beyond the power of any other rule to produce. In conclusion an original and curious 8 × 8 square is submitted in

1	14	7	12	B			
15	4	9	6				
10	5	16	3				
8	11	2	13				C
A				4	15	6	9
				14	1	12	7
				11	8	13	2
			D	5	10	3	16

Fig. 446.

1	14	7	12	3	16	5	10
15	4	9	6	13	2	11	8
10	5	16	3	12	7	14	1
8	11	2	13	6	9	4	15
2	13	8	11	4	15	6	9
16	3	10	5	14	1	12	7
9	6	15	4	11	8	13	2
7	12	1	14	5	10	3	16

Fig. 447.

Fig. 449. This square is both associated and continuous or Nasik, inasmuch as all constructive diagonals give the correct summation.

The theory upon which the writer proceeded in the construction of this square was to consider it as a compound square composed of four 4 × 4 squares, the latter being in themselves continuous but not associated. That the latter quality might obtain in the 8 × 8 square, each *quarter* of the 4 × 4 square is made the exact counterpart of the similar *quarter* in the diagonally opposite 4 × 4 square, but turned on its axis 180 degrees.

Having in this manner made an associated and continuous 8 × 8 square composed of four 4 × 4 squares, each containing the series 1 to 16 inclusive, another 8 × 8 square, made with similar properties, with a proper number series and added to the first square term to term will necessarily yield the desired result.

Practically, the work was done as follows: In one quarter of an 8×8 square, a continuous (but not associated) 4×4 square was inscribed, and in the diagonally opposite quarter another 4×4 square was written in the manner heretofore described and now illustrated in Fig. 446. A simple computation will show that in the unfilled parts of Fig. 446, if it is to be continuous, the contents of the cells C and D must be 29 and A and B must equal 5. Hence A and B may contain respectively 1 and 4, or else 2 and 3. Choosing 2 and 3 for A and B, and 14 and 15 for D and C, they were located

0	48	32	16
48	0	16	32
16	32	48	0
32	16	0	48

Fig. 448.

1	14	55	60	35	48	21	26
15	4	57	54	45	34	27	24
58	53	16	3	28	23	46	33
56	59	2	13	22	25	36	47
18	29	40	43	52	63	6	9
32	19	42	37	62	49	12	7
41	38	31	20	11	8	61	50
39	44	17	30	5	10	51	64

Fig. 449.

as marked by circles in Fig. 447, the associated or centrally balanced idea being thus preserved.

The other two quarters of the 8×8 square were then completed in the usual way of making Nasik 4×4 squares, thus producing the A primary square shown in Fig. 447, which, in accordance with our theory must be both associated and continuous which inspection confirms.

As only the numbers in the series 1 to 16 inclusive appear in this square, it is evident that they must be considered term by term with another square made with the series 0-16-32-48 in order that the final square may contain the series 1 to 64 inclusive. This is accomplished in Fig. 448, which shows a 4×4 square both associated and continuous, composed of the numbers in the above mentioned series.

At this point, two courses of operation seemed to be open, the first being to expand Fig. 448 into an 8×8 square, as in the case of the A primary square, Fig. 447, and the second being to consider Fig. 447 as a *4×4 square,* built up of sixteen subsquares of 2×2 regarded as units.

The latter course was chosen as the easier one, and each individual term in Fig. 448 was added to each of the four numbers in the corresponding quadruple cells of Fig. 447, thus giving four terms in the complete square as shown in Fg. 449. For example 0 being the term in the upper left-hand cell of Fig. 448, this term was added to 1-14-15-4 in the first quadruple cell of Fig. 447, leaving these numbers unchanged in their value, so they were simply transferred to the complete magic square Fig. 449. The second quadruple cell in Fig. 447 contains the numbers 7-12-9-6, and as the second cell in Fig. 448 contains the number 48, this number was added to each of the last mentioned four terms, converting them respectively into 55-60-57 and 54, which numbers were inscribed into the corresponding cells of Fig. 449, and so on throughout.

Attention may here be called to the "figure of equilibrium" shown in Fig. 448 by circles and its quadruple reappearance in Fig. 449 which is a complete associated and continuous 8×8 magic square, having many unique summations. L. S. F.

THE CONSTRUCTION OF MAGIC SQUARES AND RECTANGLES BY THE METHOD OF "COMPLEMENTARY DIFFERENCES."*

We are indebted to Dr. C. Planck for a new and powerful method for producing magic squares, rectangles etc. This method is especally attractive and valuable in furnishing a *general* or *universal* rule covering the construction of all conceivable types of squares and rectangles, both odd and even. It is not indeed the easiest and best method for making *all kinds* of squares, as in many cases much simpler rules can be used to advantage, but it will be found exceedingly helpful in the production of new variants, which

* This article has been compiled almost entirely from correspondence received by the writer from Dr. Planck, and in a large part of it the text of his letters has been copied almost verbatim. Its publication in present form has naturally received his sanction and endorsement. W. S. A.

might otherwise remain undiscovered, seeing that they may be non-La Hireian and ungoverned by any obvious constructive plan.

When a series of numbers is arranged in two associated columns, as shown in Fig. 450, each pair of numbers has its distinctive difference, and these "complementary differences," as they are termed by Dr. Planck, may be used very effectively in the construction of magic squares and rectangles. In practice it is often quite as efficient and simpler to use half the differences, as given in Fig. 450.

In illustrating this method we will first apply it to the con-

1	15	7
2	14	6
3	13	5
4	12	4
5	11	3
6	10	2
7	9	1
8		

Fig. 450.

a				b
	c		c	
b				a

Fig. 451.

2	5	4	3	1
7	6	8	10	9
15	13	12	11	14

Fig. 452.

struction of an associated or regular 3×5 magic rectangle, in which the natural numbers 1 to 15 inclusive are to be so arranged that every long row sums 40, and every short column sums 24. The center cell must necessarily be occupied by 8, which is the middle number of the series, and the complementary numbers must lie in associated cells, such as *a a*—*b b*—*c c* in Fig. 451.

The first operation is to lay out a 3×5 rectangle and fill it with such numbers that all the short columns shall sum 24, but in which the numbers in the columns will not be placed in any particular order. When two columns of this rectangle are filled three pairs of complementary numbers will have been used, and their differences will have disappeared, as these two columns must each sum 24. Hence, *one complementary difference must equal the sum of the other two.*

We have therefore (neglecting the middle column) to make two equations of the forms $a = b + c$ from the complementary differences, without using the same difference twice. Thus:

$$\left.\begin{array}{l} 7 = 6 + 1 \\ 5 = 3 + 2 \end{array}\right\} \quad \cdots\cdots\cdots\cdots\cdots\cdots \quad \text{(I)}$$

is such a pair of equations.

The first equation indicates that the greater of the two complements whose half difference is 7 can lie in the same column with the lesser members of the pairs whose half differences are 6 and 1. In other words, the numbers 15, 7 and 2 can lie in one column, and their complements 14, 9 and 1 in the associated column. The second equation $(5 = 3 + 2)$ gives similar information regarding the other pair of associated columns, and the three remaining numbers must then be placed in the middle column, thus producing the rectangle shown in Fig. 452.

These equations determine nothing as to the placing of the numbers in the rows, since in Fig. 452 the numbers in the columns have no definite order.

The rows may now be attacked in a similar manner. Two of the complementary differences in the upper or lower row must equal the other three, and the equation will therefore be of the order $a + b = c + d + e$.

In order that the disposition of numbers in the columns shall not be disturbed, the numbers used in this equation must be so chosen that any *two* numbers which appear together on the *same side* of an equality sign in the short column equation, must not so appear in a long row equation, also if two numbers appear on the *opposite sides* of an equality sign in a short column equation, *they must not so appear* in the long row equation.

There is only one such equation which will conform to the above rules, viz.,

$$6 + 2 = 4 + 3 + 1.$$

Interpreting this as before we have the rectangle given in Fig. 453, in which each of the three rows sums 40. We have now two rectangles, Fig. 452 showing the correct numbers in the columns, and Fig. 453 showing the proper disposition of the numbers in the rows. By combining them we get the associated or regular magic rectangle given in Fig. 454.

1	5	7	10	14
1	3	8	13	15
2	6	9	11	12

Fig. 453.

7	5	4	10	14
15	13	8	3	1
2	6	12	11	9

Fig. 454.

If a mere shuffling of pairs of complementary rows or columns is ignored, this is the *only* solution of the problem.*

There are two pairs of equations of the form

$$a = b + c$$
$$d = e + f$$

namely, the one given in (I) and

$$\left.\begin{array}{l} 7 = 5 + 2 \\ 4 = 3 + 1 \end{array}\right\} \quad \ldots\ldots\ldots\ldots\ldots\ldots\ldots\ldots \quad \text{(II)}$$

and there are nine equations of the form

$$a + b = c + d + e$$

but of these nine equations only one will go with (I) and none will go with (II) so as to conform with the above rules.

If the condition of association is relaxed there are thirty-nine different 3×5 magic rectangles.

This method can naturally be used for constructing all sizes of magic rectangles which are possible,† but we will only consider one of 5×7 as a final example.

* The solution of this problem of the associated rectangle is the first step in the construction of the higher ornate magics of composite odd orders. For example, if the above single solution for the 3×5 rectangle did not exist it would be impossible to construct a magic, pan-diagonal, associated (= regular) square of order 15, which shall be both 9-ply and 25-ply, i. e., *any* square bunch of 9 cells to sum up 9 times the mean, and *any* square bunch of 25 cells 25 times the mean. C. P.

† A magic rectangle with an odd number of cells in one side and an even number in the other, is impossible with consecutive numbers. C. P.

Fig. 455 shows the associated series of natural numbers from 1 to 35 with their half differences, from which the numbers must be chosen in accordance with the above rules. In this case three will be three equations of the order

1	35	17
2	34	16
3	33	15
4	32	14
5	31	13
6	30	12
7	29	11
8	28	10
9	27	9
10	26	8
11	25	7
12	24	6
13	23	5
14	22	4
15	21	3
16	20	2
17	19	1
18		

Fig. 455.

19	22	33	29	23	21	20
35	31	34	28	30	24	25
9	10	4	18	32	26	27
11	12	6	7	2	5	1
16	15	13	8	3	14	17

Fig. 456.

30	31	34	1	7	9	14
25	26	28	16	15	13	3
32	24	19	18	17	12	4
33	23	21	20	8	10	11
22	27	29	35	2	5	6

Fig. 457.

9	31	34	7	30	14	1
16	15	13	28	3	26	25
19	12	4	18	32	24	17
11	10	33	8	23	21	20
35	22	6	29	2	5	27

Fig. 458.

$$a + b = c + d + e$$

for the columns, and two equations of the order

$$a + b + c = d + e + f + g$$

for the rows. The following selection of numbers will satisfy the conditions:

$$\left.\begin{aligned} 1+17 &= 9+7+2 \\ 4+13 &= 8+6+3 \\ 15+16 &= 14+12+5 \end{aligned}\right\} \quad \ldots\ldots\ldots\ldots \quad \text{(III)}$$

for the columns, and

$$\left.\begin{aligned} 12+13+16 &= 17+11+9+4 \\ 7+8+10 &= 2+3+5+15 \end{aligned}\right\} \quad \ldots\ldots\ldots\ldots \quad \text{(IV)}$$

for the rows.

Fig. 456 is a rectangle made from (III) in which all the columns sum 90, and Fig. 457 is a rectangle made from (IV) in which all the rows sum 126. Combining these two rectangles produces Fig. 458 which is magic and associated.

We will now consider this method in connection with magic squares and will apply it to the construction of a square of order 5 as a first example. In this case two equations of the order

$$a+b=c+d+e$$

will be required for the rows and two more similar equations for the columns.

The following will be found suitable for the rows:

$$\left.\begin{aligned} 12+11 &= 10+9+4 \\ 8+6 &= 7+5+2 \end{aligned}\right\} \quad \ldots\ldots\ldots\ldots \quad \text{(V)}$$

and

$$\left.\begin{aligned} 11+8 &= 12+6+1 \\ 10+7 &= 9+5+3 \end{aligned}\right\} \quad \ldots\ldots\ldots\ldots \quad \text{(VI)}$$

for the columns.

It will be seen that the rule for pairs of numbers in the same equation is fulfilled in the above selection. In (V) 12 and 11 are on the same side of an equality sign, but in (VI) these numbers are on opposite sides, also, 10 and 9 are on the same side in (V) and on opposite sides in (VI) and so on.

The resulting magic square is given in Fig. 459, it is non-La Hireian, and could not easily be made in any way other than as above described.

The construction of a square of order 6 under this method presents more difficulties than previous examples, on account of the inherent disabilities natural to this square and we will consider it as a final example. The method to be employed is precisely the same as that previously discussed.

For the columns three equations should be made of the form:

$$a+b+c=d+e+f$$

or

$$a+b \quad =c+d+e+f$$

and three similar equations are required for the rows, all being subject to the rule for "pairs and equality sign" as above described.

24	3	9	4	25
21	6	11	8	19
12	16	13	10	14
7	18	15	20	5
1	22	17	23	2

Fig. 459.

On trial, however, this will be found to be impossible,* but if for one of the row- or column-equations we substitute an *inequality* whose difference is 2 we shall obtain a square of 6, which will be "associated," but in which two lines or columns will be erratic, one showing a correct summation -1 and the other a correct summation $+1$. The following equations (VII) may be used for the columns:

$$\left.\begin{array}{r} 11+7=9+5+3+1 \\ 25+17+13=21+19+15 \\ 35+31+23=33+29+27 \end{array}\right\} \quad \ldots\ldots\ldots\ldots\ldots\ldots \quad \text{(VII)}$$

and for the rows:

* It is demonstrably impossible for all orders $=4p+2$, i. e., 6, 10, 14,, etc. C.P.

$$\left.\begin{array}{r} 29+25=33+13+\ \ 7+1 \\ 35+19+\ \ 3=31+21+\ \ 5\quad\ \ \\ 27+23\neq 17+15+11+9 \end{array}\right\} \quad \ldots\ldots\ldots\ldots\ldots\ldots \text{(VIII)}$$

the last being an *inequality*. Fig. 460 shows the complementary

1	36	35
2	35	33
3	34	31
4	33	29
5	32	27
6	31	25
7	30	23
8	29	21
9	28	19
10	27	17
11	26	15
12	25	13
13	24	11
14	23	9
15	22	7
16	21	5
17	20	3
18	19	1

Fig. 460.

24	31	36	35	29	21
22	27	34	33	28	20
14	25	30	32	26	19
16	8	2	1	6	13
17	9	4	3	10	15
18	11	5	7	12	23

Fig. 461.

33	31	2	12	15	18
36	28	20	3	8	16
32	30	10	11	13	14
26	24	23	5	7	27
34	29	21	1	9	17
25	22	19	4	6	35

Fig. 462.

18	31	2	33	12	15
16	8	36	3	28	20
14	11	30	32	10	13
24	27	5	7	26	23
17	9	34	1	29	21
22	25	4	35	6	19

Fig. 463.

pairs of natural numbers 1 to 36 with their whole differences, which in this case are used in the equations (VII) and (VIII) instead of the half differences, because these differences cannot be halved without involving fractions. Fig. 461 is the square derived from equations (VII) and will be found correct in the columns. Fig. 462 is the square formed from equations (VIII) and is correct in the 1st, 2d, 5th, and 6th rows, but erratic in the 3d and 4th rows. The finished six-square made by combining Figs. 461 and 462 is shown in Fig. 463 which is associated or regular, and which gives

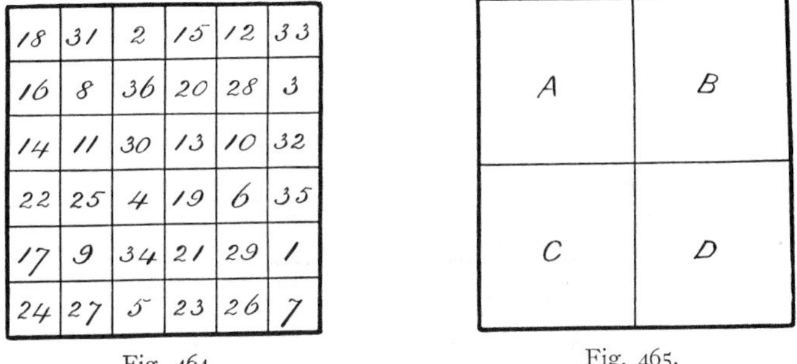

18	31	2	15	12	33
16	8	36	20	28	3
14	11	30	13	10	32
22	25	4	19	6	35
17	9	34	21	29	1
24	27	5	23	26	7

Fig. 464.

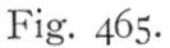

Fig. 465.

correct summations in all the columns and rows excepting the 3d and 4th rows which show − 1 and + 1 inequalities respectively.

Fig. 463, like Fig. 459, could not probably be produced by any other method than the one herein employed, and both of these squares therefore demonstrate the value of the methods for constructing new variants. Fig. 463 can be readily converted into a

7	12	1	14
2	13	8	11
16	3	10	5
9	6	15	4

Fig. 466.

7	12	14	1
2	13	11	8
9	6	4	15
16	3	5	10

Fig. 467.

continuous or pan-diagonal square by first interchanging the 4th and 6th columns and then, in the square so formed, interchanging the 4th and 6th rows. The result of these changes is given in Fig. 464 which shows correct summations in all columns and rows, excepting in the 3d and 6th rows which carry the inequalities shown in Fig. 463. This square has lost its property of association by the above change but has now correct summation in all its diagonals. It is a demonstrable fact that squares of orders $4p+2$, (i. e., 6, 10, 14 etc.) cannot be made perfectly magic in columns and rows and at the same time *either* associated or pandiagonal when constructed with consecutive numbers.

Dr. Planck also points out that the change which converts all even associated squares into pan-diagonal squares may be tersely expressed as follows:

Divide the square into four quarters as shown in Fig. 465.

Leave A untouched.

Reflect B.

Invert C.

Reflect and invert D.

1	44	32	53	2	43	31	54
58	19	39	14	57	20	40	13
38	15	59	18	37	16	60	17
29	56	4	41	30	55	3	42
23	62	10	35	24	61	9	36
48	5	49	28	47	6	50	27
52	25	45	8	51	26	46	7
11	34	22	63	12	33	21	64

Fig. 468.

The inverse change from pan-diagonal to association is not necessarily effective, but it may be demonstrated with the "Jaina" square given by Dr. Carus on p. 125, which is here repeated in Fig. 466. This is a continuous or pan-diagonal square, but after making the above mentioned changes it becomes an associated or regular square as shown in Fig. 467.

Magic squares of the 8th order can however be made to combine the pan-diagonal and associated features as shown in Fig. 468 which is contributed by Mr. Frierson, and this is true also of all larger squares of orders $4p$.

W. S. A.

NOTES ON THE CONSTRUCTION OF MAGIC SQUARES OF ORDERS IN WHICH n IS OF THE GENERAL FORM $4p+2$.

It is well known that magic squares of the above orders, i. e., 6^2, 10^2, 14^2, 18^2, etc., cannot be made perfectly pandiagonal and ornate with the natural series of numbers.

Dr. C. Planck has however pointed out that this disability is purely arithmetical, seeing that these magics can be readily constructed as perfect and ornate as any others with a properly selected series of numbers.

In all of these squares n is of the general form $4p+2$, but they can be divided into two classes:

Class I. Where n is of the form $8p-2$, as 6^2, 14^2, 22^2 etc.

Class II. Where n is of the form $8p+2$, as 10^2, 18^2, 26^2 etc.

The series for all magics of Class I may be derived by making a square of the natural series 1 to $(n+1)^2$ and discarding the numbers in the middle row and column.

Thus, for a 6^2 magic the series will be:

1	2	3	—	5	6	7
8	9	10	—	12	13	14
15	16	17	—	19	20	21
—	—	—	—	—	—	—
29	30	31	—	33	34	35
36	37	38	—	40	41	42
43	44	45	—	47	48	49

The series for all magics of Class II may be made by writing a square of the natural numbers 1 to $(n+3)^2$ and discarding the numbers in the *three* middle rows and columns. The series for a 10^2 magic, for example, will be:

1	2	3	4	5	.	.	.	9	10	11	12	13
14	15	16	17	18	.	.	.	22	23	24	25	26
27	28	29	30	31	.	.	.	35	36	37	38	39
40	41	42	43	44	.	.	.	48	49	50	51	52
53	54	55	56	57	.	.	.	61	62	63	64	65

—	—	—	—	—	. . .	—	—	—	—	—
—	—	—	—	—	. . .	—	—	—	—	—
—	—	—	—	—	. . .	—	—	—	—	—
105	106	107	108	109	. . .	113	114	115	116	117
118	119	120	121	122	. . .	126	127	128	129	130
131	132	133	134	135	. . .	139	140	141	142	143
144	145	146	147	148	. . .	152	153	154	155	156
157	158	159	160	161	. . .	165	166	167	168	169

By using series as above described, pandiagonal magics with double-ply properties, or associated magics may be readily made either by the La Hireian method with magic rectangles, or by the path method as developed by Dr. C. Planck.

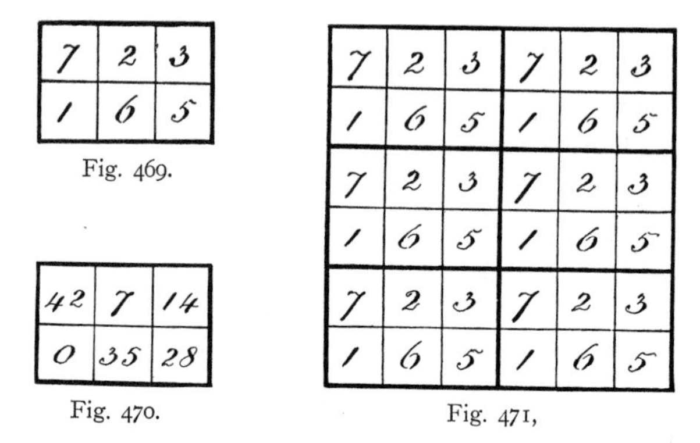

Fig. 469.

Fig. 470.

Fig. 471,

Referring now to the La Hireian method and using the 6^2 magic as a first example, the rectangles required for making the two auxiliary squares will necessarily be 2×3, and the numbers used therein will be those commonly employed for squares of the seventh order, i. e., $(6+1)^2$, with the middle numbers omitted thus:

1	2	3	—	5	6	7
0	7	14	—	28	35	42

It may be shown that a magic rectangle having an odd number of cells in one side, and an even number of cells in the other side is impossible with consecutive numbers, but with a series made as above it can be constructed without any difficulty, as shown in Figs. 469 and 470.

Two auxiliary squares may now be made by filling them with their respective rectangles. If this is done without forethought, a plain pandiagonal magic of the sixth order may result, but if attention is given to ornate qualities in the two auxiliaries, these features will naturally be carried into the final square. For example, by the arrangement of rectangles shown in Figs. 471 and 472 both auxiliaries are made magic in their six rows, six columns and twelve

0	42	0	42	0	42
35	7	35	7	35	7
28	14	28	14	28	14
0	42	0	42	0	42
35	7	35	7	35	7
28	14	28	14	28	14

Fig. 472.

7	44	3	49	2	45
36	13	40	8	41	12
35	16	31	21	30	17
1	48	5	43	6	47
42	9	38	14	37	10
29	20	33	15	34	19

Fig. 473.

7	2	3	3	2	7
1	6	5	5	6	1
7	2	3	3	2	7
1	6	5	5	6	1
7	2	3	3	2	7
1	6	5	5	6	1

Fig. 474.

0	42	0	42	0	42
35	7	35	7	35	7
28	14	28	14	28	14
28	14	28	14	28	14
35	7	35	7	35	7
0	42	0	42	0	42

Fig. 475.

diagonals, and they are also 4-ply and 9-ply. Their complementary couplets are also harmoniously connected throughout in steps of 3, 3. These ornate features are therefore transmitted into the finished 6^2 magic shown in Fig. 473. If it is desired to make this square associated, that is with its complementary couplets evenly balanced around its center, it is only necessary to introduce the feature of association into the two auxiliary squares by a rearrangement of their magic rectangles as shown in Figs. 474, 475 and 476. the last figure being a pandiagonal associated magic.

The next larger square of Class I is 14^2, and it can be made with the natural series 1 to $(14+1)^2$ arranged in a square, discarding, as before, all the numbers in the central row and column.

The rectangles for this square will necessarily be 2×7 and the numbers written therein will be those ordinarily used for a square

7	44	3	45	2	49
36	13	40	12	41	8
35	16	31	17	30	21
29	20	33	19	34	15
42	9	38	10	37	14
1	48	5	47	6	43

Fig. 476.

of the fifteenth order, $(14+1)^2$, with the middle numbers omitted, thus:

1	2	3	4	5	6	7	—	9	10	11	12	13	14	15
0	15	30	45	60	75	90	—	120	135	150	165	180	195	210

Simple forms of magic rectangles for the auxiliaries are shown in Figs. 477 and 478 but many other arrangements of the couplets will work equally well.

15	2	3	12	11	6	7
1	14	13	4	5	10	9

Fig. 477.

210	15	30	165	150	75	90
0	195	180	45	60	135	120

Fig. 478.

The smallest magic of Class II is 10^2, the series for which is given below. The rectangles used for filling the two auxiliaries of this square are 2×5, and they can be made with the numbers which would be commonly used for a square of the thirteenth order $(10+3)^2$ omitting the three middle numbers in each row thus:

1	2	3	4	5	. . .	9	10	11	12	13
0	13	26	39	52	. . .	104	117	130	143	156

Figs. 479 and 480 show these two rectangles with a simple arrangement of the numbers. The two auxiliaries and the finished 10^2 magic are given in Figs. 481, 482 and 483. Fig. 483 is magic in its

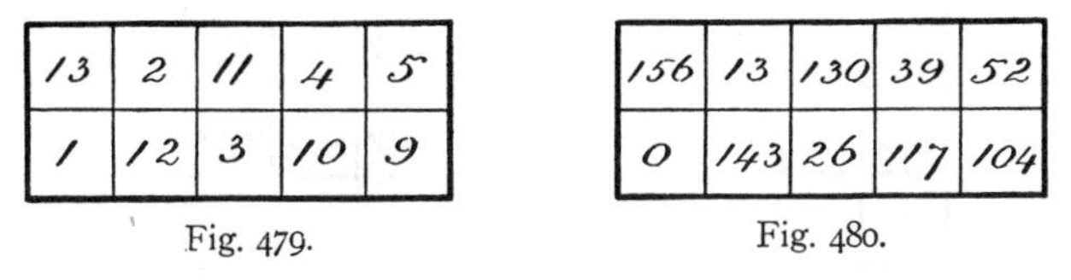

13	2	11	4	5
1	12	3	10	9

Fig. 479.

156	13	130	39	52
0	143	26	117	104

Fig. 480.

ten rows, ten columns and twenty diagonals. It is also 4-ply and 25-ply. Like the 6^2 magic, this square can also be associated by changing the disposition of the magic rectangles in the auxiliaries.

The above examples will suffice to explain the general con-

13	2	11	4	5	13	2	11	4	5
1	12	3	10	9	1	12	3	10	9
13	2	11	4	5	13	2	11	4	5
1	12	3	10	9	1	12	3	10	9
13	2	11	4	5	13	2	11	4	5
1	12	3	10	9	1	12	3	10	9
13	2	11	4	5	13	2	11	4	5
1	12	3	10	9	1	12	3	10	9
13	2	11	4	5	13	2	11	4	5
1	12	3	10	9	1	12	3	10	9

Fig. 481.

struction of these squares by the La Hireian method with magic rectangles. It may however be stated that although the series previously described for use in building these squares include the lower numerical values, there are other series of higher numbers which will produce equivalent magic results.

0	156	0	156	0	156	0	156	0	156
143	13	143	13	143	13	143	13	143	13
26	130	26	130	26	130	26	130	26	130
117	39	117	39	117	39	117	39	117	39
104	52	104	52	104	52	104	52	104	52
0	156	0	156	0	156	0	156	0	156
143	13	143	13	143	13	143	13	143	13
26	130	26	130	26	130	26	130	26	130
117	39	117	39	117	39	117	39	117	39
104	52	104	52	104	52	104	52	104	52

Fig. 482.

13	158	11	160	5	169	2	167	4	161
144	25	146	23	152	14	155	16	153	22
39	132	37	134	31	143	28	141	30	135
118	51	120	49	126	40	129	42	127	48
117	54	115	56	109	65	106	63	108	57
1	168	3	166	9	157	12	159	10	165
156	15	154	17	148	26	145	24	147	18
27	142	29	140	35	131	38	133	36	139
130	41	128	43	122	52	119	50	121	44
105	64	107	62	113	53	116	55	114	61

Fig. 483.

The following table illustrates another rule covering the selection of numbers for all magic squares of these orders.

ORDER OF SQUARE	NATURAL SERIES	DISCARDING NUMBERS IN
6th	1 to $(6+1)^2$	the middle row and column.
10th	1 to $(10+3)^2$	the 3 middle rows and columns.
14th	1 to $(14+5)^2$	the 5 middle rows and columns.
18th	1 to $(18+7)^2$	the 7 middle rows and columns.
22nd	1 to $(22+9)^2$	the 9 middle rows and columns.
26th	1 to $(26+11)^2$	the 11 middle rows and columns.
		and so forth.

These figures show that this rule is equivalent to taking the numbers of the natural series $\left(\frac{3n-4}{2}\right)^2$ and omitting the central $\frac{n-4}{2}$ rows and columns. In comparing the above with the rules previously given, for which we are indebted to Dr. C. Planck, it will be seen that in cases of magics larger than 10^2 it involves the use of unnecessarily large numbers.

The numerical values of the ply properties of these squares are naturally governed by the dimensions of the magic rectangles used in their construction. Thus the rectangle of the 6^2 magic (Fig. 473) is 2×3, and this square is 2^2-ply and 3^2-ply. The rectangle of the 10^2 magic being 2×5, the square may be made 2^2-ply and 5^2-ply, and so forth.

The formation of these squares by the "path" method which has been so ably developed by Dr. C. Planck* may now be considered. The first step is to rearrange the numbers of the given series in such a cyclic order or sequence, that each number being written consecutively into the square by a well defined rule or path, the resulting magic will be identical with that made by the La Hireian method, or equivalent thereto in magic qualities. Starting, as before, with the 6^2 magic, the proper sequence of the first six numbers is found in what may be termed the "continuous diagonal" of its magic rectangle. Referring to Fig. 469, this sequence is seen to be 1, 2, 5,

* *The Theory of Path Nasiks,* by C. Planck, M.A., M.R.C.S., published by A. T. Lawrence, Rugby, England.

7, 6, 3, but it is obvious that there may be as many different sequences as there are variations in the magic rectangles.

The complete series given on page 267 must now be rearranged in its *lines and columns* in accordance with the numerical sequence

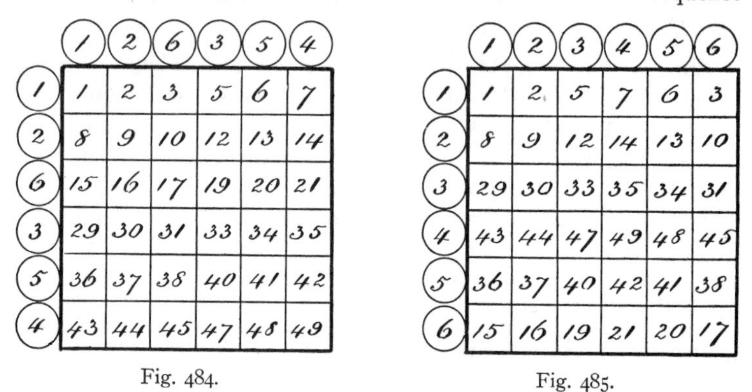

	(1)	(2)	(6)	(3)	(5)	(4)
(1)	1	2	3	5	6	7
(2)	8	9	10	12	13	14
(6)	15	16	17	19	20	21
(3)	29	30	31	33	34	35
(5)	36	37	38	40	41	42
(4)	43	44	45	47	48	49

Fig. 484.

	(1)	(2)	(3)	(4)	(5)	(6)
(1)	1	2	5	7	6	3
(2)	8	9	12	14	13	10
(3)	29	30	33	35	34	31
(4)	43	44	47	49	48	45
(5)	36	37	40	42	41	38
(6)	15	16	19	21	20	17

Fig. 485.

of the first six numbers as above indicated. To make this arrangement quite clear, the series given on p. 267 is reproduced in Fig. 484, the numbers written in circles outside the square showing the numerical order of lines and columns under rearrangement. Fig. 485 shows the complete series in new cyclic order, and to construct a square directly therefrom, it is only necessary to write these numbers con-

7		3		2	
			8		
1		5		6	

Fig. 486.

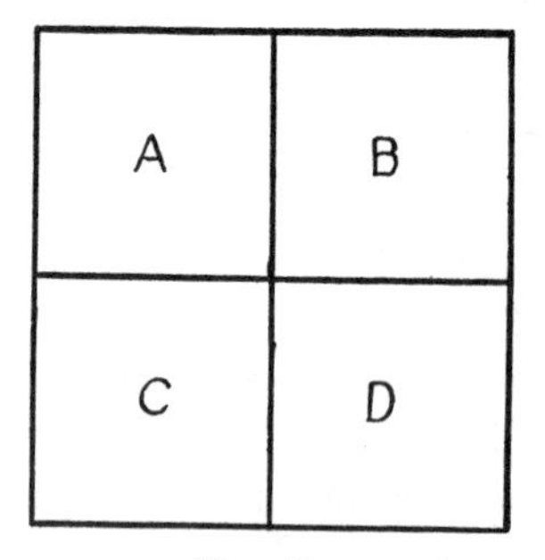

Fig. 487.

secutively along the proper paths. Since the square will be pandiagonal *it may be commenced anywhere,* so in the present example we will place 1 in the fourth cell from the top in the first column, and will use the paths followed in Fig. 473 so as to reproduce that square. The paths may be written $\begin{vmatrix} 3, 2 \\ 4, 3 \end{vmatrix}$ and since we can always write

1	2	3	4	9	13	12	11	10	5
14	15	16	17	22	26	25	24	23	18
27	28	29	30	35	39	38	37	36	31
40	41	42	43	48	52	51	50	49	44
105	106	107	108	113	117	116	115	114	109
157	158	159	160	165	169	168	167	166	161
144	145	146	147	152	156	155	154	153	148
131	132	133	134	139	143	142	141	140	135
118	119	120	121	126	130	129	128	127	122
53	54	55	56	61	65	64	63	62	57

Fig. 488.

13	160	2	161	11	169	4	158	5	157
27	140	38	139	29	131	36	142	35	133
117	56	106	57	115	65	108	54	109	63
144	23	155	22	146	14	153	25	152	16
130	43	119	44	128	52	121	41	122	50
1	166	12	165	3	157	10	168	9	159
39	134	28	135	37	143	30	132	31	141
105	62	116	61	107	53	114	64	113	55
156	17	145	18	154	26	147	15	148	24
118	49	129	48	120	40	127	51	126	42

Fig. 489.

$-(n-a)$ instead of a, we may write this $\begin{vmatrix} 3, 2 \\ -2, 3 \end{vmatrix}$. This only means that the numbers in the first column of Fig. 485 (which may be termed the *leading numbers*) are to be placed in order along the path (3, 2), as in the numbers enclosed in circles in Fig. 473; and then starting from each cell thus occupied, the remaining five numbers in each of the six rows of Fig. 485 are to be written along the path (−2, 3). It will be seen that this is equivalent to writing the successive rows of Fig. 485 intact along the path (−2, 3), or (3, −2) and using a "break-step" (1, −1), as in Fig. 486 where the first break-step is shown with an arrow. The break-step is always given

21	2	3	4	17	16	15	8	13
1	20	19	18	5	6	7	14	9

Fig. 490.

23	2	21	4	19	6	17	8	9	10	13
1	22	3	20	5	18	7	16	15	14	11

Fig. 491.

29	2	27	4	25	6	23	8	9	20	11	18	13
1	28	3	26	5	24	7	22	21	10	19	12	17

Fig. 492.

by summing up the coordinates; thus, the paths here being $\begin{vmatrix} 3, 2 \\ -2, 3 \end{vmatrix}$, by summing the columns we get (1, 5), that is (1, −1). The resulting square is, of course, identical with Fig. 473.

As previously stated, this square being pandiagonal, it may be commenced in any of its thirty-six cells, and by using the same methods as before, different aspects of Fig. 473 will be produced. Also, since by this method complementary pairs are always separated by a step $(n/2, n/2)$, any of the thirty-six squares thus formed may be made associated by the method described under the heading "Magic Squares by Complementary Differences," viz., Divide the square into four quarters as shown in Fig. 487; leave A untouched, reflect B, invert C and reflect and invert D. For this concise and elegant method of changing the relative positions of the complementary couplets in a square we are indebted to Dr. Planck.

The next square in order is 10^2. The series of numbers used is given on page 267 and their rearrangement in proper cyclic order for direct entry may be found as before in the continuous diagonal of its magic rectangle. The sequence shown in Fig. 479 is, 1, 2, 3, 4, 9, 13, 12, 11, 10, 5, and the complete rearrangement of the series in accordance therewith is given in Fig. 488. Various 10^2 magics may be made by using this series with different paths. The paths $\begin{vmatrix} 5, 4 \\ -4, 5 \end{vmatrix}$ will produce Fig. 483, and $\begin{vmatrix} 5, 2 \\ 2, 5 \end{vmatrix}$ will make Fig. 489, which is equivalent to Fig. 483 in its ornate features.

These squares and all similarly constructed larger ones of these orders may be changed to the form of association wherein the complementary couplets are evenly balanced around the center of the square, by the method previously explained. It will be unnecessary to prolong the present article by giving any examples of larger squares of this class, but the simple forms of magic rectangles for 18^2 and 22^2 and 26^2 magics, shown in Figs. 490, 491, and 492, may be of some assistance to those who desire to devote further study to these interesting squares.* W. S. A. L. S. F.

NOTES ON THE CONSTRUCTION OF MAGIC SQUARES OF ORDERS IN WHICH n IS OF THE GENERAL FORM $8p+2$.

It has just been shown that the minimum series to be used in constructing this class of squares is selected from the series 1, 2,

* More generally, if p, q are relative primes, the square of order pq will be magic on its pq rows, pq columns and $2pq$ diagonals, and at the same time p^2-ply and q^2-ply, if it be constructed with the paths $\begin{vmatrix} p, q \\ q, p \end{vmatrix}$, and the period be taken from the continuous diagonal of the magic rectangle $p \times q$. The limitations are dictated by the magic rectangle. Evidently p and q must both be >1, and consecutive numbers must fail if the order is $\equiv 2 \pmod{4}$; in all other cases consecutive numbers will suffice. C. P.

3, ... $(n+3)^2$, by discarding 3 rows and columns from the natural square of the order $n+3$.

It is not necessary, however, to discard the three central rows and columns, as was therein explained, there being numerous variations, the total number of which is always equal to $\left(\frac{n+2}{4}\right)^2$

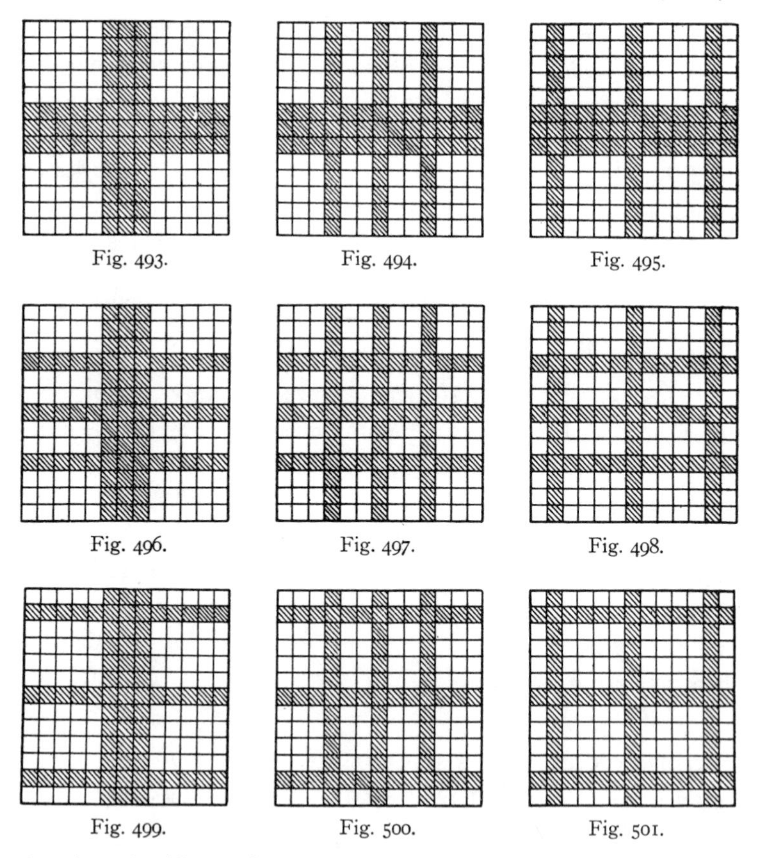

Fig. 493. Fig. 494. Fig. 495.

Fig. 496. Fig. 497. Fig. 498.

Fig. 499. Fig. 500. Fig. 501.

therefore the 10^2 can be constructed with 9 different series, the 18^2 with 25 different series, the 26^2 with 49 different series, and so on.

In Figs. 493 to 501 are shown all the possible variations of discarding rows and columns for the 10^2, Fig. 493 representing the series explained in the foregoing article.

The central row and column must always be discarded, the

remaining two rows and columns can be cast out symmetrically in relation to their parallel central row or column and should be an

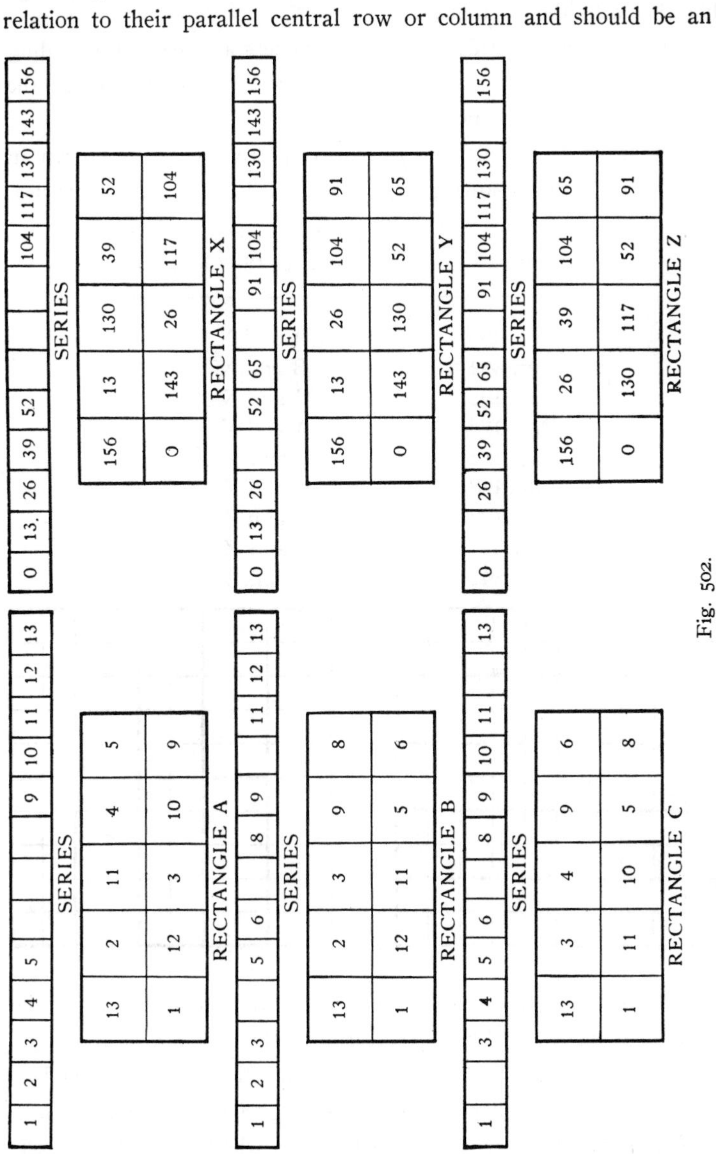

Fig. 502.

odd number of rows or columns from it. In other words, we cast out the central row, then on each side of it we cast out the 1st, 3d,

5th, or 7th, etc. rows from it, and irrespective of the rows, we do likewise with the columns.

In a manner already explained, numbers are selected according to the series desired and arranged in rectangles with which the magic square is constructed.

A set of rectangles with their respective series is shown in Fig. 502, and the following table will give directions for their use.

SERIES	RECTANGLES (SEE Fig. 502)
Fig. 493	A and X
Fig. 494	B and X
Fig. 495	C and X
Fig. 496	A and Y
Fig. 497	B and Y
Fig. 498	C and Y
Fig. 499	A and Z
Fig. 500	B and Z
Fig. 501	C and Z

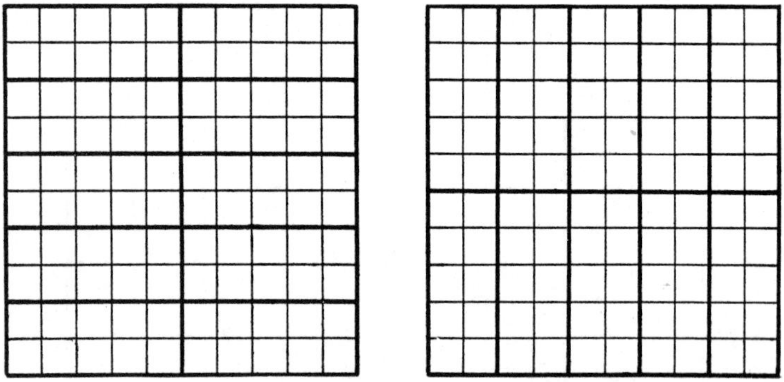

Fig. 503.

For example, suppose we were to construct a square, using the series denoted in Fig. 495. By referring to the table it is seen that we must employ rectangles C and X. By using the La Hireian method these rectangles are placed as shown in Fig. 503, care being taken to arrange them in respect to the final square, whether it is to be associated or non-associated.*

* See preceding article.

65	107	56	113	58	117	55	108	61	110
40	128	49	122	47	118	50	127	44	125
143	29	134	35	136	39	133	30	139	32
14	154	23	148	21	144	24	153	18	157
169	3	160	9	162	13	159	4	165	6
53	115	62	109	60	105	63	114	57	112
52	120	43	126	45	130	42	121	48	123
131	37	140	31	138	27	141	36	135	34
26	146	17	152	19	156	16	147	22	149
157	11	166	5	164	1	167	10	161	8

Fig. 504.

1	2	3	5	6	8	9	11	12	13
27	28	29	31	32	34	35	37	38	39
40	41	42	44	45	47	48	50	51	52
53	54	55	57	58	60	61	63	64	65
66	67	68	70	71	73	74	76	77	78
92	93	94	96	97	99	100	102	103	104
105	106	107	109	110	112	113	115	116	117
118	119	120	122	123	125	126	128	129	130
131	132	133	135	136	138	139	141	142	143
157	158	159	161	162	164	165	167	168	169

Fig. 505.

1	2	11	9	6	13	12	3	5	8
27	28	37	35	32	39	38	29	31	34
118	119	128	126	123	130	129	120	122	125
105	106	115	113	110	117	116	107	109	112
92	93	102	100	97	104	103	94	96	99
157	158	167	165	162	169	168	159	161	164
131	132	141	139	136	143	142	133	135	138
40	41	50	48	45	52	51	42	44	47
53	54	63	61	58	65	64	55	57	60
66	67	76	74	71	78	77	68	70	73

Fig. 506.

5	162	1	168	11	161	6	157	12	167
100	73	104	67	94	74	99	78	93	68
57	110	53	116	63	109	58	105	64	115
126	47	130	41	120	48	125	52	119	42
135	32	131	38	141	31	136	27	142	37
9	164	13	158	3	165	8	169	2	159
96	71	92	77	102	70	97	66	103	76
61	112	65	106	55	113	60	117	54	107
122	45	118	51	128	44	123	40	129	50
139	34	143	28	133	35	138	39	132	29

Fig. 507.

A non-associated square resulting from rectangles C and X is shown in Fig. 504. Another example is shown in Figs. 505, 506 and 507. Here a series corresponding to Fig. 500 has been selected and the natural square is shown in Fig. 505, the heavy lines indicating the discarded rows and columns. The rows and columns are re-arranged according to the numerical sequence of the continuous diagonals* of rectangles B and Z of Fig. 502, this re-arrangement being shown in Fig. 506.

In constructing the final square, Fig. 507, an advance move -4, -5 and a break move 1, 1 was used.

It will be unnecessary to show examples of higher orders of these squares, as their methods of construction are only extensions of what has been already described. It may be mentioned that these squares when non-associated can be transformed into associated squares by the method given in the preceding article. H. A. S.

GEOMETRIC MAGIC SQUARES AND CUBES.

The term "geometric" has been applied to that class of magic squares wherein the numbers in the different rows, columns, and diagonals being multiplied together give similar products. They are analogous in all respects to arithmetical magic squares.

Any feature produced in an arithmetical square can likewise be produced in a geometric square, the only difference being that the features of the former are shown by summations while those of the latter are shown by products. Where we use an arithmetical series for one, we use a geometric series for the other, and where one is constructed by a method of differences the other is constructed by ratios.

These geometric squares may be considered unattractive because of the large numbers involved, but they are interesting to study, even though the actual squares are not constructed. The absurdity of constructing large geometric squares can be easily shown. For example, suppose we were to construct an 8th order square using the series $2^0, 2^1, 2^2, 2^3, \ldots 2^{63}$, the lowest number would be 1 and

* See preceding article.

the highest number would be 9,223,372,036,854,775,808. Who would be willing to test the accuracy of such a square by multiplying together the numbers in any of its rows or columns?

Analogous to the arithmetical squares the geometric squares may be constructed with a straight geometric series, a broken geometric series, or a series which has no regular progression.

I have divided the methods of construction into four groups, namely: the "Exponential method," the "Exponential La Hireian method," the "Ratio method," and the "Factorial method."

The Exponential Method.

The most common way of constructing these squares is with a straight geometric series, arranged in the same order as a straight arithmetical series would be in any summation square. This is equivalent to the following.

Form any magic with a straight arithmetical series as in Fig. 508. Consider these numbers as exponents by repeating any number

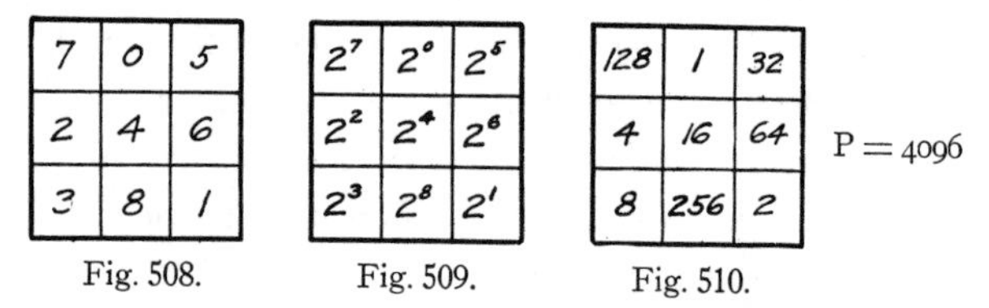

7	0	5
2	4	6
3	8	1

Fig. 508.

2^7	2^0	2^5
2^2	2^4	2^6
2^3	2^8	2^1

Fig. 509.

128	1	32
4	16	64
8	256	2

Fig. 510.

P = 4096

(in this case 2) before each of them, which will give us a square as shown in Fig. 509. It may be noticed that 2 is taken 12 times as a factor in each of the rows, columns, and diagonals, therefore forming a geometric square with constant products of 4096. The square transposed in natural numbers is shown in Fig. 510.

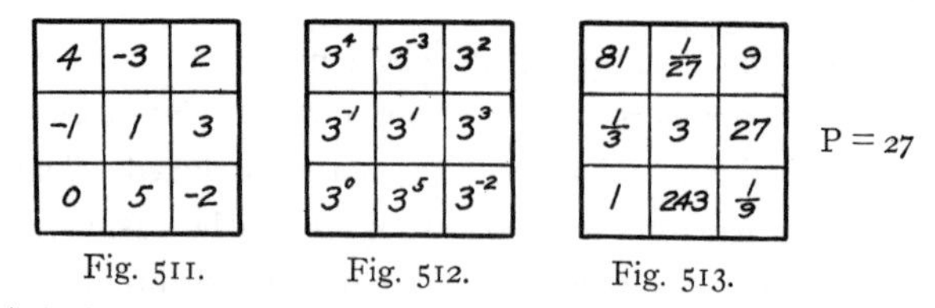

4	-3	2
-1	1	3
0	5	-2

Fig. 511.

3^4	3^{-3}	3^2
3^{-1}	3^1	3^3
3^0	3^5	3^{-2}

Fig. 512.

81	$\frac{1}{27}$	9
$\frac{1}{3}$	3	27
1	243	$\frac{1}{9}$

Fig. 513.

P = 27

Fig. 511, 512 and 513 show the same process involving negative exponents.

Figs. 514, 515 and 516 show how fractional exponents may be used; and the use of both fractional and negative exponents is shown in Figs. 517, 518 and 519.

Figs. 520 and 521 show the exponential method applied to a fourth order square. The exponents in Fig. 520 taken alone, obviously form an arithmetical magic.

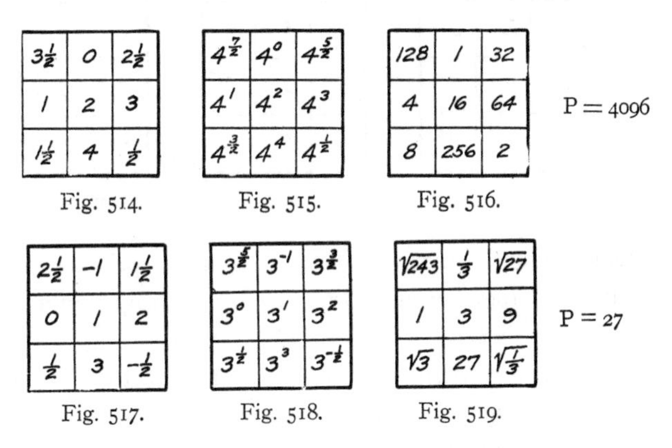

Fig. 514. Fig. 515. Fig. 516.

Fig. 517. Fig. 518. Fig. 519.

This square is an associated square with the products of each complementary pair equaling 32.

2^{-5}	2^{9}	2^{8}	2^{-2}
2^{6}	2^{0}	2^{1}	2^{3}
2^{2}	2^{4}	2^{5}	2^{-1}
2^{7}	2^{-3}	2^{-4}	2^{10}

Fig. 520.

$\frac{1}{32}$	512	256	$\frac{1}{4}$
64	1	2	8
4	16	32	$\frac{1}{2}$
128	$\frac{1}{8}$	$\frac{1}{16}$	1024

P = 1024

Fig. 521.

The Exponential La Hireian Method.

Two primary squares are shown in Figs. 522 and 523. One is filled with the powers 0, 1 and 2 of the factor 2, and the other with the powers 0, 1 and 2 of the factor 5. Each primary square in itself is a geometric magic with triplicate numbers. Figs. 522 and 523 multiplied together, cell by cell, will produce the magic shown in Fig. 524.

The factor numbers, in this case 2 and 5, are not necessarily different, but when they are alike the exponents must suit the condition, to avoid duplicate numbers in the final square. To make this clearer: if we form two primary squares that will add together and form an arithmetical magic, the same factor number may be added to each of these primary squares, using the former numbers as exponents, and the two will become geometric primary squares that will multiply together and form a geometric magic without duplicate numbers.

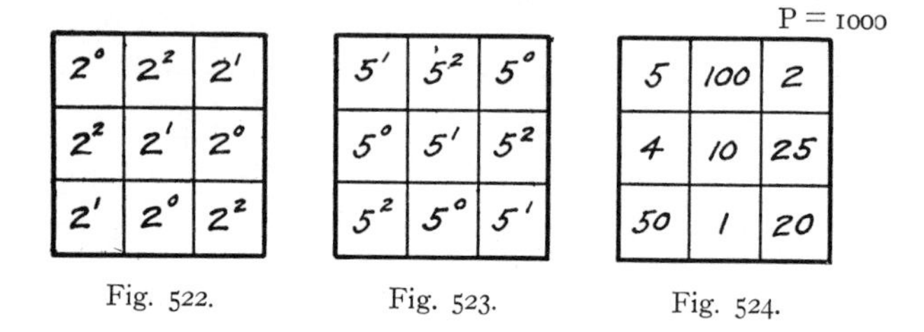

2^0	2^2	2^1
2^2	2^1	2^0
2^1	2^0	2^2

Fig. 522.

5^1	5^2	5^0
5^0	5^1	5^2
5^2	5^0	5^1

Fig. 523.

P = 1000

5	100	2
4	10	25
50	1	20

Fig. 524.

Figs. 525, 526 and 527 show the same methods applied to the fourth order squares. This is a Jaina square, and is consequently pandiagonal and also contains the other Jaina features.

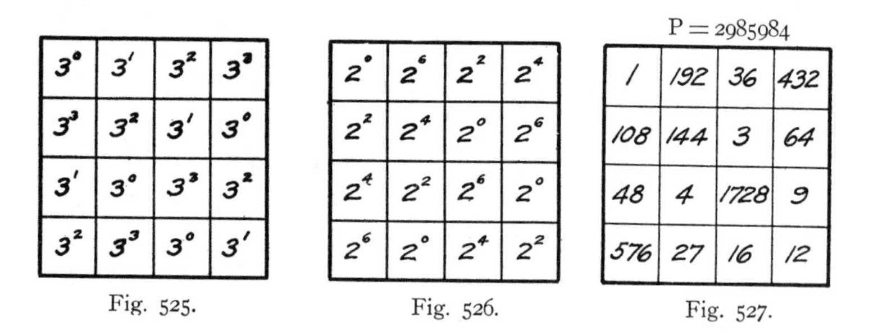

3^0	3^1	3^2	3^3
3^3	3^2	3^1	3^0
3^1	3^0	3^3	3^2
3^2	3^3	3^0	3^1

Fig. 525.

2^0	2^6	2^2	2^4
2^2	2^4	2^0	2^6
2^4	2^2	2^6	2^0
2^6	2^0	2^4	2^2

Fig. 526.

P = 2985984

1	192	36	432
108	144	3	64
48	4	1728	9
576	27	16	12

Fig. 527.

Figs. 528, 529, 530 show the application of a double set of factors to the primary squares. The constants of Fig. 528 are 3×5^3 and those of Fig. 529 are $2^3 \times 7$. This is also a Jaina square.

The Ratio Method.

If we fill a square with numbers as in Fig. 531, such that the ratios between all horizontally adjacent cells are equal, and the ratios between all vertically adjacent cells are equal, we have a natural square which can be formed into a geometric magic by any of the well-known methods.

The horizontal ratios in Fig. 531 are 2 as represented by the figure at the end of the division line, and the vertical ratios are 3 as indicated, and Fig. 532 shows the magic arrangement of this series.

In a fourth order square, as in Fig. 533, the horizontal ratios

3^0	5^1	5^2	3^1
3^1	5^2	5^1	3^0
5^1	3^0	3^1	5^2
5^2	3^1	3^0	5^1

Fig. 528.

2^1	2^0	2^2	7^1
2^2	7^1	2^1	2^0
7^1	2^2	2^0	2^1
2^0	2^1	7^1	2^2

Fig. 529.

P = 21000

2	5	100	21
12	175	10	1
35	4	3	50
25	6	7	20

Fig. 530.

are not necessarily equal, and neither are the vertical ratios. A magic may be made from this natural square by forming the numbers in the upper row into a primary square as in Fig. 534. The numbers in the left-hand column are then formed into another primary square as in Fig. 535. These two primary squares will then produce the magic shown in Fig. 536.

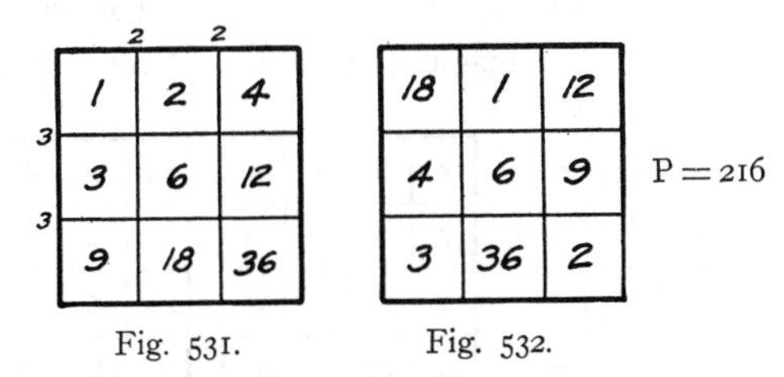

1	2	4
3	6	12
9	18	36

Fig. 531.

18	1	12
4	6	9
3	36	2

P = 216

Fig. 532.

Fig. 537 is a balanced natural square. This series will produce a perfect Jaina, or Nasik,* or an associated square. Figs. 538, 539 and 540 show it arranged in a Nasik formation.

Mr. L. S. Frierson's arithmetical equation squares also have their geometric brothers. Where he applies the equation $a - b =$

* A concise description of Nasik squares is given in *Enc. Brit.*

$c-d$, we use the proportion $a:b::c:d$. Fig. 542 shows a natural equation square, and besides the proportions there shown, the diagonals of the magic depend on the necessary proportion $a:b::c:d$ as indicated in the respective cells of Fig. 544*a*.

P = 7560

Column multipliers: 2, $\frac{3}{2}$, 3; row multipliers: 4, $\frac{5}{4}$, $\frac{7}{5}$

1	2	3	9
4	8	12	36
5	10	15	45
7	14	21	63

Fig. 533.

1	2	3	9
9	3	2	1
2	1	9	3
3	9	1	2

Fig. 534.

1	7	4	5
4	5	1	7
5	4	7	1
7	1	5	4

Fig. 535.

1	14	12	45
36	15	2	7
10	4	63	3
21	9	5	8

Fig. 536.

The magic is then formed by revolving the diagonals 180° as is shown in Fig. 543, or by interchanging the numbers represented by like letters in Fig. 541.

P = 14400

Column multipliers: 2, $\frac{5}{2}$, 2; row multipliers: 3, $\frac{4}{3}$, 3

1	2	5	10
3	6	15	30
4	8	20	40
12	24	60	120

Fig. 537.

1	2	10	5
10	5	1	2
1	2	10	5
10	5	1	2

Fig. 538.

1	12	1	12
3	4	3	4
12	1	12	1
4	3	4	3

Fig. 539.

1	24	10	60
30	20	3	8
12	2	120	5
40	15	4	6

Fig. 540.

Another form of natural equation square is shown in Fig. 546. The diagonals in this square depend on the equation $a \times b = c \times d$ (see Fig. 544*b*). The magic is made by interchanging the numbers

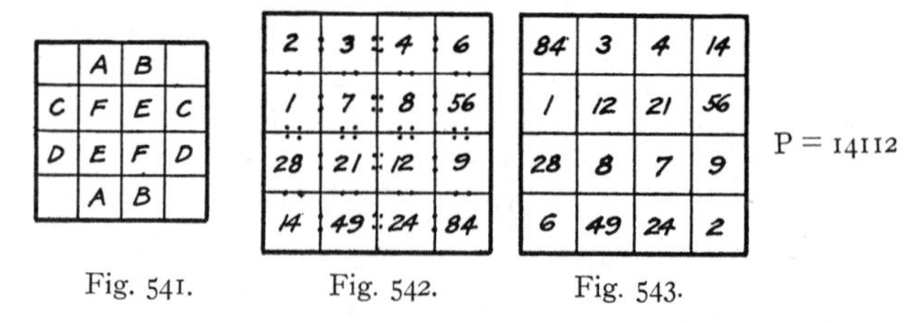

	A	B	
C	F	E	C
D	E	F	D
	A	B	

Fig. 541.

2	3	4	6
1	7	8	56
28	21	12	9
14	49	24	84

Fig. 542.

84	3	4	14
1	12	21	56
28	8	7	9
6	49	24	2

Fig. 543.

P = 14112

represented by like letters in Fig. 545, producing Fig. 547 and then adjusting to bring the numbers represented by the A's and D's in Fig. 545, in one diagonal and the numbers represented by the B's and C's in the other diagonal, or in other words, shifting the left-hand column of Fig. 547 so as to make it the right-hand column, and then shifting the bottom line of the square thus formed to the top. The result of these changes is shown in Fig. 548.

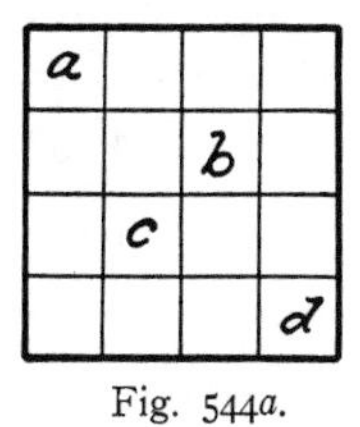

Fig. 544*a*.

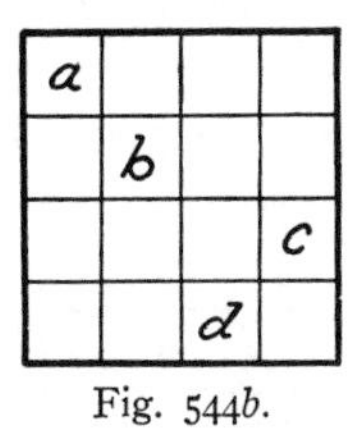

Fig. 544*b*.

		B	A
		A	B
C	D		
D	C		

Fig. 545.

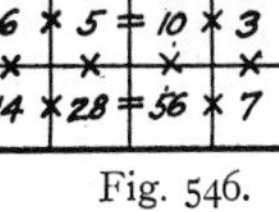

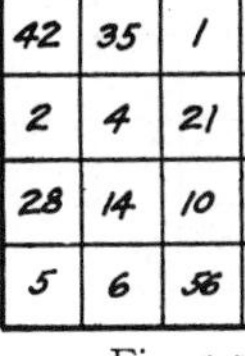

Fig. 546.

42	35	1	8
2	4	21	70
28	14	10	3
5	6	56	7

Fig. 547.

P = 11760

6	56	7	5
35	1	8	42
4	21	70	2
14	10	3	28

Fig. 548.

	2	$\frac{3}{2}$	$\frac{4}{3}$	$\frac{5}{4}$	
	1	2	3	4	5
6	6	12	18	24	30
$\frac{7}{6}$	7	14	21	28	35
$\frac{11}{7}$	11	22	33	44	55
$\frac{13}{11}$	13	26	39	52	65

Fig. 549.

1	2	3	4	5
3	4	5	1	2
5	1	2	3	4
2	3	4	5	1
4	5	1	2	3

Fig. 550.

1	7	13	6	11
6	11	1	7	13
7	13	6	11	1
11	1	7	13	6
13	6	11	1	7

Fig. 551.

P = 720720

1	14	39	24	55
18	44	5	7	26
35	13	12	33	4
22	3	28	65	6
52	30	11	2	21

Fig. 552.

Fig. 549 is a fifth order natural square, and Figs. 550, 551 and 552 clearly show the method of forming the magic, which is pan-diagonal.

In the same manner Dr. Planck constructed his arithmetical Nasik squares* of orders $4p+2$, we can likewise construct geometric squares.

Fig. 553 shows a natural 7×7 square with the central row and column cast out. This is formed by path method into the Nasik square, rearranging the columns in this order 1, 4, 32, 64, 16, 2

$P > 22 \times 10^{42}$

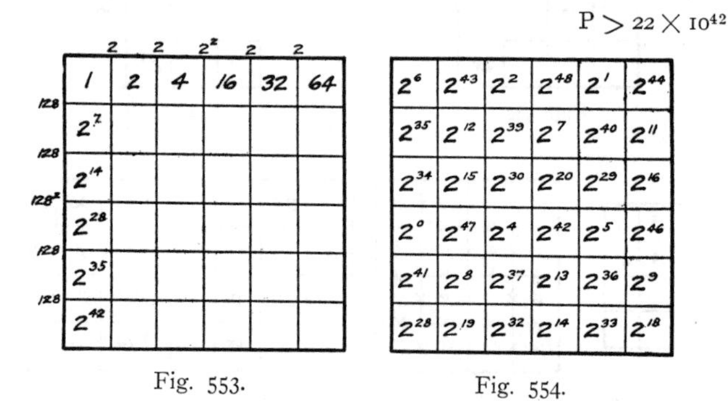

Fig. 553. Fig. 554.

and the rows in this order 1, 2^7, 2^{28}, 2^{42}, 2^{35}, 2^{14} and using advance move 2, 3 and a break-move $-1, -1$.

The Factorial Method.

In this method we fill two primary squares, each with n sets of any n different numbers, such that each row, column, and diagonal contains each of the n different numbers.

To avoid duplicates in the magic, the primary squares should have only one number in common, or they may not have any number in common. Also, no two numbers in one primary square should have the same ratio as two numbers in the other primary square.

This may be more clearly explained by an example. Suppose we select two sets of numbers as follows for constructing a fourth order square.

1 2 4 7
1 3 5 6

Four sets of the upper row of numbers are to fill one primary

* See "Notes on the Construction of Magic Squares" (n in the form of $4p+2$), p. 267.

square and four sets of the lower row are to fill the other. These two groups contain only one number in common, but the magic would contain duplicate numbers due to the duplicate ratios 2:4 as 3:6. Therefore $2\times6=4\times3$, consequently the duplicate numbers would be 12. But if we interchange the numbers 2 and 5, the fault will be corrected and the square can then be constructed without duplicate numbers.

The square in Fig. 555 is constructed with the two groups

1 2 3 4
1 5 6 7

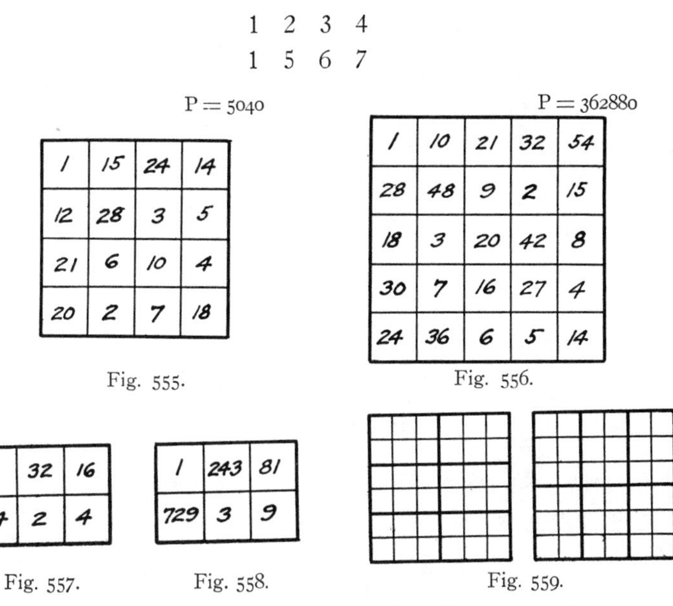

P = 5040

1	15	24	14
12	28	3	5
21	6	10	4
20	2	7	18

Fig. 555.

P = 362880

1	10	21	32	54
28	48	9	2	15
18	3	20	42	8
30	7	16	27	4
24	36	6	5	14

Fig. 556.

1	32	16
64	2	4

Fig. 557.

1	243	81
729	3	9

Fig. 558.

Fig. 559.

A fifth order square is shown in Fig. 556 and in this case the following groups are used:

1 2 3 4 6
1 5 7 8 9

This square is pan-diagonally magic.

I will now show how a Nasik sixth order square may be made by a method derived from Dr. Planck's method of constructing Nasik squares with arithmetical series.

Fill two six-celled rectangles, each with six different numbers, the two rectangles to have no more than one number in common.

The numbers in each rectangle should be arranged so that the products of its horizontal rows are equal, and the products of its vertical rows are equal.

Two of such sets of numbers that will suit the above conditions will not be found so readily as in Dr. Planck's examples above mentioned.

729	192	9	46656	3	576
32	486	2592	2	7776	162
11664	12	144	2916	48	36
1	15552	81	64	243	5184
23328	6	288	1458	96	18
16	972	1296	4	3888	324

P = 101,559,956,668,416.

Fig. 560.

The two sets forming the magic rectangles in Figs. 557 and 558 are taken from the following groups:

$$2^0 \quad 2^1 \quad 2^2 \quad 2^3 \quad 2^4 \quad 2^5 \quad 2^6$$
$$3^0 \quad 3^1 \quad 3^2 \quad 3^3 \quad 3^4 \quad 3^5 \quad 3^6$$

Each group is a geometrical series of seven numbers, and in forming the rectangle, the central number in each group is omitted.

1	2	4
5	10	20
25	50	100

3	6	12
15	30	60
75	150	300

9	18	36
45	90	180
225	450	900

Fig. 561.

The rectangles are arranged in primary squares as shown in Fig. 559, and the two rectangles in Figs. 557 and 558 so arranged will produce the square in Fig. 560. This square is pan-diagonal, 2^2-ply and 3^2-ply.*

* A square is said to be m^2-ply when the numbers in any m^2 group of contiguous cells give a constant product in geometric squares, or a constant sum in arithmetical squares.

Geometric Magic Cubes.

I will here briefly describe the analogy between the series which may be used in constructing cubes, and those used in constructing squares.

It is obvious that an unbroken geometric series of any sort may

1	90	300
60	25	18
450	12	5

150	4	45
9	30	100
20	225	6

180	75	2
50	36	15
3	10	900

P = 27000

Fig. 562.

be arranged in a cube of any order, by placing the numbers in the cube in the same progression as the numbers of an arithmetical series would be placed in forming an arithmetical cube. This may be accomplished by an extension of the method exemplified in Figs. 508 to 521 inclusive.

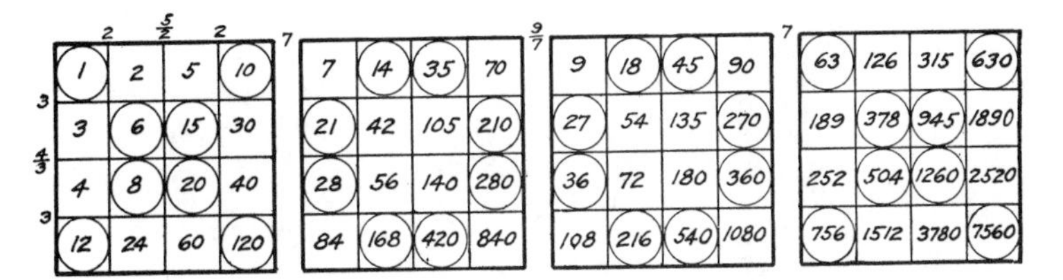

1	2	5	10
3	6	15	30
4	8	20	40
12	24	60	120

7	14	35	70
21	42	105	210
28	56	140	280
84	168	420	840

9	18	45	90
27	54	135	270
36	72	180	360
108	216	540	1080

63	126	315	630
189	378	945	1890
252	504	1260	2520
756	1512	3780	7560

Fig. 563.

P = 57,153,600

7560	2	5	756
3	1260	504	30
4	945	378	40
630	24	60	63

7	540	216	70
360	42	105	36
270	56	140	27
84	45	18	840

9	420	168	90
280	54	135	28
210	72	180	21
108	35	14	1080

120	126	315	12
189	20	8	1890
252	15	6	2520
10	1512	3780	1

Fig. 564.

In using the Exponential La Hireian method, the same process is followed in cubes as in squares, the main difference being that three primary cubes are necessarily used.

Fig. 561 shows a natural cubic series, obtained by the ratio method. The three squares represent the three planes of the cube.

The numbers 5 at the left of the first square represent the ratio between vertically adjacent cells in each of the planes. The numbers 2 above represent the ratio between horizontally adjacent cells in each of the planes, and the numbers 3 between the squares represent the ratio between adjacent cells from plane to plane.

By rearranging this series into a cube according to the path methods as in arithmetical cubes many results may be obtained, one of which is shown in Fig. 562.

A fourth order balanced or associated series is shown in Fig. 563. This series is analogous to the plane series in Fig. 537, and may be transformed into a magic cube by the following well-known method:

Interchange the numbers in all associated pairs of cells which are inclosed in circles, producing the result shown in Fig. 564.

The possibilities in using the Factorial method in constructing cubes, have not been investigated by the writer. H. A. S.

CHAPTER XII.

THE THEORY OF REVERSIONS.

SQUARES like those shown in Figs. 565 and 566, in which the numbers occur in their natural order, are known as *natural squares.* In such squares, it will be noticed that the numbers in associated cells are complementary, i. e., their sum is twice the mean number. It follows that any two columns equally distant from the central bar of the lattice are complementary columns, that is, the magic sum will be the mean of their sums. Further any two numbers in these complementary columns which lie in the same

1	2	3	4
5	6	7	8
9	10	11	12
13	14	15	16

Fig. 565.

1	2	3	4	5	6
7	8	9	10	11	12
13	14	15	16	17	18
19	20	21	22	23	24
25	26	27	28	29	30
31	32	33	34	35	36

Fig. 566.

row have a constant difference, and therefore the sums of the two columns differ by n times this difference. If then we raise the lighter column and depress the heavier column by $n/2$ times this difference we shall bring both to the mean value. Now we can effect this change by interchanging half the numbers in the one column with the numbers in the other column lying in their respective rows. The same is true with regard to rows, so that if we can make $n/2$ horizontal interchanges between every pair of complementary columns and the same number of vertical interchanges between every pair of complementary rows, we shall have the magic sum in all rows and columns. It is easy to see that we can do this by reversing half the rows and half the columns, provided the two operations are so arranged as not to interfere with one another. This last condition can be assured by always turning over columns and rows in associated pairs, for then we shall have made horizontal interchanges only between pairs of numbers previously untouched between pairs, each of whose constituents has already received an equal vertical displacement; and similarly with the vertical interchanges. By this method, it will be noticed, we always secure magic central diagonals, for however we choose our rows and columns we only alter the central diagonals of the natural

1	58	59	4	5	62	63	8
16	55	54	13	12	51	50	9
17	42	43	20	21	46	47	24
32	39	38	29	28	35	34	25
40	31	30	37	36	27	26	33
41	18	19	44	45	22	23	48
56	15	14	53	52	11	10	49
57	2	3	60	61	6	7	64

Fig. 567.

square (which are already magic) by interchanging pairs of complementaries with other pairs of complementaries.

Since the $n/2$ columns have to be arranged in pairs on either side of the central vertical bar of the lattice, $n/2$ must be even, and so the method, *in its simplest form,* applies only to orders $\equiv 0 \pmod 4$. We may formulate the rule thus: *For orders of form $4m$, reverse m pairs of complementary columns and m pairs of complementary rows, and the crude magic is completed.*

In the following example the curved lines indicate the rows and columns which have been reversed (Fig. 567).

We have said that this method applies only when $n/2$ is even but we shall now show that by a slight modification it can be applied to all even orders. For suppose n is double-of-odd; we cannot then arrange half the columns in pairs about the center since their number is odd, but we can so arrange $n/2-1$ rows and $n/2-1$ columns, and if we reverse all these rows and columns we shall have made $n/2-1$ interchanges between every pair of complementary rows and columns. We now require only to make the

16	2	3	13
5	11	10	8
9	7	6	12
4	14	15	1

Fig. 568.

one further interchange between every pair of rows and columns, without interfering with the previous changes or with the central diagonals. To effect this is always easy with any orders $\equiv 2 \pmod 4$, (6, 10, 14 etc.), excepting the first. In the case of 6^2 an artifice is necessary. If we reverse the two central diagonals of a square it will be found, on examination, that this is equivalent to reversing two rows and two columns; in fact, this gives us a

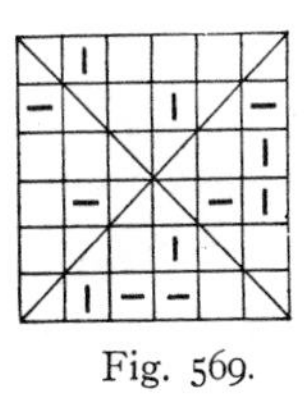

Fig. 569.

36	32	3	4	5	31
12	29	9	28	26	7
13	14	22	21	17	24
19	23	16	15	20	18
25	11	27	10	8	30
6	2	34	33	35	1

Fig. 570.

method of forming the magic 4^2 from the natural square with the least number of displacements, thus:

Applying this idea, we can complete the crude magic 6^2 from the scheme shown in Fig. 569 where horizontal lines indicate horizontal interchanges, and vertical lines vertical interchanges; the lines through the diagonals implying that the diagonals are to be reversed. The resulting magic is shown in Fig. 570.

The general method here described is known as the *method of reversions,* and the artifice used in the double-of-odd orders is called the *broken reversion.* The method of reversions, as applied to all even orders, both in squares and cubes, was first(?) investigated by the late W. Firth, Scholar of Emmanuel, Cambridge.*

The broken reversion for 6^2 may, of course, be made in various ways, but the above scheme is one of the most symmetrical, and may be memorialized thus: *For horizontal changes commence at the two middle cells of the bottom row, and progress upward and divergently along two knight's paths. For vertical changes turn the square on one of its sides and proceed as before.*

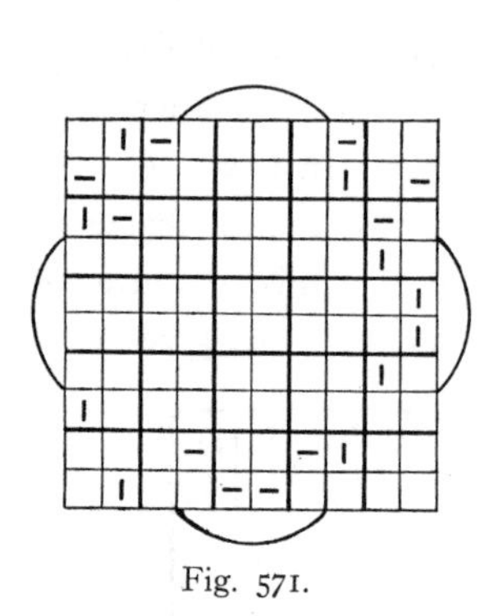

Fig. 571.

1	92	8	94	95	96	97	3	9	10
20	12	13	84	85	86	87	88	19	11
71	29	23	74	75	76	77	28	22	30
40	39	38	67	66	65	64	33	62	31
50	49	48	57	56	55	54	43	42	51
60	59	58	47	46	45	44	53	52	41
70	69	68	37	36	35	34	63	32	61
21	72	73	24	25	26	27	78	79	80
81	82	83	17	15	16	14	18	89	90
91	2	93	4	6	5	7	98	99	100

Fig. 572.

In dealing with larger double-of-odd orders we may leave the central diagonals "intact" and invert $n/2 - 1$ rows and $n/2 - 1$ columns. The broken reversion can then always be effected in a multitude of ways. It must be kept in mind, however, that in making horizontal changes we must not touch numbers which have been already moved horizontally, and if we use a number which has received a vertical displacement we can only change it with a number which has received an equal vertical displacement, and similarly with vertical interchanges. Lastly we must not touch the central diagonals.

* Died 1889. For historical notice see pp. 304-305.

Fig. 571 is such a scheme for 10^2, with the four central rows and columns reversed, and Fig. 572 shows the completed magic.

It is unnecessary to formulate a rule for making the reversions in these cases, because we are about to consider the method from a broader standpoint which will lead up to a general rule.

If the reader will consider the method used in forming the magic 6^2 by reversing the central diagonals, he will find that this artifice amounts to taking in every column two numbers equally distant from the central horizontal bar and interchanging each of them with its complementary in the associated cell, the operation being so arranged that two and only two numbers are moved in each row. This, as we have already pointed out, is equivalent to reversing two rows and two columns. Now these skew interchanges need not be made on the central diagonals—they can be made in any part of the lattice, provided the conditions just laid down are attended to. If then we make a second series of skew changes of like kind, we shall have, in effect, reversed 4 rows and 4 columns, and so on, each complete skew reversion representing two rows and columns. Now if $n \equiv 2 \pmod 4$ we have to reverse $n/2 - 1$ rows and columns before making the broken reversion, therefore the same result is attained by making $(n-2)/4$ complete sets of skew reversions and one broken reversion. In like manner, if $n \equiv 0 \pmod 4$, instead of reversing $n/2$ rows and columns we need only to make $n/4$ sets of skew reversions.

We shall define the symbol $[\times]$ as implying that skew interchanges are to be made between opposed pairs of the four numbers symmetrically situated with regard to the central horizontal and vertical bars, one of which numbers occupies the cell in which the symbol is placed. In other words we shall assume that Fig. 573*a* indicates what we have hitherto represented as in Fig. 573*b*. Further, it is quite unnecessary to use two symbols for a vertical or horizontal change, for Fig. 573*c* sufficiently indicates the same as Fig. 573*d*. If these abbreviations are granted, a scheme like Fig. 569 may be replaced by a small square like Fig. 574, which is to be applied to the top left-hand corner of the natural 6^2.

Fig. 575 is the extended scheme from Fig. 574, and Fig. 576

is the resulting magic. The small squares of symbols like Fig. 574 may be called *index squares.*

The law of formation for index squares is sufficiently obvious. To secure magic rows and columns in the resulting square, the symbols — and | must occur once on each row and column of the index, and the symbol × an equal number of times on each row and column; that is, if there are two series × ×. . . . × the symbol × must appear twice in every row and twice in every column, and

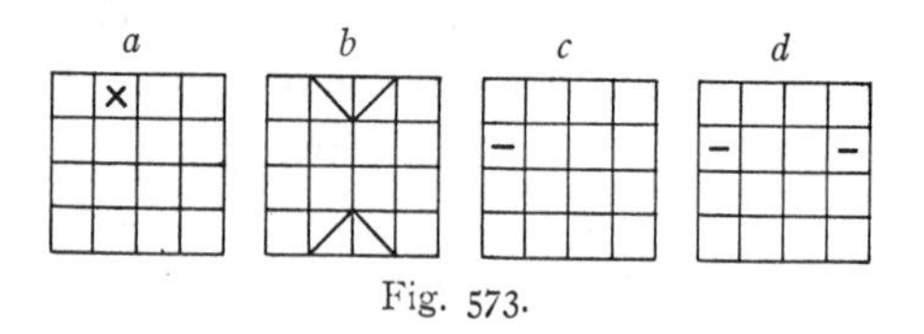

Fig. 573.

so on. But we already know by the theory of paths that these conditions can be assured by laying the successive symbolic periods along parallel paths of the index, whose coordinates are prime to the order of the index. If we decide always to use parallel diagonal paths and always to apply the index to the top left-hand corner of the natural square, the index square will be completely repre-

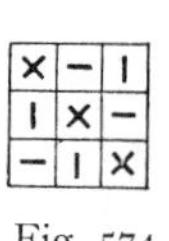

Fig. 574.

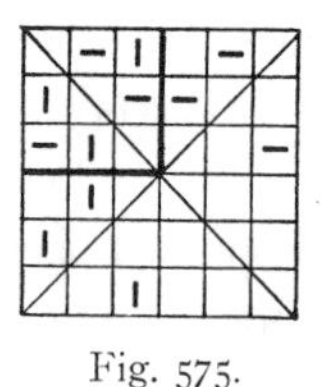

Fig. 575.

36	5	33	4	2	31
25	29	10	9	26	12
18	20	22	21	17	13
19	14	16	15	23	24
7	11	27	28	8	30
6	32	3	34	35	1

Fig. 576.

sented by its top row. In Fig 574 this is [× | — | |], which we may call the index-rod of the square, or we may simply call Fig. 576 the magic [× | — | |]. Remembering that we require $(n-2)/4$ sets of skew reversions when $n \equiv 2 \pmod 4$ and $n/4$ when $n \equiv 0$, it is obvious that the following rule will give crude magic squares of any even order n:

Take a rod of $n/2$ cells, $n/4$ symbols of the form ×, (using the integral part of $n/4$ only), and if there is a remainder when n

is divided by 4, add the symbols | and —. Place one of the symbols × in the left-hand cell of the rod, and the other symbols in any cell, but not more than one in each cell. The result is an index-rod for the magic n^2.

Take a square lattice of order $n/2$, and lay the rod along the top row of the lattice. Fill up every diagonal slanting downward and to the right which has a symbol in its highest cell with repetitions of that symbol. The resulting index-square if applied to the top left-hand corner of the natural n^2, with the symbols allowed the operative powers already defined, will produce the magic n^2.

The following are index-rods for squares of even orders:

4^2 [× ·]
6^2 [× · — · |]
8^2 [× · · × ·]
10^2 [× · · | · × · —]
12^2 [× · · · × · × ·]
14^2 [× · — · × · · · × · |]

When the number of cells in the rod exceeds the number of symbols, as it always does excepting with 6^2, the first cell may be left blank. Also, if there are sufficient blank cells, a × may

[× · | · | · — · × · —]

×	|	|	—	×	—
—	×	|	|	—	×
×	—	×	|	|	—
—	×	—	×	|	|
|	—	×	—	×	|
|	|	—	×	—	×

Fig. 577.

144	134	135	9	140	7	6	137	4	10	11	133
24	131	123	124	20	127	126	17	21	22	122	13
120	35	118	112	113	31	30	32	33	111	26	109
48	107	46	105	101	102	43	44	100	39	98	37
85	59	94	57	92	90	55	89	52	87	50	60
73	74	70	81	68	79	78	65	76	63	71	72
61	62	75	69	77	67	66	80	64	82	83	84
49	86	58	88	56	54	91	53	93	51	95	96
97	47	99	45	41	42	103	104	40	106	38	108
36	110	34	28	29	114	115	116	117	27	119	25
121	23	15	16	125	19	18	128	129	130	14	132
12	2	3	136	8	138	139	5	141	142	143	1

Fig. 578.

be replaced by two vertical and two horizontal symbols. Thus 12^2 might be given so [× · | · | · — · × · —]. This presentation of

12^2 is shown in Figs. 577, 578, and 14^2 from the index-rod given above, in Figs. 579 and 580.

Of course the employment of diagonal paths in the construction of the index is purely a matter of convenience. In the following index for 10^2, (Fig. 581) the skew-symbols are placed along two

×	—	×			×	I

×	—	×			×	I
I	×	—	×			×
×	I	×	—	×		
	×	I	×	—	×	
		×	I	×	—	×
×			×	I	×	—
—	×			×	I	×

Fig. 579.

196	13	194	4	5	191	189	8	188	10	11	185	2	183
169	181	26	179	19	20	176	175	23	24	172	17	170	28
168	156	166	39	164	34	35	36	37	159	32	157	41	155
43	153	143	151	52	149	49	50	146	47	144	54	142	56
57	58	138	130	136	65	134	133	62	131	67	129	69	70
126	72	73	123	117	121	78	77	118	80	116	82	83	113
98	111	87	88	108	104	106	105	93	103	95	96	100	85
99	97	101	102	94	90	92	91	107	89	109	110	86	112
84	114	115	81	75	79	119	120	76	122	74	124	125	71
127	128	68	60	66	132	64	63	135	61	137	59	139	140
141	55	45	53	145	51	147	148	48	150	46	152	44	154
42	30	40	158	38	160	161	162	163	33	165	31	167	29
15	27	171	25	173	174	22	21	177	178	18	180	16	182
14	184	12	186	187	9	7	190	6	192	193	3	195	1

Fig. 580.

parallel paths (2, 1) and the symbols — and | are then added so that each shall appear once in each row and once in each column, but neither of them on the diagonal of the index slanting upward and to the left.

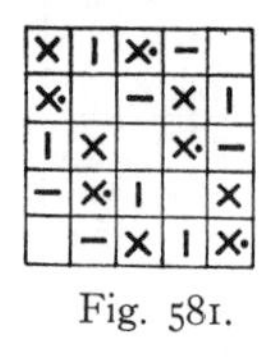

Fig. 581.

Crude cubes of even orders we shall treat by the index-rod as in the section on squares. The reader will remember that we constructed squares of orders ≡ 0 (mod 4) by reversing half the rows and half the columns, and it is easy to obtain an analogous method for the cubes of the same family. Suppose we reverse the V-planes* in associated pairs; that is, turn each through an angle of 180° round a horizontal axis parallel to the paper-plane so that the associated columns in each plane are interchanged and reversed. We evidently give to every row of the cube the magic sum, for half the numbers in each row will be exchanged for their complemen-

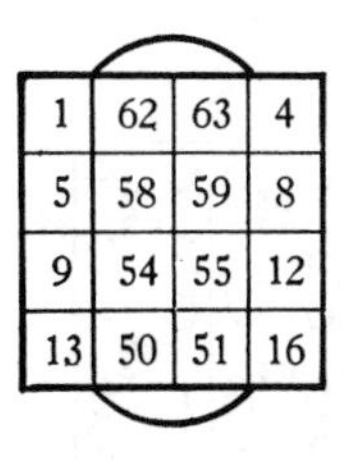

1	62	63	4
5	58	59	8
9	54	55	12
13	50	51	16

17	46	47	20
21	42	43	24
25	38	39	28
29	34	35	32

33	30	31	36
37	26	27	40
41	22	23	44
45	18	19	48

49	14	15	52
53	10	11	56
57	6	7	60
61	2	3	64

Magic in rows only.
Fig. 582. The natural 4^3 with V-planes reversed.

1	62	63	4
56	11	10	53
60	7	6	57
13	50	51	16

17	46	47	20
40	27	26	37
44	23	22	41
29	34	35	32

33	30	31	36
24	43	42	21
28	39	38	25
45	18	19	48

49	14	15	52
8	59	58	5
12	55	54	9
61	2	3	64

Magic in rows and columns.
Fi. 583. Being Fig. 583 with H-planes reversed.

1	62	63	4
56	11	10	53
60	7	6	57
13	50	51	16

32	35	34	29
41	22	23	44
37	26	27	40
20	47	46	17

48	19	18	45
25	38	39	28
21	42	43	24
36	31	30	33

49	14	15	52
8	59	58	5
12	55	54	9
61	2	3	64

Magic in rows, columns and lines.
Fig. 584. Being Fig. 19, with P-planes reversed.
CRUDE MAGIC 4^3.

taries. If we do likewise with H-planes and P-planes the rows and lines† will become magic. But as with the square, and for like reasons, these three operations can be performed without mutual

* P-plane = Presentation-, or Paper-plane; H-plane = Horizontal plane; V-plane = Vertical plane.

† "Line" = a contiguous series of cells measured at right angles to the paper-plane.

interference. Hence the simple general rule for all cubes of the double-of-even orders:

Reverse, in associated pairs, half the V-planes, half the H-planes and half the P-planes.

With this method the central great diagonals, of course, maintain their magic properties, as they must do for the cube to be considered even a crude magic. To make the operation clear to

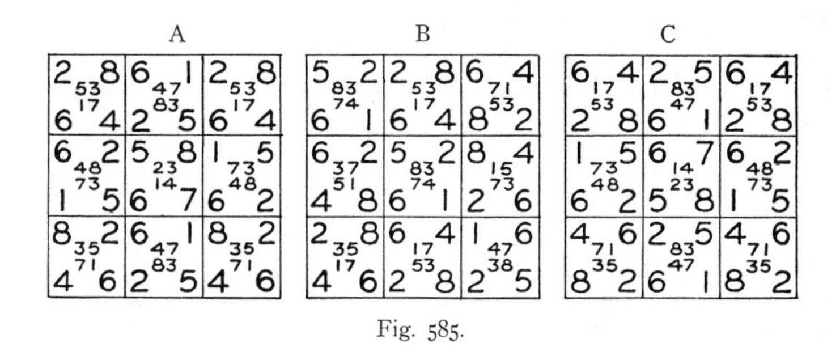

Fig. 585.

the reader we append views of 4^3 at each separate stage, the central pair of planes being used at each reversion.

By this method the reader can make any crude magic cube of order $4m$. With orders of form $4m + 2$ we find the same difficulties as with squares of like orders. So far as we are aware no magic cube of this family had been constructed until Firth suc-

1	17	24
23	3	16
18	22	2

15	19	8
7	14	21
20	9	13

26	6	10
12	25	5
4	11	27

Fig. 586.

ceeded with 6^3 in 1889. Firth's original cube was built up by the method of "pseudo-cubes," being an extension to solid magics of Thompson's method. The cube of 216 cells was divided into 27 subsidiary cubes each containing 2 cells in an edge. The 8 cells of each subsidiary were filled with the numbers 1 to 8 in such a way that each row, column, line, *and central great diagonal* of the large cube summed 27. The cube was then completed by using the magic 3^3 in the same way that 6^2 is constructed from 3^2. Firth formulated no rule for arrangement of the numbers in the pseudo-cubes, and great difficulty was encountered in balancing the central great diagonals. His pseudo-skeleton is shown in Fig. 585, where each plate represents two P-planes of 6^3, each plate containing 9 pseudo-cubes. The numbers in the subsidiaries are shown in diagrammatic perspective, the four "larger" numbers lying in the anterior layer, and the four "smaller" numbers, grouped in the center, in the posterior layer.

I

2	8	134	129	186	192
6	4	130	133	190	188
182	178	21	24	121	125
177	181	22	23	126	122
144	138	174	169	16	10
140	142	170	173	12	14

II

5	3	132	135	189	187
1	7	136	131	185	191
180	184	18	19	127	123
183	179	17	20	124	128
139	141	172	175	11	13
143	137	176	171	15	9

III.

117	114	146	152	62	60
118	113	150	148	64	58
54	50	109	106	168	164
52	56	110	105	162	166
154	160	70	68	97	102
156	158	66	72	98	101

120	115	149	147	63	59
119	116	145	151	61	59
51	55	112	107	161	165
53	49	111	108	167	163
155	157	65	71	100	103
153	159	69	67	99	104

IV

206	204	42	45	78	76
202	208	46	41	74	80
89	93	198	199	38	34
94	90	197	200	33	37
28	30	82	85	212	214
32	26	86	81	216	210

V

201	207	48	43	73	79
205	203	44	47	77	75
95	91	193	196	36	40
92	96	194	195	39	35
31	25	88	83	215	209
27	29	84	87	211	213

VI

Fig. 587.

If we use this with the magic of Fig. 586 we obtain the magic 6^3 shown in Fig. 587.

This cube is non-La Hireian, as is frequently the case with magics constructed by this method.

The scheme of pseudo-cubes for 6^3 once found, we can easily extend the method to any double-of-odd order in the following manner. Take the pseudo-scheme of next lower order [e. g., 6^3 to make 10^3, 10^3 to make 14^3 etc.]. To each of three outside plates of cubes, which meet at any corner of the skeleton, apply a replica-plate, and to each of the other three faces a complementary to the plate opposed to it, that is a plate in which each number replaces its complementary number (1 for 8, 2 for 7, etc.). We now have a properly balanced skeleton for the next double-of-odd order, wanting only its 12 edges. Consider any three edges that meet at a corner of the cube; they can be completed (wanting their corner-cubes) by placing in each of them any row of cubes from the original skeleton. Each of these three edges has three other edges parallel to it, two lying in the same square planes with it and the third diagonally opposed to it. In the former we may place edges complementary to the edge to which they are parallel, and in the latter a replica of the same. The skeleton wants now only its 8 corner pseudo-cubes. Take any cube and place it in four corners, no two of which are in the same row, line, column, or great diagonal (e. g., B, C, E, H in Fig. 602), and in the four remaining corners place its complementary cube. The skeleton is now complete, and the cube may be formed from the odd magic of half its order.

This method we shall not follow further, but shall now turn to the consideration of index-cubes, an artifice far preferable.

Before proceeding, the reader should carefully study the method of the index-rod as used for magic squares (pp. 299-302).

The reversion of a pair of planes in each of the three aspects, as previously employed for 4^3, is evidently equivalent to interchanging two numbers with their complementaries in every row, line, and column of the natural cube. If therefore we define the symbol $\times$ as implying that such an interchange is to be made not only from the cell in which it is placed, but also from the three other cells with which it is symmetrically situated in regard to the central horizontal and vertical bars of its P-plane, and can make one such symbol operate in every row, line and column of an index-cube whose edge is half that of the great cube, we shall have secured the equivalent of the above-mentioned reversion. For example, a $\times$ placed in the second cell of the top row of any P-plane of 4^3, will denote that the four numbers marked a in Fig. 588 are each to be interchanged with its complement, which lies in the associated cell in the associated P-plane.

From this it follows that we shall have a complete reversion scheme for any order $4m$, by placing in every row, line and column of the index $(2m)^3$, m of the symbols $\times$. In the case of orders $4m \times 2$, after placing m such symbols in the cube $(2m+1)^3$, we have still to make the equivalent of one reversed plane in each of the three aspects. This amounts to making one symmetrical vertical interchange, one symmetrical horizontal interchange, and one symmetrical interchange at right angles to the paper-plane in every

	a	a	
	a	a	

Fig. 588.

row, line and column. If we use the symbol | to denote such a vertical interchange, not only for the cell in which it stands, but also for the associated cell, and give like meanings to — and ·, for horizontal changes and changes along lines, we shall have made the broken reversion when we allow each of these symbols to operate once in every row, column and line of the index. For example, *a* in Fig. 589 means *b* in its own P-plane, and *c* in the associated P-plane; while *d* indicates that the numbers lying in its own

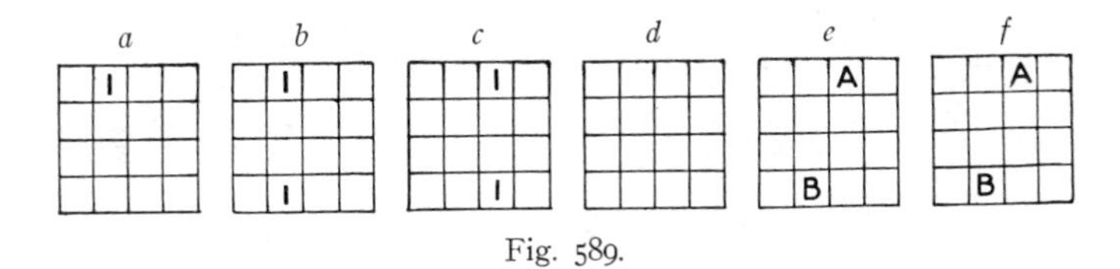

Fig. 589.

P-plane as in *e* are to be interchanged, A with A and B with B, with the numbers lying in the associated plane *f*. We can always prepare the index, provided the rod does not contain a less number of cells than the number of symbols, by the following rule, n being the order.

Take an index-rod of $n/2$ cells, $n/4$ symbols of the form $\times$, (using the integral part of $n/4$ only), and if there is any remainder when n is divided by 4 add the three symbols |, —, ·. Now prepare an index square in the way described on page 300, but using the diagonals upward and to the right instead of upward to the left,* and take this square as the first P-plane of an index-cube. Fill

I

64	2	3	61
5	59	58	8
9	55	54	12
52	14	15	49

II

48	18	19	45
21	43	42	24
25	39	38	28
36	30	31	33

III

32	34	35	29
37	27	26	40
41	23	22	44
20	46	47	17

IV

16	50	51	13
53	11	10	56
57	7	6	60
4	62	63	1

Fig. 590.

every *great* diagonal of the cube, running to the *right, down* and *away,* which has a symbol in this P-plane cell, with repetitions of that symbol.† This index-cube applied to the near, left-hand, top

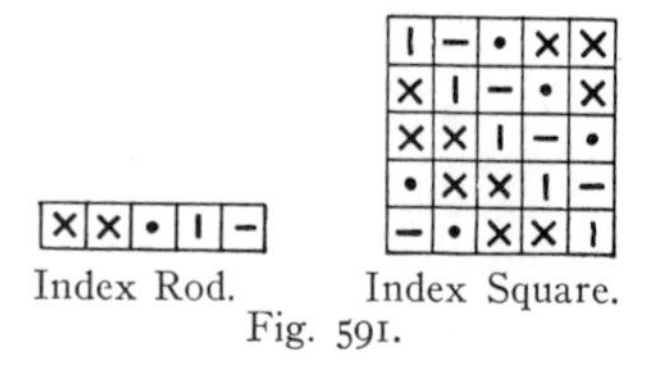

Index Rod. Index Square.
Fig. 591.

corner of the natural n^3, with the symbols allowed the operative powers already defined, will make the magic n^3.

This method for even orders applies universally with the single

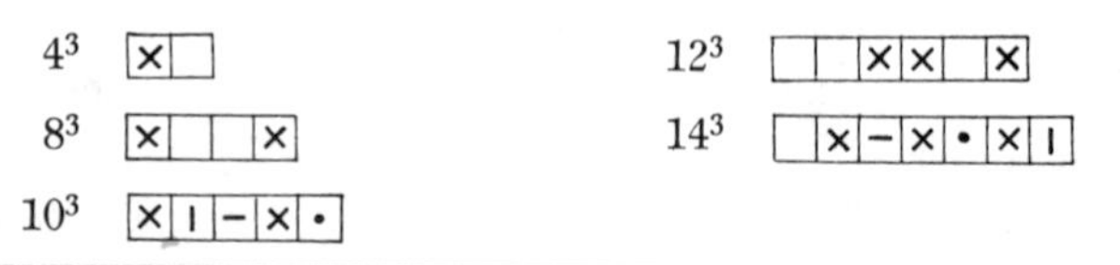

Fig. 592. Index Cube.

exception of 6^3, and in the case of 6^3 we shall presently show that the broken reversion can still be made by scattering the symbols over the whole cube. The following are index-rods for various cubes.

4^3	×						
8^3	×			×			
10^3	×	I	—	×	•		
12^3			×	×		×	
14^3		×	—	×	•	×	I

* Either way will do, but it happens that the former has been used in the examples which follow.

† More briefly, in the language of Paths, the symbols are laid, in the square, on (1, 1); their repetitions in the cube, on (1, —1, 1).

As in the case of index-rods for squares, the first cell may be left blank, otherwise it must contain a $\times$.

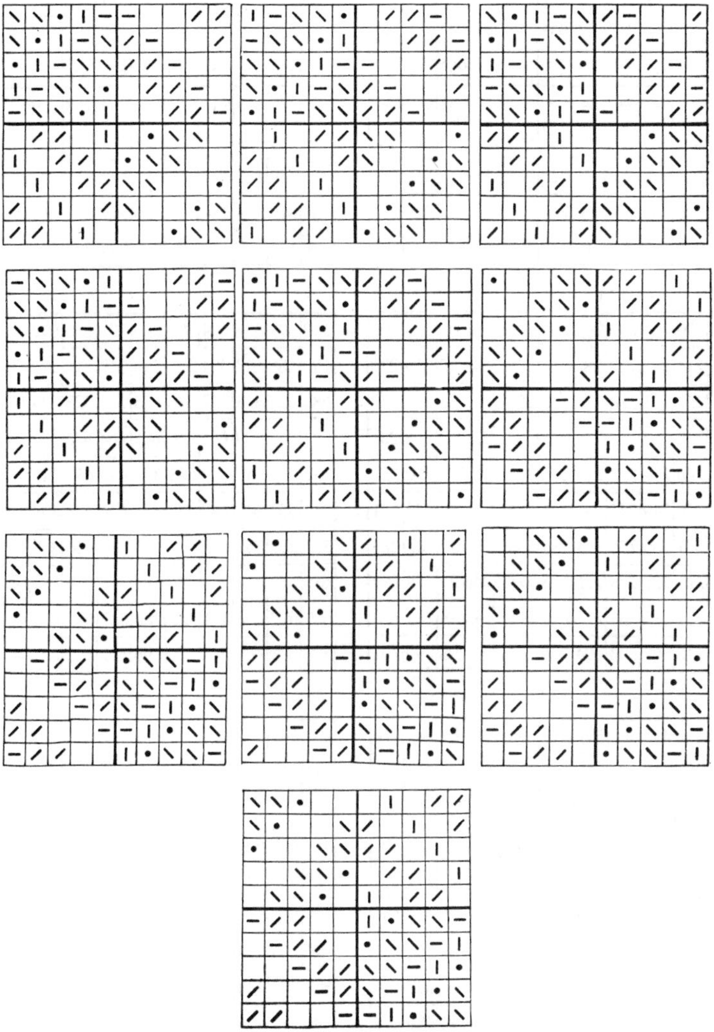

Fig. 593. Extended Reversion Scheme for 10^3.

Fig. 590 is a 4^3, made with the index-rod given above. It has only half the numbers removed from their natural places. Figs.

1000	999	903	94	6	5	7	8	992	991
990	912	83	17	986	985	14	18	19	981
921	72	28	977	976	975	974	23	29	30
61	39	968	967	935	36	964	963	32	40
50	959	958	944	55	46	47	953	952	41
51	949	948	54	45	56	957	943	942	60
31	62	938	937	65	966	934	933	69	70
71	22	73	927	926	925	924	78	79	980
920	82	13	84	916	915	87	88	989	911
910	909	93	4	95	96	97	998	902	901

191	109	898	897	805	106	894	893	102	110
120	889	888	814	185	116	117	883	882	111
880	879	823	174	126	125	127	128	872	871
870	832	163	137	866	865	134	138	139	861
841	152	148	857	856	855	854	143	149	150
151	142	153	847	846	845	844	158	159	860
840	162	133	164	836	835	167	168	869	831
830	829	173	124	175	176	177	878	822	821
181	819	818	184	115	186	887	813	812	190
101	192	808	807	195	896	804	803	199	200

800	702	293	207	796	795	204	208	209	791
711	282	218	787	786	785	784	213	219	220
271	229	778	777	725	226	774	773	222	230
240	769	768	734	265	236	237	763	762	231
760	759	743	254	246	245	247	248	752	751
750	749	253	244	255	256	257	758	742	741
261	739	738	264	235	266	767	733	732	270
221	272	728	727	275	776	724	723	279	280
281	212	283	717	716	715	714	288	289	790
710	292	203	294	706	705	297	298	799	701

310	699	698	604	395	306	307	693	692	301
690	689	613	384	316	315	317	318	682	681
680	622	373	327	676	675	324	328	329	671
631	362	338	667	666	665	664	333	339	340
351	349	658	657	645	346	654	653	342	350
341	352	648	647	355	656	644	643	359	360
361	332	363	637	636	635	634	368	369	670
630	372	323	374	626	625	377	378	679	621
620	619	383	314	385	386	387	688	612	611
391	609	608	394	305	396	697	603	602	400

501	492	408	597	596	595	594	403	409	410
481	419	588	587	515	416	584	583	412	420
430	579	578	524	475	426	427	573	572	421
570	569	533	464	436	435	437	438	562	561
560	542	453	447	556	555	444	448	449	551
550	452	443	454	546	545	457	458	559	541
540	539	463	434	465	466	467	568	532	531
471	529	528	474	425	476	577	523	522	480
411	482	518	517	485	586	514	513	489	490
491	402	493	507	506	505	504	498	499	600

401	502	503	497	496	495	494	508	599	510
511	512	488	487	415	516	484	483	519	590
521	479	478	424	525	576	527	473	472	530
470	469	433	534	535	536	567	538	462	461
460	442	543	544	456	455	547	558	549	451
450	552	553	557	446	445	554	548	459	441
440	439	563	564	566	565	537	468	432	431
580	429	428	574	575	526	477	423	422	571
581	589	418	417	585	486	414	413	582	520
591	592	598	407	406	405	404	593	509	500

Fig. 594. First 6 plates of 10^3, made from Fig. 593. (Sum = 5005.)

591 and 592 are the index-rod, index-square and index-cube for 10^3, and Fig. 593 is the extended reversion scheme obtained from these, in which ╲ and ╱ denote single changes between associated cells, and the symbols |, —, and ·, single changes parallel to columns, rows, and lines. Figs. 594 and 595 show the resulting cube.

If we attack 6^3 by the general rule, we find 4 symbols, ×, —, |, ·, and only 3 cells in the rod; the construction is therefore

601	399	398	304	605	696	607	393	392	610
390	389	313	614	615	616	687	618	382	381
380	322	623	624	376	375	627	678	629	371
331	632	633	367	366	365	364	638	669	640
641	642	358	357	345	646	354	353	649	660
651	659	348	347	655	356	344	343	652	650
661	662	668	337	336	335	334	663	639	370
330	672	673	677	326	325	674	628	379	321
320	319	683	684	686	685	617	388	312	311
700	309	308	694	695	606	395	303	302	691

300	202	703	704	296	295	707	798	709	291
211	712	713	287	286	285	284	718	789	720
721	722	278	277	225	726	274	273	729	780
731	269	268	234	735	766	737	263	262	740
260	259	243	744	745	746	757	748	252	251
250	249	753	754	756	755	747	258	242	241
770	239	238	764	765	736	267	233	232	761
771	779	228	227	775	276	224	223	772	730
781	782	788	217	216	215	214	783	719	290
210	792	793	797	206	205	794	708	299	201

801	802	198	197	105	806	194	193	809	900
811	189	188	114	815	886	817	183	182	820
180	179	123	824	825	826	877	828	172	171
170	132	833	834	166	165	837	868	839	161
141	842	843	157	156	155	154	848	859	850
851	852	858	147	146	145	144	853	849	160
140	862	863	867	136	135	864	838	169	131
130	129	873	874	876	875	827	178	122	121
890	119	118	884	885	816	187	113	112	881
891	899	108	107	895	196	104	103	892	810

100	99	3	904	905	906	997	908	92	91
90	12	913	914	86	85	917	988	919	81
21	922	923	77	76	75	74	928	979	930
931	932	68	67	35	936	64	63	939	970
941	59	58	44	945	956	947	53	52	950
960	49	48	954	955	946	57	43	42	951
961	969	38	37	965	66	34	33	962	940
971	972	978	27	26	25	24	973	929	80
20	982	983	987	16	15	984	918	89	11
10	9	993	994	996	995	907	98	2	1

Fig. 595. Last 4 plates of 10^3, made from Fig. 593. (Sum = 5005.)

impossible. Suppose we construct an index-cube from the rod | × | | | | | — |, we shall find it impossible to distribute the remaining symbol [·] in the extended reversion-scheme obtained from this index. The feat, however, is possible if we make (for this case only) a slight change in the meanings of | and —. By the general rule × operates on 4 cells in its own P-plane, where, by the rule of association, the planes are paired thus: $\left|\begin{array}{lcr} 1 & \text{with} & 6 \\ 2 & \text{"} & 5 \\ 3 & \text{"} & 4 \end{array}\right|$. In interpreting the meanings of | and —, in this special case, we must make

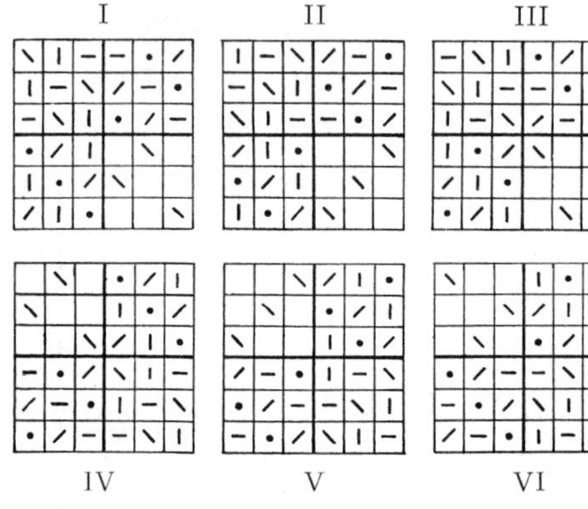

Fig. 596. Extended Reversion-Scheme for 6^3.

I

216	32	4	3	185	211
25	11	208	207	8	192
18	203	21	196	200	13
199	197	15	22	194	24
7	206	190	189	29	30
186	2	213	34	35	181

II

67	41	178	177	38	150
48	173	63	154	170	43
168	56	52	51	161	163
162	50	165	58	59	157
169	155	45	64	152	66
37	176	148	147	71	72

III

78	143	105	112	140	73
138	98	82	81	119	133
91	89	130	129	86	126
85	128	124	123	95	96
120	80	135	100	101	115
139	113	75	106	110	108

109	107	111	76	104	144
102	116	117	136	83	97
121	122	94	93	131	90
132	92	88	87	125	127
84	137	99	118	134	79
103	77	142	141	74	114

IV

145	146	70	69	179	42
151	65	153	46	62	174
60	158	159	166	53	55
54	167	57	160	164	49
61	47	172	171	44	156
180	68	40	39	149	175

V

36	182	183	214	5	31
187	188	28	27	209	12
193	23	195	16	20	204
19	17	202	201	14	198
210	26	10	9	191	205
6	215	33	184	212	1

VI

Fig. 597, made from Fig. 596. Sum = 651.

a cyclic change in the right-hand column of this little table. Thus for "|" $\left|\begin{array}{lcr} 1 & \text{with} & 5 \\ 2 & \text{"} & 4 \\ 3 & \text{"} & 6 \end{array}\right|$, and for "—" $\left|\begin{array}{lcr} 1 & \text{with} & 4 \\ 2 & \text{"} & 6 \\ 3 & \text{"} & 5 \end{array}\right|$. This means that a [|], for example, in the second P-plane has its usual meaning in that plane, and also acts on the two cells which would be the associated cells if the 4th plane were to become the 5th, etc. If we extend this scheme, there will be just room to properly distribute the [·]'s in the two parallelopipeds which form the right-

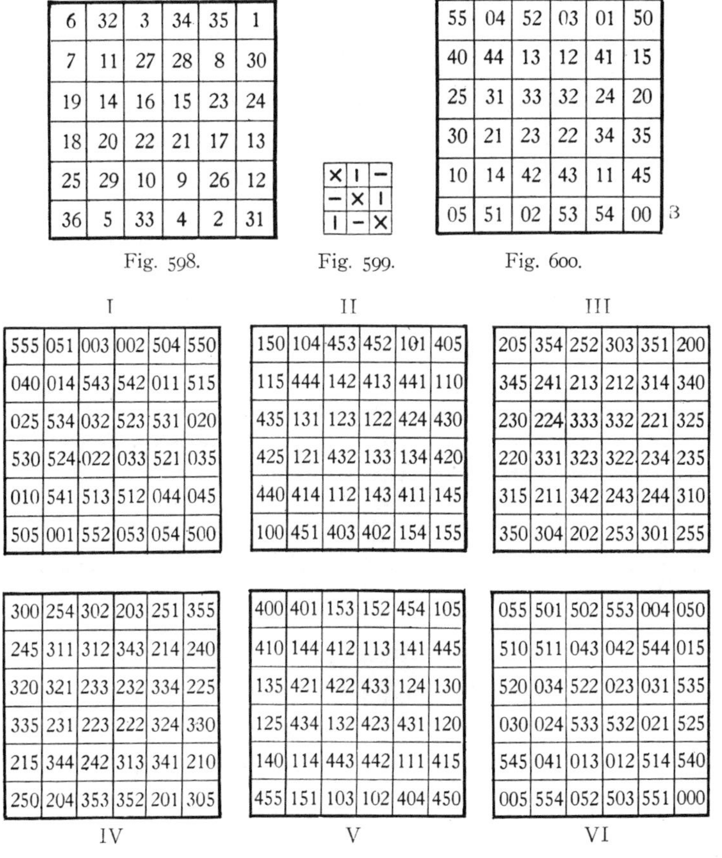

6	32	3	34	35	1
7	11	27	28	8	30
19	14	16	15	23	24
18	20	22	21	17	13
25	29	10	9	26	12
36	5	33	4	2	31

Fig. 598.

<table>
<tr><td>×</td><td>|</td><td>−</td></tr>
<tr><td>−</td><td>×</td><td>|</td></tr>
<tr><td>|</td><td>−</td><td>×</td></tr>
</table>

Fig. 599.

55	04	52	03	01	50
40	44	13	12	41	15
25	31	33	32	24	20
30	21	23	22	34	35
10	14	42	43	11	45
05	51	02	53	54	00

Fig. 600.

I

555	051	003	002	504	550
040	014	543	542	011	515
025	534	032	523	531	020
530	524	022	033	521	035
010	541	513	512	044	045
505	001	552	053	054	500

II

150	104	453	452	101	405
115	444	142	413	441	110
435	131	123	122	424	430
425	121	432	133	134	420
440	414	112	143	411	145
100	451	403	402	154	155

III

205	354	252	303	351	200
345	241	213	212	314	340
230	224	333	332	221	325
220	331	323	322	234	235
315	211	342	243	244	310
350	304	202	253	301	255

300	254	302	203	251	355
245	311	312	343	214	240
320	321	233	232	334	225
335	231	223	222	324	330
215	344	242	313	341	210
250	204	353	352	201	305

IV

400	401	153	152	454	105
410	144	412	113	141	445
135	421	422	433	124	130
125	434	132	423	431	120
140	114	443	442	111	415
455	151	103	102	404	450

V

055	501	502	553	004	050
510	511	043	042	544	015
520	034	522	023	031	535
030	024	533	532	021	525
545	041	013	012	514	540
005	554	052	503	551	000

VI

Fig. 601.

hand upper and left-hand lower quarters of the cube, as shown in Fig. 596.

This scheme produces the cube shown in Fig. 597, which is magic on its 36 rows, 36 columns, 36 lines, *and on its 4 central great diagonals.*

Fig. 596 is the identical scheme discovered by Firth in 1889, and was obtained a few months later than the pseudo-skeleton shown in Fig. 585. A year or two earlier he had discovered the broken reversion for squares of even order, but he never generalized the method, or conceived the idea of an index-cube. The development of the method as here described was worked out by the present writer in 1894. About the same time Rouse Ball, of Trinity College, Cambridge, independently arrived at the method of reversions for squares (compare the earlier editions of his

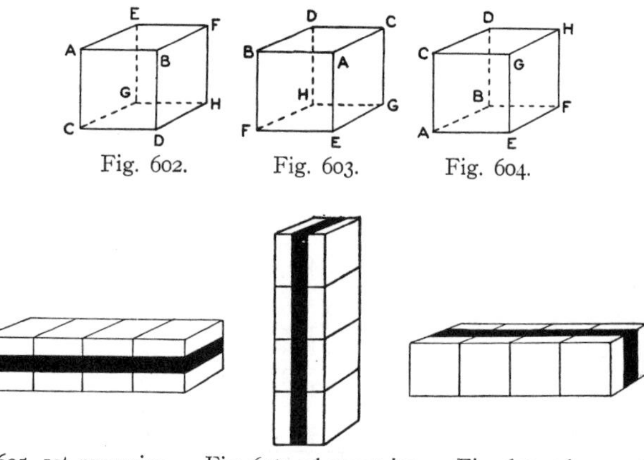

Fig. 602. Fig. 603. Fig. 604.

Fig. 605, 1st reversion. Fig. 606, 2d reversion. Fig. 607, 3d reversion.

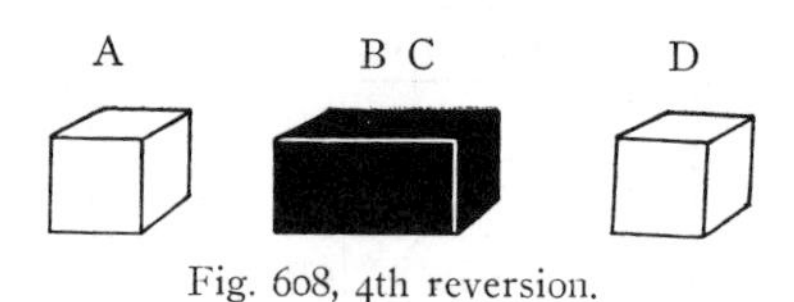

Fig. 608, 4th reversion.

Mathematical Recreations, Macmillan), and in the last edition, 1905, he adopts the idea of an index-square; but he makes no application to cubes or higher dimensions. There is reason to believe, however, that the idea of reversions by means of an index-square was known to Fermat. In his letter to Mersenne of April 1, 1640, (*Œuvres de Fermat,* Vol. II, p. 193), he gives the square of order 6 shown in Fig. 598. This is obtained by applying the index (Fig. 599) to the *bottom* left-hand corner of the natural square written from below upward, i. e., with the numbers 1 to 6 in the bottom row, 7 to 12 in the row above this, etc. There is nothing surprising in this method of writing the natural square, in fact it is suggested by the conventions of Cartesian geometry, with which Fermat was familar. There is a much later similar instance: Cayley, in 1890, dealing with "Latin squares," writes from below upward, although Euler, in his original Memoire (1782), wrote

1	2	3	4
248	247	246	245
252	251	250	249
13	14	15	16

65	66	67	68
184	183	182	181
188	187	186	185
77	78	79	80

129	130	131	132
120	119	118	117
124	123	122	121
141	142	143	144

193	194	195	196
56	55	54	53
60	59	58	57
205	206	207	208

17	18	19	20
232	231	230	229
236	235	234	233
29	30	31	32

81	82	83	84
168	167	166	165
172	171	170	169
93	94	95	96

145	146	147	148
104	103	102	101
108	107	106	105
157	158	159	160

209	210	211	212
40	39	38	37
44	43	42	41
221	222	223	224

33	34	35	36
216	215	214	213
220	219	218	217
45	46	47	48

97	98	99	100
152	151	150	149
156	155	154	153
109	110	111	112

161	162	163	164
88	87	86	85
92	91	90	89
173	174	175	176

225	226	227	228
24	23	22	21
28	27	26	25
237	238	239	240

49	50	51	52
200	199	198	197
204	203	202	201
61	62	63	64

113	114	115	116
136	135	134	133
140	139	138	137
125	126	127	128

177	178	179	180
72	71	70	69
76	75	74	73
189	190	191	192

241	242	243	244
8	7	6	5
12	11	10	9
253	254	255	256

Fig. 609.

from above downward. Another square of order 6, given by Fermat, in the same place, is made from the same index, but is disguised because he uses a "deformed" natural square.

It is interesting to note that all these reversion magics (unlike those made by Thompson's method), are La Hireian, and also that the La Hireian scheme can be obtained by turning a single outline on itself. To explain this statement we will translate the square in Fig. 576 into the scale whose radix is 6, first decreasing every number by unity. This last artifice is merely equivalent to using the n^2 consecutive numbers from 0 to $n^2 - 1$, instead of from 1 to n^2, and is convenient because it brings the scheme of units and the scheme of 6's digits into uniformity.

1	254	255	4
248	11	10	245
252	7	6	249
13	242	243	16

65	190	191	68
184	75	74	181
188	71	70	185
77	178	179	80

129	126	127	132
120	139	138	117
124	135	134	121
141	114	115	144

193	62	63	196
56	203	202	53
60	199	198	57
205	50	51	208

17	238	239	20
232	27	26	229
236	23	22	233
29	226	227	32

81	174	175	84
168	91	90	165
172	87	86	169
93	162	163	96

145	110	111	148
104	155	154	101
108	151	150	105
157	98	99	160

209	46	47	212
40	219	218	37
44	215	214	41
221	34	35	224

33	222	223	36
216	43	42	213
220	39	38	217
45	210	211	48

97	158	159	100
152	107	106	149
156	103	102	153
109	146	147	112

161	94	95	164
88	171	170	85
92	167	166	89
173	82	83	176

225	30	31	228
24	235	234	21
28	231	230	25
237	18	19	240

49	206	207	52
200	59	58	197
204	55	54	201
61	194	195	64

113	142	143	116
136	123	122	133
140	119	118	137
125	130	131	128

177	78	79	180
72	187	186	69
76	183	182	73
189	66	67	192

241	14	15	244
8	251	250	5
12	247	246	9
253	2	3	256

Fig. 610.

If we examine this result as shown in Fig. 600 we find that the scheme for units can be converted into that for the 6's, by turning the skeleton through 180° about the axis AB; that is to say, a single outline turned upon itself will produce the magic.

The same is true of the cube; that is, just as we can obtain a La Hireian scheme for a square by turning a single square outline once upon itself, so a similar scheme for a cube can be obtained by turning a cubic outline twice upon itself. If we reduce all the numbers in Fig. 597 by unity and then "unroll" the cube, we get the La Hireian scheme of Fig. 601 in the scale radix 6.

If now we represent the skeleton of the 6^2's: (left-hand) digits by Fig. 602, and give this cube the "twist" indicated by Fig. 603, we

1	254	255	4
248	11	10	245
252	7	6	249
13	242	243	16

65	190	191	68
184	75	74	181
188	71	70	185
77	178	179	80

129	126	127	132
120	139	138	117
124	135	134	121
141	114	115	144

193	62	63	196
56	203	202	53
60	199	198	57
205	50	51	208

224	35	34	221
41	214	215	44
37	218	219	40
212	47	46	209

160	99	98	157
105	150	151	108
101	154	155	104
148	111	110	145

96	163	162	93
169	86	87	172
165	90	91	168
84	175	174	81

32	227	226	29
233	22	23	236
229	26	27	232
20	239	238	17

240	19	18	237
25	230	231	28
21	234	235	24
228	31	30	225

176	83	82	173
89	166	167	92
85	170	171	88
164	95	94	161

112	147	146	109
153	102	103	156
149	106	107	152
100	159	158	97

48	211	210	45
217	38	39	220
213	42	43	216
36	223	222	33

49	206	207	52
200	59	58	197
204	55	54	201
61	194	195	64

113	142	143	116
136	123	122	133
140	119	118	137
125	130	131	128

177	78	79	180
72	187	186	69
76	183	182	73
189	66	67	192

241	14	15	244
8	251	250	5
12	247	246	9
253	2	3	256

Fig. 611.

shall get the skeleon of the 6's (middle) digits, and the turn suggested by Fig. 604 gives that of the units (right-hand) digits. Thus a single outline turned twice upon itself gives the scheme.

We can construct any crude magic octahedroid* of double-

* DIMENSIONS	REGULAR FIGURE	BOUNDARIES
2	*Tetra*gon (or square	*4* one-dimensional straight lines
3	*Hexa*hedron (cube)	*6* two-dimensional squares
4	*Octa*hedroid	*8* three-dimensional cubes
etc.	etc.	etc.

of-even order, by the method of reversions, as shown with 4^4 in Figs. 605 to 608.

The first three reversions will be easily understood from the figures, but the fourth requires some explanation. It actually amounts to an interchange between every pair of numbers in associated cells of the parallelopiped formed by the two central cubical

1	254	255	4
248	11	10	245
252	7	6	249
13	242	243	16

192	67	66	189
73	182	183	76
69	186	187	72
180	79	78	177

128	131	130	125
137	118	119	140
133	122	123	136
116	143	142	113

193	62	63	196
56	203	202	53
60	199	198	57
205	50	51	208

224	35	34	221
41	214	215	44
37	218	219	40
212	47	46	209

97	158	159	100
152	107	106	149
156	103	102	153
109	146	147	112

161	94	95	164
88	171	170	85
92	167	166	89
173	82	83	176

32	227	226	29
233	22	23	236
229	26	27	232
20	239	238	17

240	19	18	237
25	230	231	28
21	234	235	24
228	31	30	225

81	174	175	84
168	91	90	165
172	87	86	169
93	162	163	96

145	110	111	148
104	155	154	101
108	151	150	105
157	98	99	160

48	211	210	45
217	38	39	220
213	42	43	216
36	223	222	33

49	206	207	52
200	59	58	197
204	55	54	201
61	194	195	64

144	115	114	141
121	134	135	124
117	138	139	120
132	127	126	129

80	179	178	77
185	70	71	188
181	74	75	184
68	191	190	65

241	14	15	244
8	251	250	5
12	247	246	9
253	2	3	256

Fig. 612.

selections. If the reader will use a box or some other "rectangular" solid as a model, and numbers the 8 corners, he will find that such a change cannot be effected in three-dimensional space by turning the parallelopiped as a whole, on the same principle that a right hand cannot, by any turn, be converted into a left hand. But such a change can be produced by a single turn in 4-dimensional space;

in fact this last reversion is made with regard to an axis in the 4th, or imaginary direction. The following four figures (609-612) show each stage of the process, and if the reader will compare them with the results of a like series of reversions made from a different aspect of the natural octahedroid, he will find that the "imaginary" reversion then becomes a real reversion, while one of the reversions which was real becomes imaginary. Fig. 609 is the natural 4^4 after the first reversion, magic in columns only; Fig. 610 is Fig. 609 after the second reversion, magic in rows and columns; Fig. 611 is Fig. 610 after the third reversion, magic in rows, columns and lines; and Fig. 612 is Fig. 611 after the fourth reversion, magic

	×	×		×			×	×			×		×	×	
×			×		×	×			×	×		×			×
×			×		×	×			×	×		×			×
	×	×		×			×	×			×		×	×	
×			×		×	×			×	×		×			×
	×	×		×			×	×			×		×	×	
	×	×		×			×	×			×		×	×	
×			×		×	×			×	×		×			×
×			×		×	×			×	×		×			×
	×	×		×			×	×			×		×	×	
	×	×		×			×	×			×		×	×	
×			×		×	×			×	×		×			×
	×	×		×			×	×			×		×	×	
×			×		×	×			×	×		×			×
×			×		×	×			×	×		×			×
	×	×		×			×	×			×		×	×	

Fig. 613. Skew Reversion for 4^4.

in rows, columns, lines and *i*'s, = crude magic 4^4. The symbol *i* denotes series of cells parallel to the imaginary edge.

Fig. 612 is magic on its 64 rows, 64 columns, 64 lines, and 64 *i*'s and on its 8 central hyperdiagonals. Throughout the above operations the columns of squares have been taken as forming the four cells of the P_1-aspect;* the rows of squares taken to form cubes, of course, show the P_2-aspect.

This construction has been introduced merely to accentuate the analogy between magics of various dimensions; we might have

* Since the 4th dimension is the square of the second, two aspects of the octahedroid are shown in the presentation plane. The 3d and 4th aspects are in H-planes and V-planes. Since there are two P-plane aspects it might appear that each would produce a different H-plane and V-plane aspect; but this is a delusion.

obtained the magic 4^4 much more rapidly by a method analogous to that used for 4^3 (Fig. 590). We have simply to interchange each number in the natural octahedroid occupying a cell marked [×] in Fig. 613, with its complementary number lying in the associated cell of the associated cube. Fig. 613 is the extended skew-reversion scheme from the index-rod [|×].

All magic octahedroids of double-of-odd order $> 10^4$ can be constructed by the index-rod, for just as we construct an index-square from the rod, and an index-cube from the square, so we can construct an index-octahedroid from the cube. The magics 6^4 and 10^4 have not the capacity for construction by the general rule, but they may be obtained by scattering the symbols over the whole figure as we did with 6^3.

C. P.

CHAPTER XIII.

MAGIC CIRCLES, SPHERES AND STARS.

MAGIC circles, spheres and stars have been apparently much less studied than magic squares and cubes. We cannot say that this is because their range of variety and development is limited; but it may be that our interest in them has been discouraged, owing to the difficulty of showing them clearly on paper, which is especially the case with those of three dimensions.

It is the aim of the present chapter to give a few examples of what might be done in this line, and to explain certain methods of construction which are similar in some respects to the methods used in constructing magic squares.

MAGIC CIRCLES.

The most simple form of magic spheres is embodied in all perfect dice. It is commonly known that the opposite faces of a die contain complementary numbers; that is, 6 is opposite to 1, 5 is opposite to 2, and 4 is opposite to 3—the complementaries in each case adding to 7—consequently, any band of four numbers encircling the die, gives a summation of 14. This is illustrated in Fig. 614, which gives a spherical representation of the die; and if we imagine this sphere flattened into a plane, we have the diagram shown in Fig. 615, which is the simplest form of magic circle.

Fig. 616 is another construction giving the same results as Fig. 615; the only difference being in the arrangement of the circles. It will be noticed in these two diagrams that any pair of complementary numbers is common to two circles, which is a rule also used in constructing many of the following diagrams.

Fig. 617 contains the series 1, 2, 3 . . . . 12 arranged in four circles of six numbers each, with totals of 39. Any one of these circles laps the other three, making six points of intersection on which are placed three pairs of complementary numbers according to the above rule. The most simple way of following this rule is to start by placing number 1 at any desired point of intersection; then by tracing

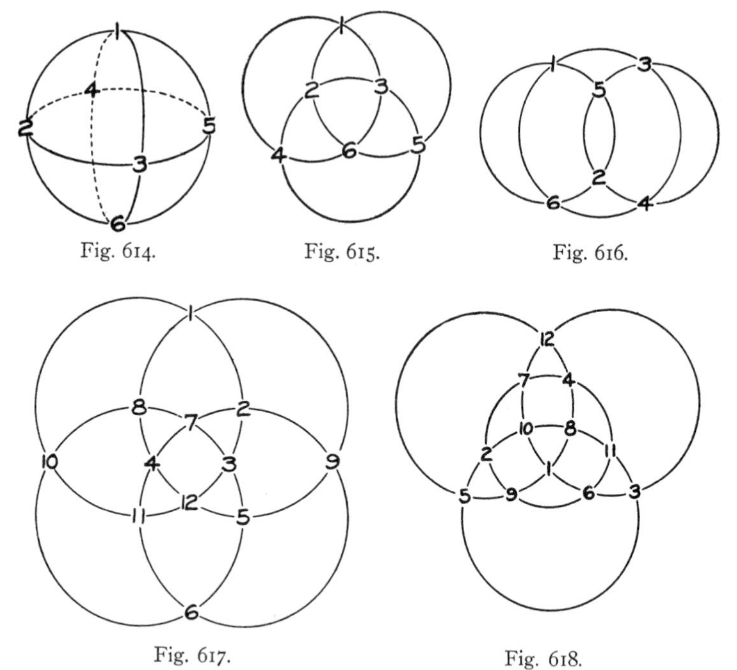

Fig. 614. Fig. 615. Fig. 616.

Fig. 617. Fig. 618.

out the two circles from this point, we find their second point of intersection, on which must be placed the complementary number of 1. Accordingly we locate 2 and its complementary, 3 and its complementary, and so on until the diagram is completed.

Fig. 618 is the same as Fig. 617, differing only in the arrangement of the circles.

Fig. 619 contains the series 1, 2, 3 . . . . 20 arranged in five circles of eight numbers each, with totals of 84.

Fig. 620 contains the series 1, 2, 3 . . . . 14 arranged in five circles of six numbers each, with totals of 45. It will be noticed in this diagram, that the 1 and 14 pair is placed at the intersections of three circles, but such intersections may exist as long as each circle contains the same number of pairs.

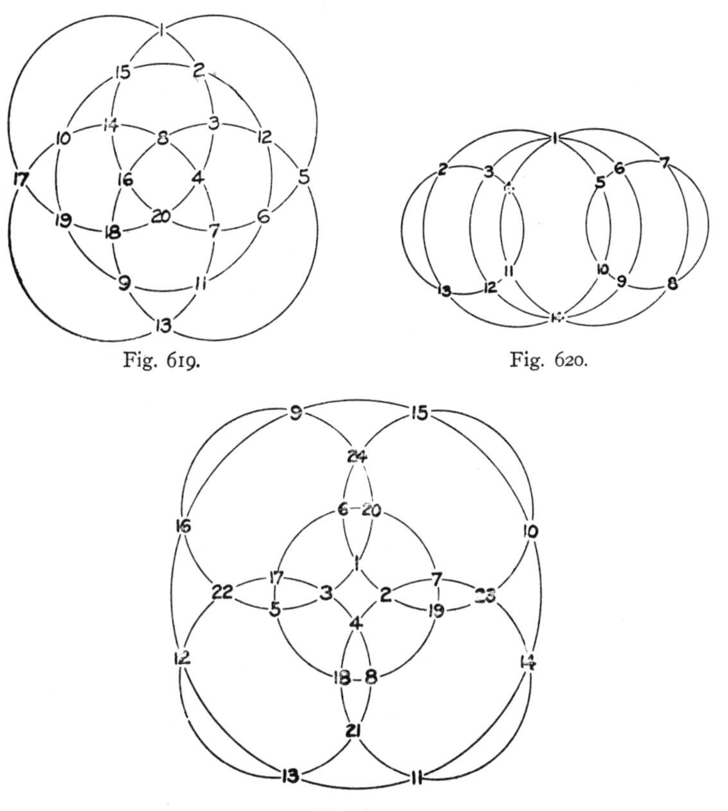

Fig. 619.

Fig. 620.

Fig. 621.

Fig. 621 contains the series 1, 2, 3 . . . . 24 arranged in six circles of eight numbers each, with totals of 100.

Fig. 622 contains the series 1, 2, 3 . . . . 30 arranged in six circles of ten numbers each, with totals of 155. Also, if we add together any two diametrical lines of four and six numbers respectively, we will get totals of 155; but this is only in consequence of the complementaries being diametrically opposite.

Fig. 623 contains the series 1, 2, 3...40 arranged in eight circles of ten numbers each, with totals of 205.

Fig. 624 contains the series 1, 2, 3....8 arranged in eight circles of four numbers each, with totals of 18. This diagram involves a feature not found in any of the foregoing examples, which is due to the arrangement of the circles. It will be noticed that each

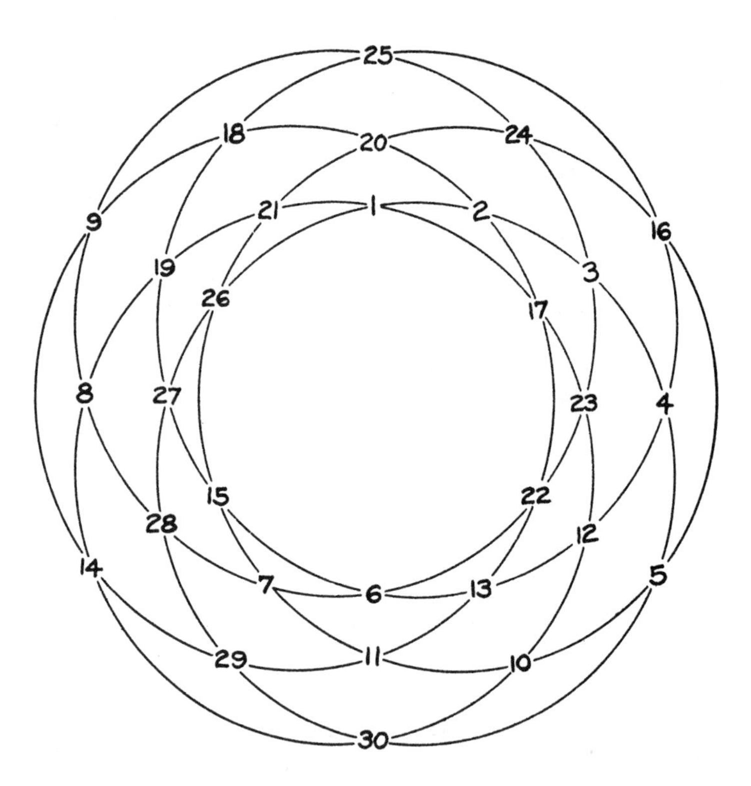

Fig. 622.

number marks the intersection of four circles, but we find that no other point is common to the same four circles, consequently we need more than the foregoing rule to meet these conditions. If we place the pairs on horizontally opposite points, all but the two large circles will contain two pairs of complementaries. The totals of the two large circles must be accomplished by adjusting the pairs. This adjustment is made in Fig. 625, which shows the two selections that will give totals of 18.

Fig. 626 contains the series 1, 2, 3.... 24 arranged in ten circles of six numbers each, with totals of 75. This is accomplished by placing the pairs on radial lines such that each of the six equal circles contains three pairs. It then only remains to adjust these

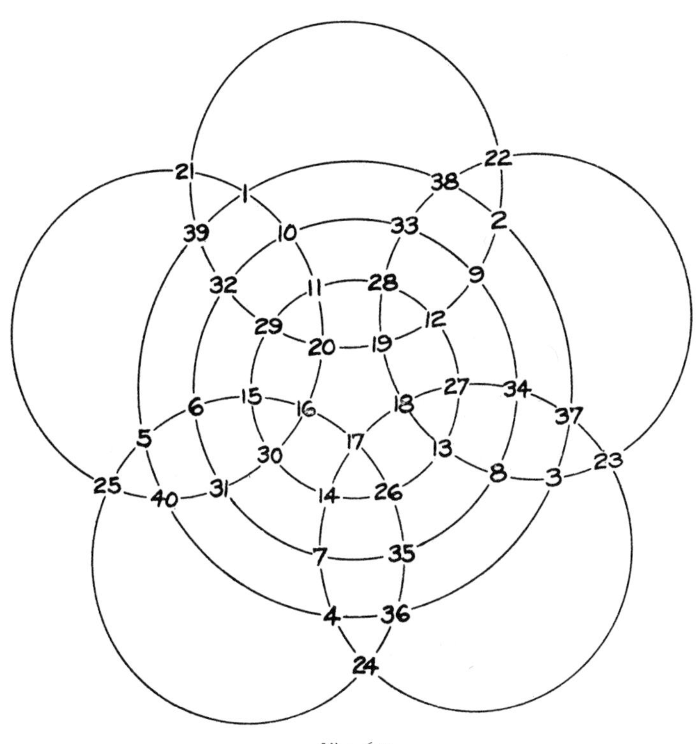

Fig. 623.

pairs to give the constant totals to each of the four concentric circles. Their adjustment is shown diagrammatically in Fig. 627, which is one of many selections that would suit this case.

Fig. 628 contains the series 1, 2, 3.... 12 arranged in seven circles and two diametrical lines of four numbers each with totals of 26.

The large number of tangential points renders this problem

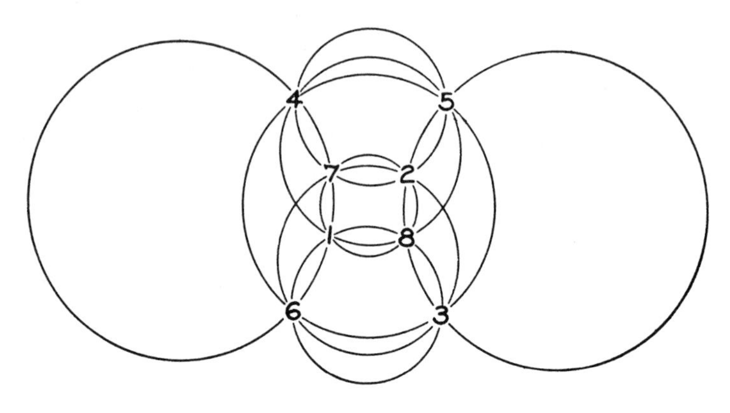

Fig. 624.

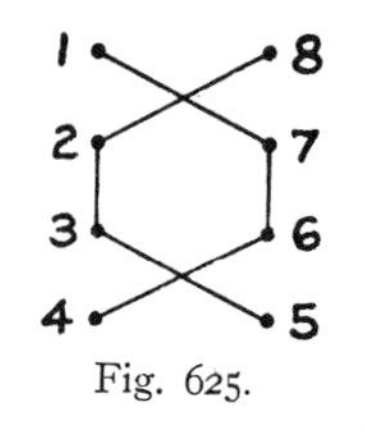

Fig. 625.

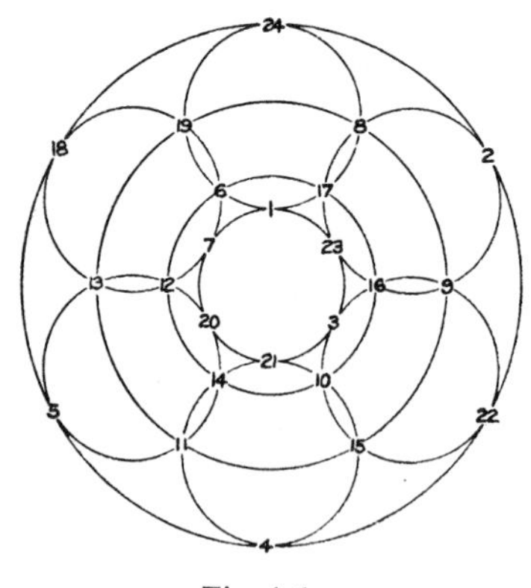

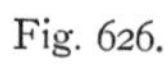

Fig. 626.

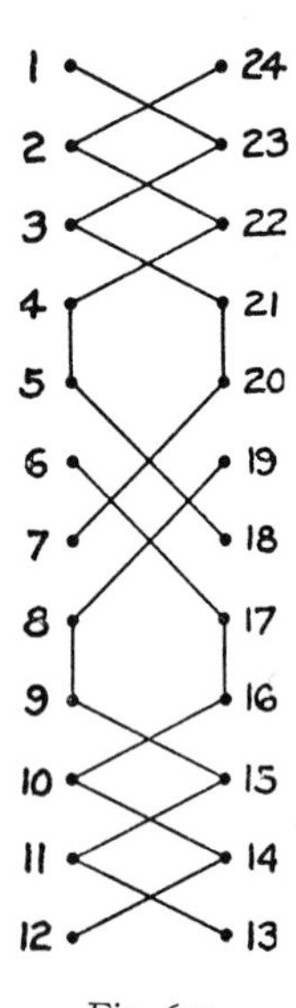

Fig. 627.

quite difficult, and it appears to be solvable only by La Hireian methods. It was derived by adding together the respective numbers of the two primary diagrams Figs. 629 and 630, and Fig. 630 was in turn derived from the two primary diagrams Figs. 631 and 632.

We begin first with Fig. 629 by placing four each of the num-

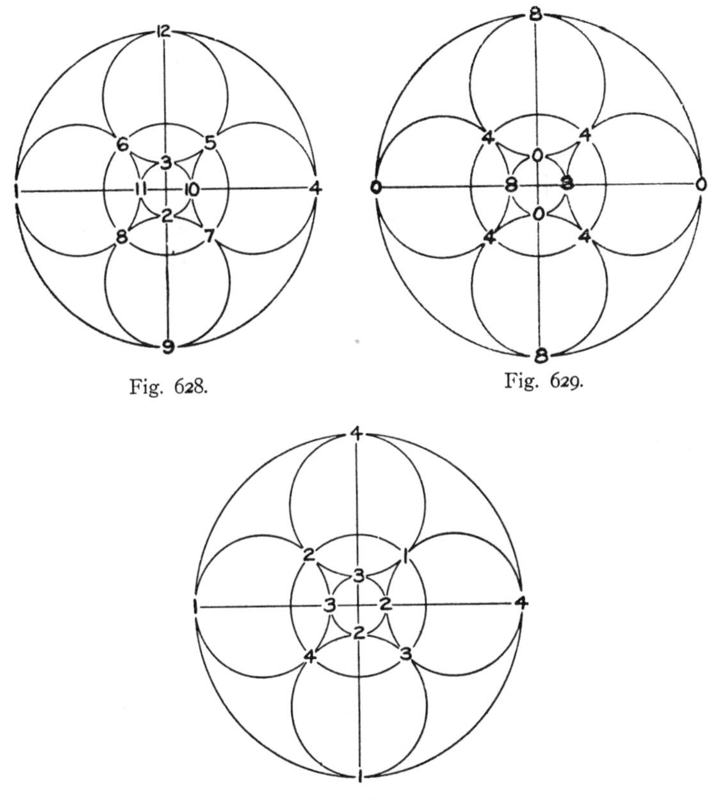

Fig. 628.

Fig. 629.

Fig. 630.

bers 0, 4, and 8 so that we get nine totals amounting to 16. This is done by placing the 4's on the non-tangential circle; which leaves it an easy matter to place the 0's and 8's in their required positions. Fig. 630 must then be constructed so as to contain three sets of the series 1, 2, 3, 4; each set to correspond in position respective to the three sets in Fig. 629, and give totals of 10. This could be done by experiment, but their positions are much easier found with the two diagrams, Figs. 631 and 632. Fig. 631 contains six 0's and six 2's giving totals of 4, while Fig. 632 contains six 1's and six 2's giving

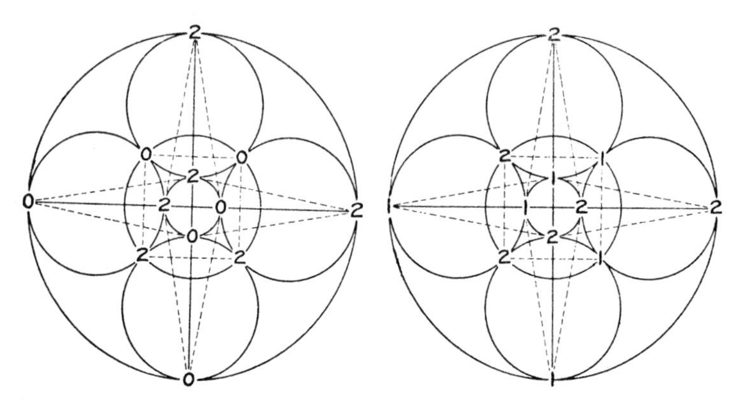

Fig. 631. Fig. 632.

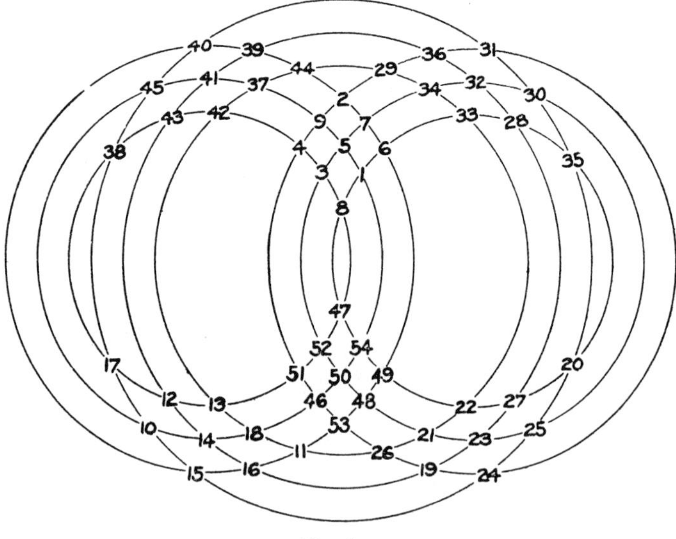

Fig. 633.

totals of 6. It will be noticed in Fig. 629 that the 0's form a horizontal diamond, the 8's a vertical diamond and the 4's a square, which three figures are shown by dotted lines in Figs. 631 and 632.

Besides giving the required totals, Figs. 631 and 632 must have their numbers so arranged, that we can add together the respective diamonds and squares, and obtain the series 1, 2, 3, 4 for each diamond and square, which is shown in Fig. 630. Figs. 630 and 629 are then added together which gives us the result as shown in Fig. 628.

This diagram was first designed for a sphere, in which case

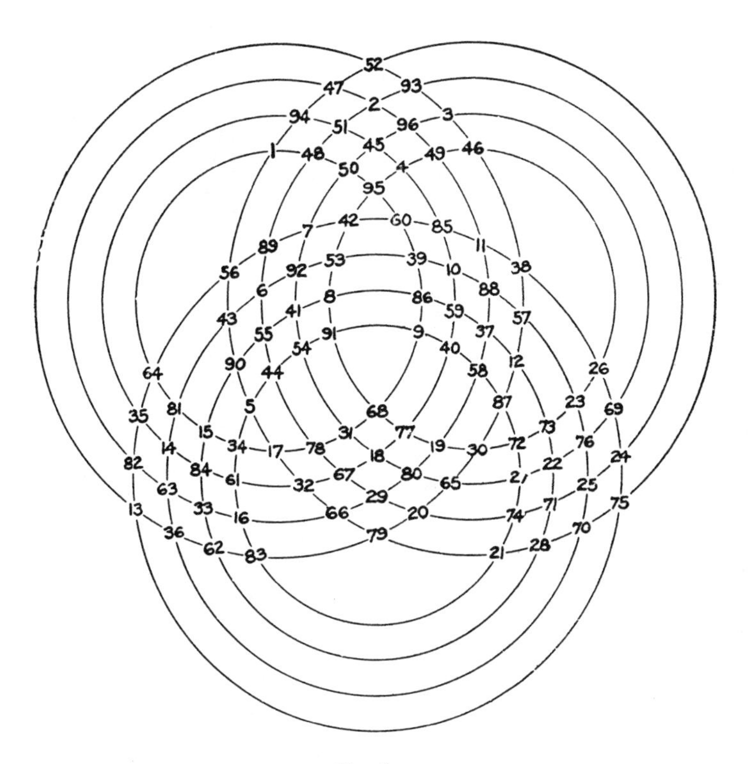

Fig. 634.

the two diametrical lines and the 5, 6, 7, 8 circle were great circles on the sphere and placed at right angles to each other as are the three circles in Fig. 614. The six remaining circles were equal and had their tangential points resting on the great circles. The diagrams used here are easier delineated and much easier to understand than the sphere would have been.

Fig. 633 contains the series 1, 2, 3 .... 54 arranged in nine circles of twelve numbers each with totals of 330. The arrangement also forms six 3×3 magic squares.

We begin this figure by placing the numbers 1 to 9 in magic square order, filling any one of the six groups of points; then,

1	2	3	4
4	3	2	1
2	1	4	3
3	4	1	2

Fig. 635.

0	92	44	48
44	48	0	92
48	44	92	0
92	0	48	44

Fig. 636.

0	44	48	92
4	40	52	88
8	36	56	84
12	32	60	80
16	28	64	76
20	24	68	72

Fig. 637.

according to the first general rule, we locate the complementaries of each of these numbers, forming a second and complementary square. We locate the remaining two pairs of squares in the same manner. The pairs of squares in the figure are located in the same

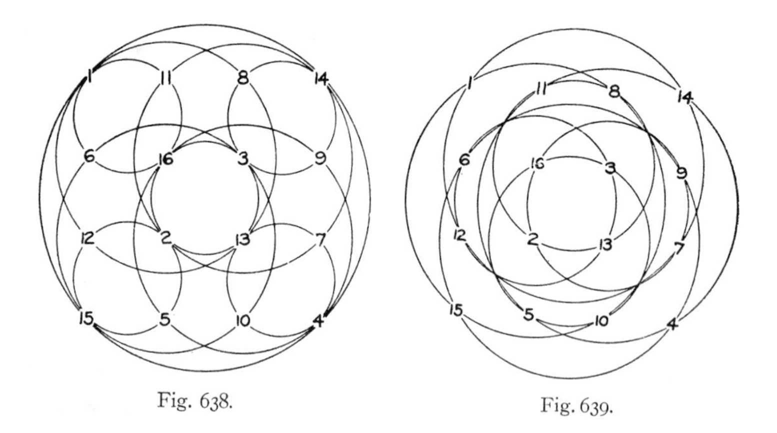

Fig. 638. Fig. 639.

relative positions as the pairs of numbers in Fig. 616, in which respect the two figures are identical.

Fig. 634 contains the series 1, 2, 3 .... 96 aranged in twelve circles of sixteen numbers each, with totals of 776. The sum of the sixteen numbers in each of the six squares is also 776. These squares possess the features of the ancient Jaina square, and are constructed by the La Hireian method as follows.

The series 0, 4, 8, 12 .... 92 are arranged in six horizontal groups of four numbers, as shown in Fig. 637, by running the series down, up, down, and up through the four respective vertical rows. The upper horizontal row of Fig. 637 is used to form the primary square Fig. 636; likewise, five other squares are formed with the remaining groups of Fig. 637. These six squares are each, in turn, added to the primary square, Fig. 635, giving the six squares in Fig. 634. There is no necessary order in the placing of these squares, since their summations are equal.

Figs. 638 and 639 show the convenience of using circles to show up the features of magic squares. The two diagrams represent the same square, and show eighteen summations amounting to 34.

H. A. S.

MAGIC SPHERES.

In constructing the following spheres, a general rule of placing complementary numbers diametrically opposite, has been followed, in which cases we would term them associated. This conforms with a characteristic of magic squares and cubes.

Fig. 640 is a sphere containing the series 1, 2, 3....26 arranged in nine circles of eight numbers each, with totals of 108.

In this example, it is only necessary to place the pairs at diametrically opposite points; because all the circles are great circles, which necessitates the diametrically opposite position of any pair common to two or more circles. Otherwise we are at liberty to place the pairs as desired; so, in this sphere it was chosen to place the series 1, 2, 3....9 in magic square form, on the front face, and in consequence, we form a complementary square on the rear face.

Fig. 641 is a sphere containing the series 1, 2, 3....26, arranged in seven circles of eight numbers each, with totals of 108.

This was accomplished by placing the two means of the series at the poles, and the eight extremes in diametrically opposite pairs on the central horizontal circle. In order to give the sphere "associated" qualities, as mentioned before, the remaining numbers should be placed as shown by diagram in Fig. 642. This shows the two selections for the upper and lower horizontal circles. The numbers for the upper circle are arranged at random, and the numbers in the lower circle are arranged in respect to their complementaries in the upper circle.

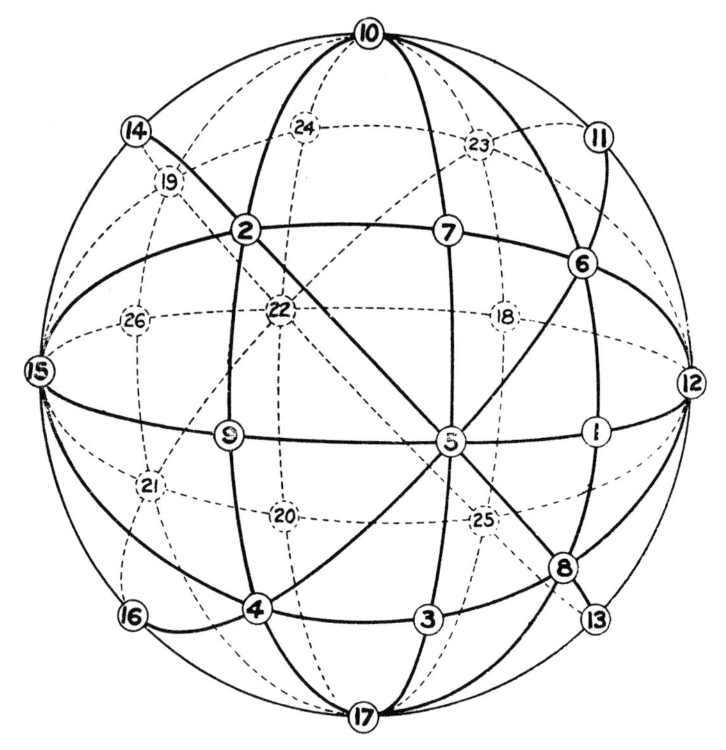

Fig. 640.

Fig. 644 is a sphere containing the series 1, 2, 3 . . . . 62 arranged in eleven circles of twelve numbers each, with totals of 378.

This is a modification of the last example and represents the parallels and meridians of the earth. Its method of construction is also similar, and the selections are clearly shown in Fig. 643.

Fig. 645 shows two concentric spheres containing the series 1, 2, 3 . . . . 12 arranged in six circles of four numbers each, with totals of 26. It also has three diametrical lines running through the spheres with totals of 26.

The method for constructing this is simple, it being only necessary to select three pairs of numbers for each sphere and place the complementaries diametrically opposite each other.

Fig. 646 is the same as the last example with the exception that

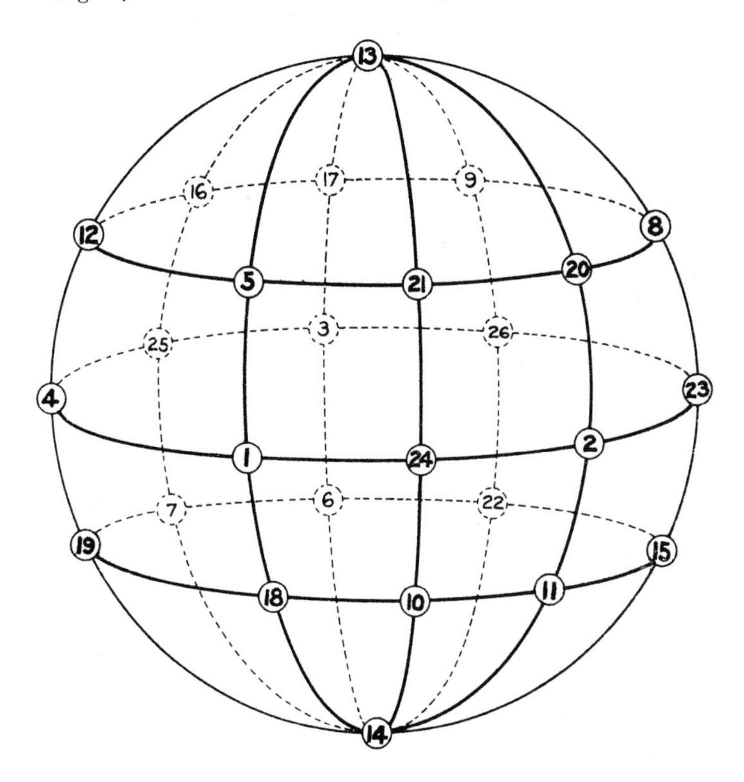

Fig. 641.

two of the circles do not give the constant total of 26; but with this sacrifice, however, we are able to get twelve additional summations of 26, which are shown by the dotted circles in Figs. 647, 648 and 649. Fig. 647 shows the vertical receding plane of eight numbers, Fig. 648, the horizontal plane; and Fig. 649, the plane parallel to the picture, the latter containing the two concentric circles that do not give totals of 26.

In this example all pairs are placed on radial lines with one number in each sphere which satisfies the summations of the twelve dotted circles. The selections for the four concentric circles are

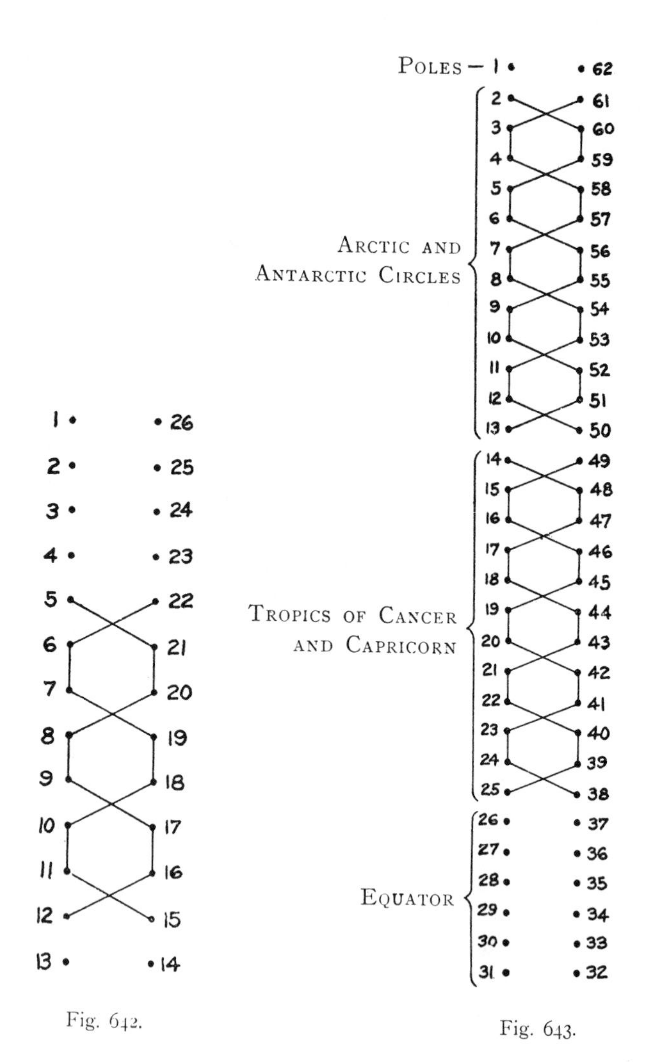

Fig. 642.

Fig. 643.

shown in Fig. 650. The full lines show the selections for Fig. 647 and the dotted lines for Fig. 648. It is impossible to get constant totals for all six concentric circles.

Fig. 651 is a sphere containing the series 1, 2, 3....98, arranged in fifteen circles of sixteen numbers each, with totals of 792. It contains six 3×3 magic squares, two of which, each form the nucleus of a 5×5 concentric square. Also, the sum of any two diametrically opposite numbers is 99.

To construct this figure, we must select two complementary

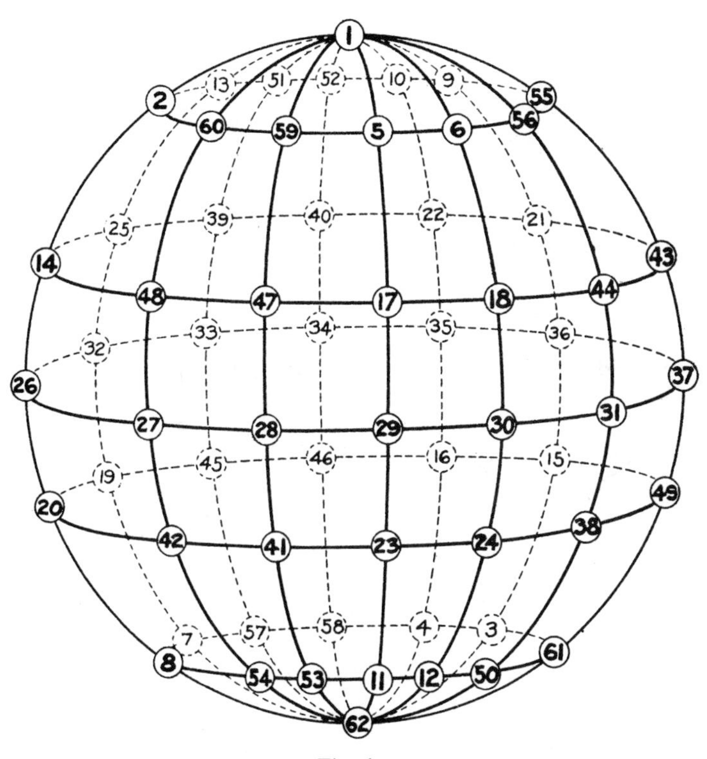

Fig. 644.

sets of 25 numbers each, that will form the two concentric squares; and four sets of 9 numbers each, to form the remaining squares, the four sets to be selected in two complementary pairs.

This selection is shown in Fig. 652, in which the numbers enclosed in full and dotted circles represent the selection for the front and back concentric squares respectively. The numbers marked with

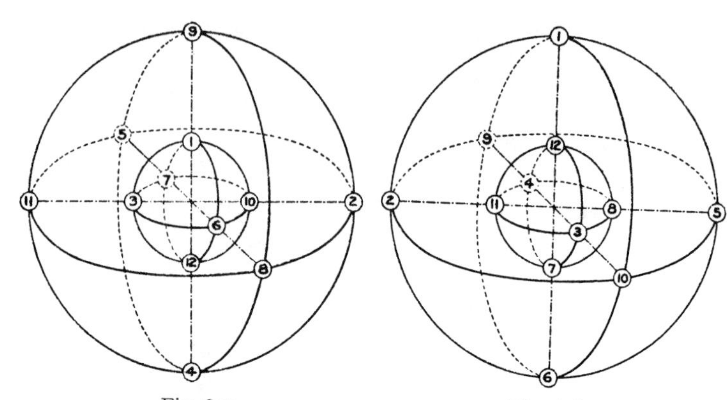

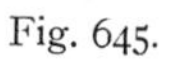
Fig. 645.

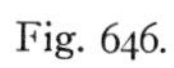
Fig. 646.

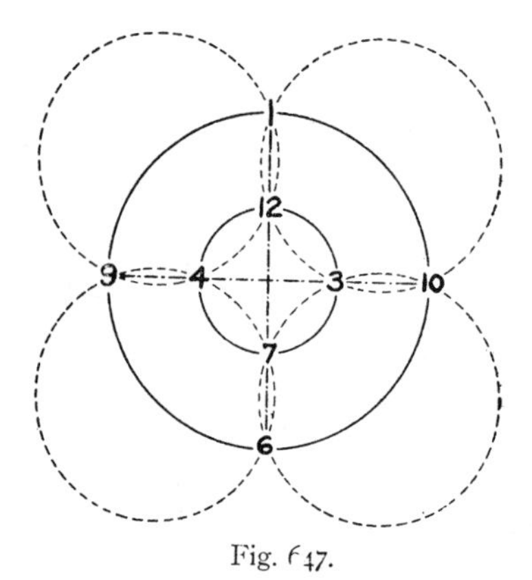

Fig. 647.

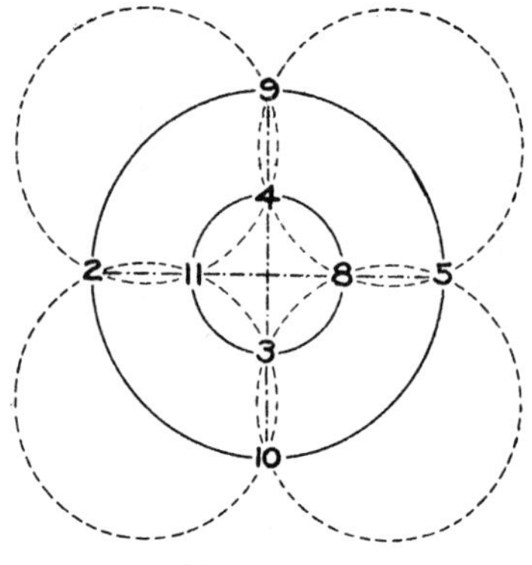

Fig. 648.

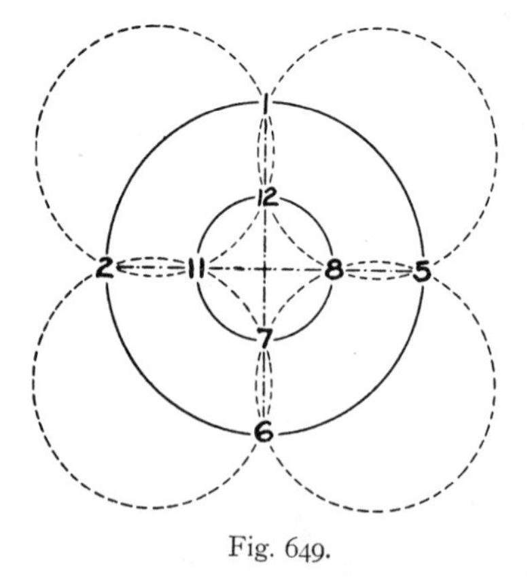

Fig. 649.

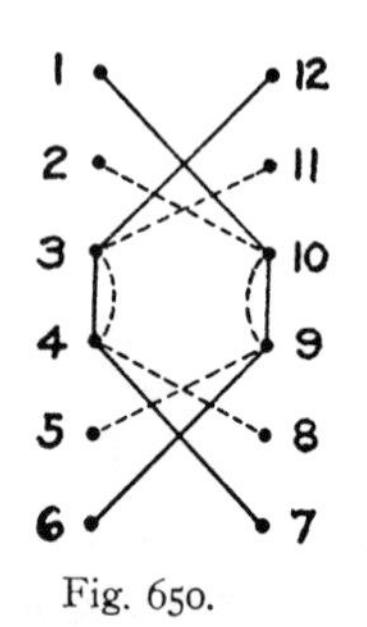

Fig. 650.

T, B, L and R represent the selections for the top, bottom, left and right horizon squares respectively.

After arranging the numbers in the top horizon square, we locate the complementary of each number, diametrically opposite and accordingly form the bottom square. The same method is used in placing the left and right square.

The numbers for the front concentric square are duplicated in

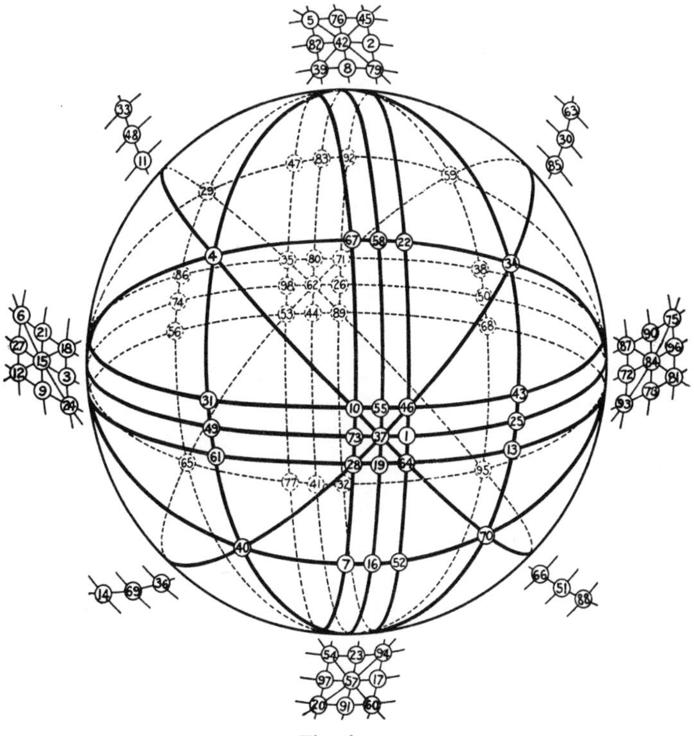

Fig. 651.

Fig. 653. The numbers marked by dot and circle represent the selection for the nucleus square, and the diagram shows the selections for the sides of the surrounding panel, the numbers 4, 70, 34 and 40 forming the corners.

By placing the complementaries of each of the above 25 numbers, diametrically opposite, we form the rear concentric square.

After forming the six squares, we find there are twelve num-

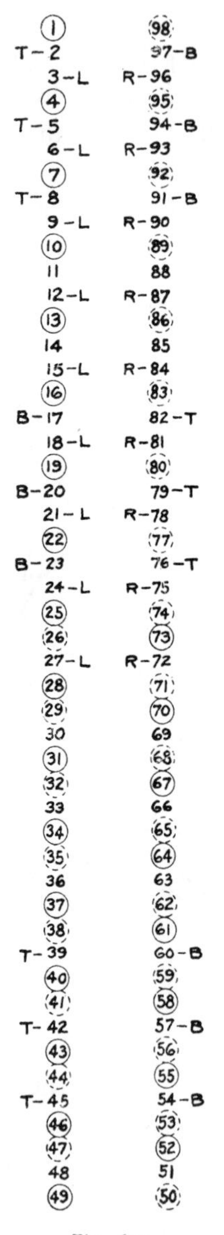

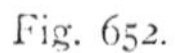
Fig. 652.

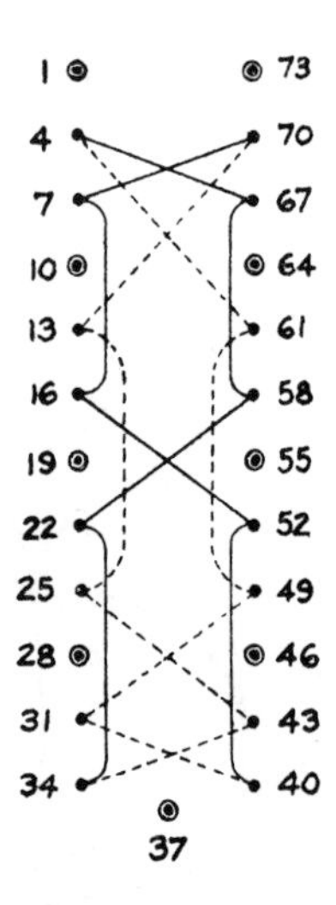

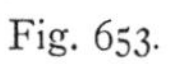
Fig. 653.

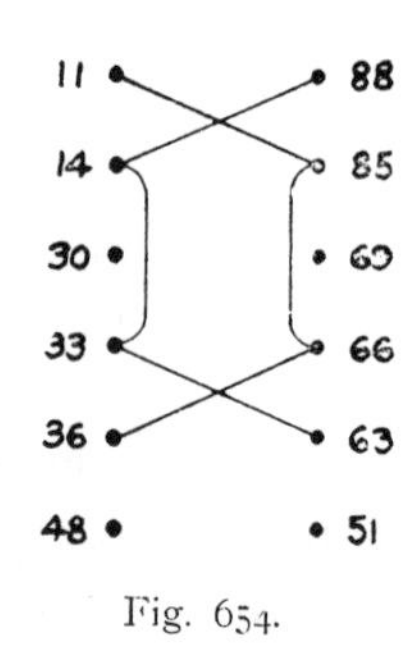

Fig. 654.

bers left, which are shown in Fig. 654. These are used to form the four horizon triads. Two pairs are placed on the central circle, and by selection, as shown in the diagram, we fill in the other two circles with complementary numbers diametrically opposite. The above selection is such that it forms two groups of numbers, each with a summation of 198; this being the amount necessary to complete the required summations of the horizon circles.

There are many selections, other than those shown in Fig. 652, which could have been taken. A much simpler one would be to select the top 25 pairs for the front and back concentric squares.

H. A. S.

MAGIC STARS.

We are indebted to Mr. Frederick A. Morton, Newark, N. J., for these plain and simple rules for constructing magic stars of all orders.

A five-pointed star being the smallest that can be made, the rules will be first applied to this one.

Choosing for its constant, or summation (S) = 48, then:

$$(5 \times 48)/2 = 120 = \text{sum of series.}$$

Divide 120 into two parts, say 80 and 40, although many other divisions will work out equally well. Next find a series of five numbers, the sum of which is one of the above two numbers. Selecting 40, the series $6+7+8+9+10=40$ can be used. These numbers must now be written in the central pentagon of the star following the direction of the dotted lines, as shown in Fig. 655. Find the sum of every pair of these numbers around the circle beginning in this case with $6+9=15$ and copy the sums in a separate column (A) as shown below:

	(A)
$6+9=15$	$17+15+16=48$
$7+10=17$	$16+17+15=48$
$8+6=14$	$15+14+19=48$
$9+7=16$	$19+16+13=48$
$10+8=18$	$13+18+17=48$

Place on each side of 15, numbers not previously used in the central pentagon, which will make the total of the three numbers = 48 or S. 17 and 16 are here selected. Copy the last number of the trio (16) under the first number (17) as shown above, and under 16 write the number required to make the sum of the second trio = 48 (in this case 15). Write 15 under 16, and proceed as before to the end. If proper numbers are selected to make the sum of the first trio = 48, it will be found that the first number of the first trio will be the same as the last number of the last trio (in this case 17) and this result will indicate that the star will sum correctly if the numbers in the first column are written in their

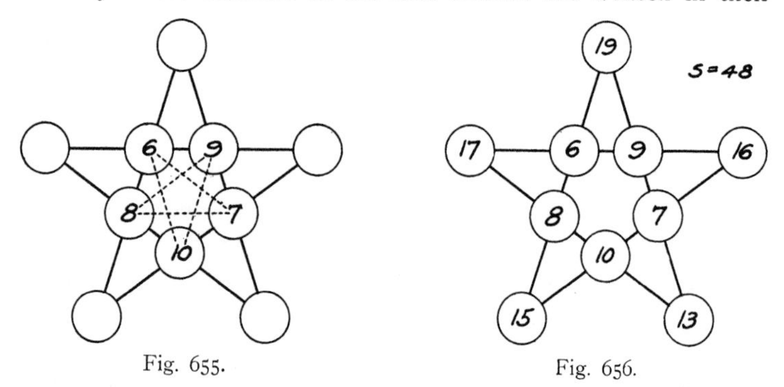

Fig. 655. Fig. 656.

proper order at the points of the star, as shown in Fig. 656. If the first and last numbers prove different, a simple operation may be used to correct the error. When the last number is *more* than the first number, add half the difference between the two numbers to the first number and proceed as before, but if the last number is *less* than the first number, then *subtract* half the difference from the first number. One or other of these operations will always correct the error.

For example, if 14 and 19 had been chosen instead of 17 and 16, the numbers would then run as follows:

$$14 + 15 + 19 = 48$$
$$19 + 17 + 12 = 48$$
$$12 + 14 + 22 = 48$$
$$22 + 16 + 10 = 48$$
$$10 + 18 + 20 = 48$$

The difference between the first and last numbers is seen to be 6 and 20 being *more* than 14, half of 6 *added* to 14 makes 17 which is the correct starting number. Again, if 21 and 12 had been selected, then:

$$21 + 15 + 12 = 48$$
$$12 + 17 + 19 = 48$$
$$19 + 14 + 15 = 48$$
$$15 + 16 + 17 = 48$$
$$17 + 18 + 13 = 48$$

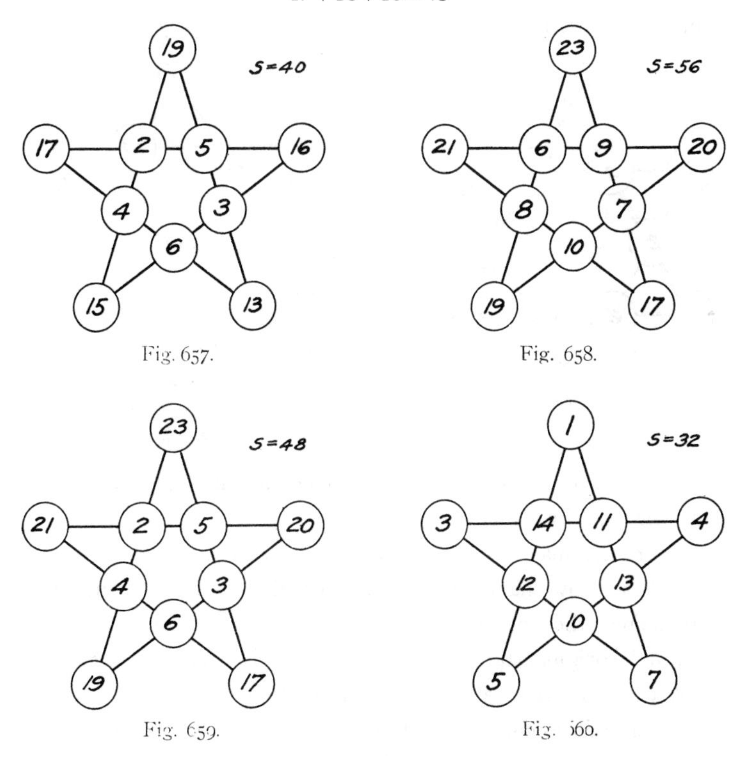

Fig. 657. Fig. 658. Fig. 659. Fig. 660.

The difference between the first and last numbers is here 8, and the last number being *less* than the first, half of this difference subtracted from 21 leaves 17 as before.

It is obvious that the constant S of a star of any order may be changed almost indefinitely by adding or subtracting a number selected so as to avoid the introduction of duplicates. Thus, the constant of the star shown in Fig. 656 may be reduced from 48 to 40 by subtracting 4 from each of the five inside numbers, or it may be increased to 56 by adding 4 to each of the five outside numbers and another variant may then be made by using the five inside numbers of S = 40, and the five outside numbers of S = 56. These three variants are shown respectively in Figs. 657, 658 and 659.

It is also obvious that any pair of five-pointed or other stars may be superposed to form a new star, and by rotating one star over the other, four other variants may be made; but in these and similar operations duplicate numbers will frequently occur, which

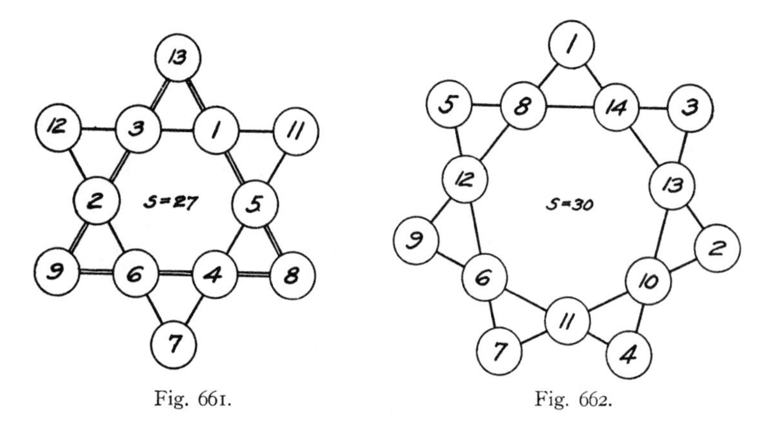

Fig. 661. Fig. 662.

of course will make the variant ineligible although its constant must necessarily remain correct.

Variants may also be made in this and all other orders of magic stars, by changing each number therein to its complement with some other number that is larger than the highest number used in the original star. The highest number in Fig. 656, for example is 19. Choosing 20 as a number on which to base the desired variant 19 in Fig. 656 is changed to 1, 17 to 3 and so on throughout, thus making the new five-pointed star shown in Fig. 660 with S = 32.

The above notes on the construction of variants are given in detail as they apply to *all orders* of magic stars and will not need repetition.

The construction of a six-pointed star may now be considered Selecting 27 as a constant:

$(6 \times 27)/2 = 81 =$ sum of the series.

Divide 81 into two parts, say 60 and 21, and let the sum of the six numbers in the inner hexagon $= 21$, leaving 60 to be divided among the outer points. Select a series of six numbers, the sum of which is 21, say 1, 2, 3, 4, 5, 6, and arrange these six numbers in hexagonal form, so that the sum of each pair of opposite numbers

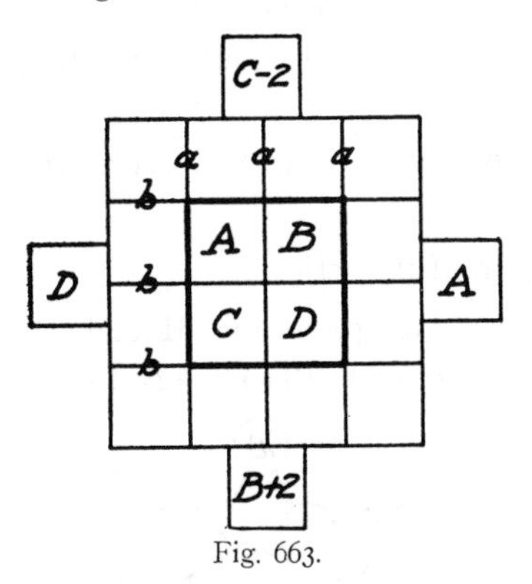

Fig. 663.

1	2	3	4
6	7	8	9
11	12	13	14
16	17	18	19

Fig. 664.

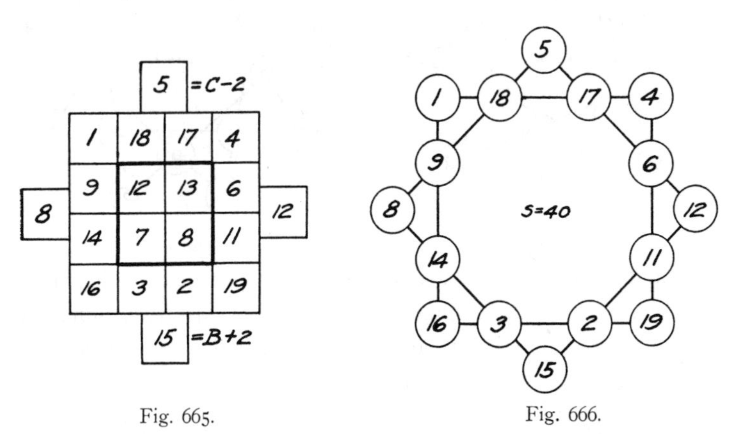

Fig. 665.

Fig. 666.

$= 7$. Fig. 661 shows that these six inside numbers form part of two triangles, made respectively with single and double lines. The outside numbers of each of these two triangles must be computed separately according to the method used in connection with the five-pointed star. Beginning with the two upper numbers in the single-lined triangle and adding the couplets together we have:

	(A)
$3+1=4$	$12+4+11=27$
$5+4=9$	$11+9+7=27$
$6+2=8$	$7+8+12=27$

Writing these sums in a separate column (A) and proceeding as before described, the numbers 12, 11, 7 are obtained for the points of the single-lined triangle, and in the same manner 13, 8, 9 are found for the points of the double-lined triangle, thus completing the six-pointed star Fig. 661.

The next larger star has seven points. Selecting 30 for a constant, which is the lowest possible:

$$(7\times 30)/2=105=\text{sum of the series.}$$

Dividing this sum as before into two parts, say 31 and 74, seven numbers are found to sum 74, say, $6+8+10+11+12+13+14$

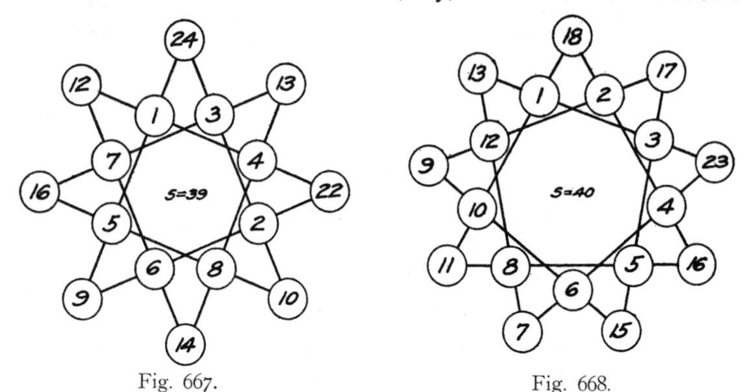

Fig. 667. Fig. 668.

$=74$, and these numbers are written around the inside heptagon as shown in Fig. 662. Adding them together in pairs, their sums are written in a column and treated as shown below, thus determining the numbers for the points of Fig. 662.

$14+13=27$	$1+27+2=30$
$10+11=21$	$2+21+7=30$
$6+12=18$	$7+18+5=30$
$8+14=22$	$5+22+3=30$
$13+10=23$	$3+23+4=30$
$11+6=17$	$4+17+9=30$
$12+8=20$	$9+20+1=30$

The next larger star has eight points and it can be made in two different ways, viz., By arranging the numbers in one continuous line throughout as in stars already described having an odd number of points, or by making it of two interlocking squares. The latter form of this star may be constructed by first making a 4^2 with one extra cell on each of its four sdes, as shown in Fig. 663. A series of sixteen numbers is then selected which will meet the

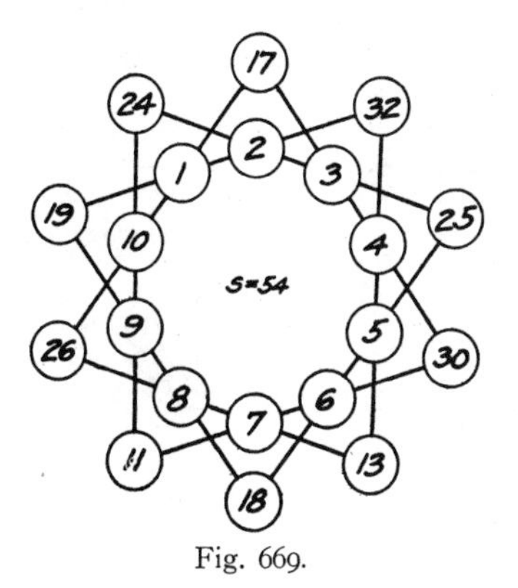

Fig. 669.

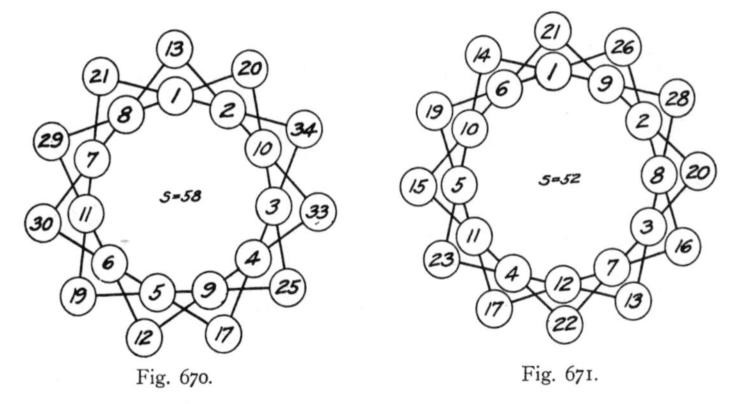

Fig. 670.

Fig. 671.

conditions shown by italics *a, a, a,* and *b, b, b,* in the figure, i. e., all differences between row numbers must be the same, and also all differences between column numbers, but the two differences must be unlike. The constant (S) of the series when the latter is arranged as a magic 4^2 must also be some multiple of 4. The series is then put into magic formation by the old and well-known rule for making magic squares of the 4th order. The central 2×2 square is now eliminated and the numbers therein transferred to the four extra outside cells as indicated by the letters A. B. C. D. Finally all numbers are transferred in their order into an eight-pointed star.

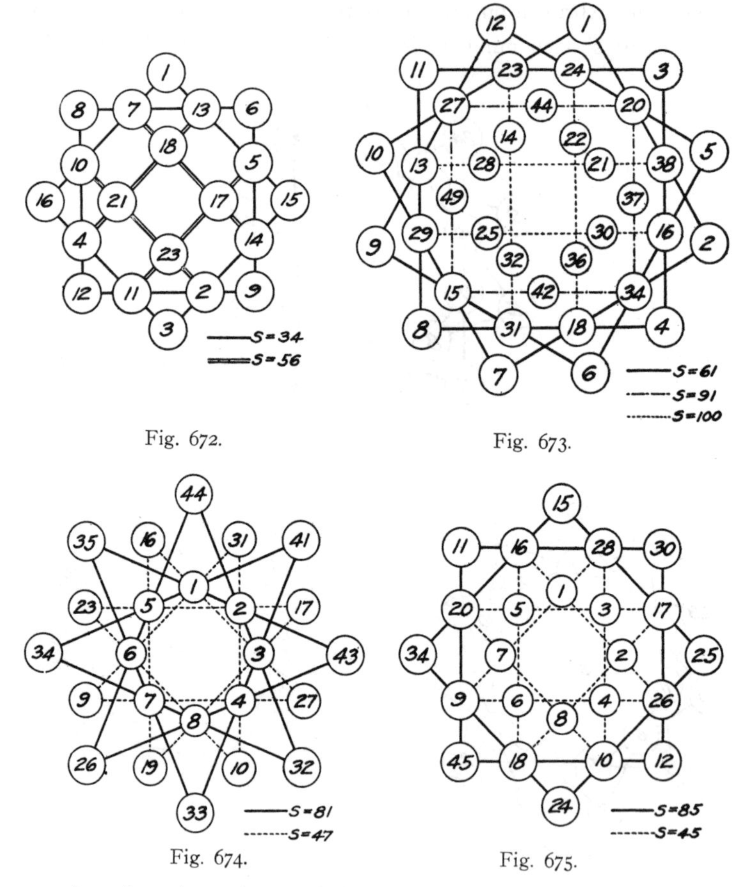

Fig. 672. Fig. 673.

Fig. 674. Fig. 675.

A series of numbers meeting the required conditions is shown in Fig. 664, and its arrangement according to the above rules is given in Fig. 665, the numbers in which, transferred to an eight-pointed star, being shown in Fig. 666, $S=40$. The 4^2 magic arrangement of the series must be made in accordance with Fig. 665, for other magic arrangements will often fail to work out, and will never do so in accordance with Fig. 663. The above instructions cover the simplest method of making this form of star but it can be constructed in many other different ways and also with constants which are not evenly divisible by 4.

Turning now to the construction of the eight-pointed star by the continuous line method, inspection of Figs. 666 and 667 will show that although the number of points is the same in each star yet the arrangement of numbers in their relation to one another in the eight quartets is entirely different.

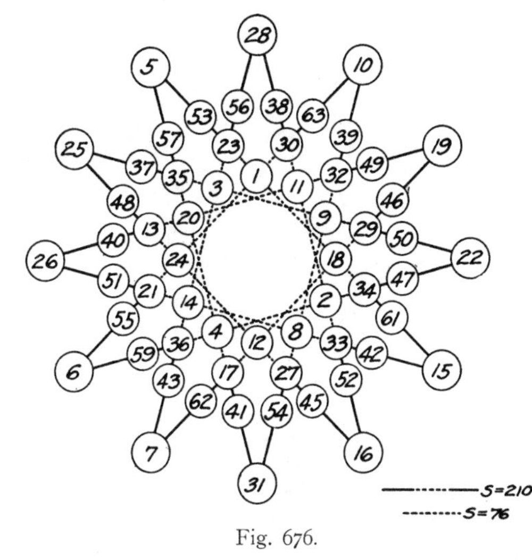

Fig. 676.

Choosing a constant of 39 for an example:

$$(39 \times 8)/2 = 156 = \text{sum of series.}$$

This sum is now divided into two parts, say 36 and 120. The sum of the first eight digits being 36, they may be placed around the inside octagon so that the sum of each opposite pair of numbers = 9, as shown in Fig. 667. Adding them together in pairs, as indicated by the connecting lines in the figure, their sums are written in a column and treated as before explained, thus giving the correct numbers to be arranged around the points of the star Fig. 667.

These rules for making magic stars of all orders are so simple that further examples are deemed unnecessary. Nine-, ten-, eleven-, and twelve-pointed stars, made by the methods described, are shown respectively in Figs. 668, 669, 670 and 671. Several other diagrams of ingenious and more intricate star patterns made by Mr. Morton are also appended for the interest of the reader in Figs. 672 to 681 inclusive.

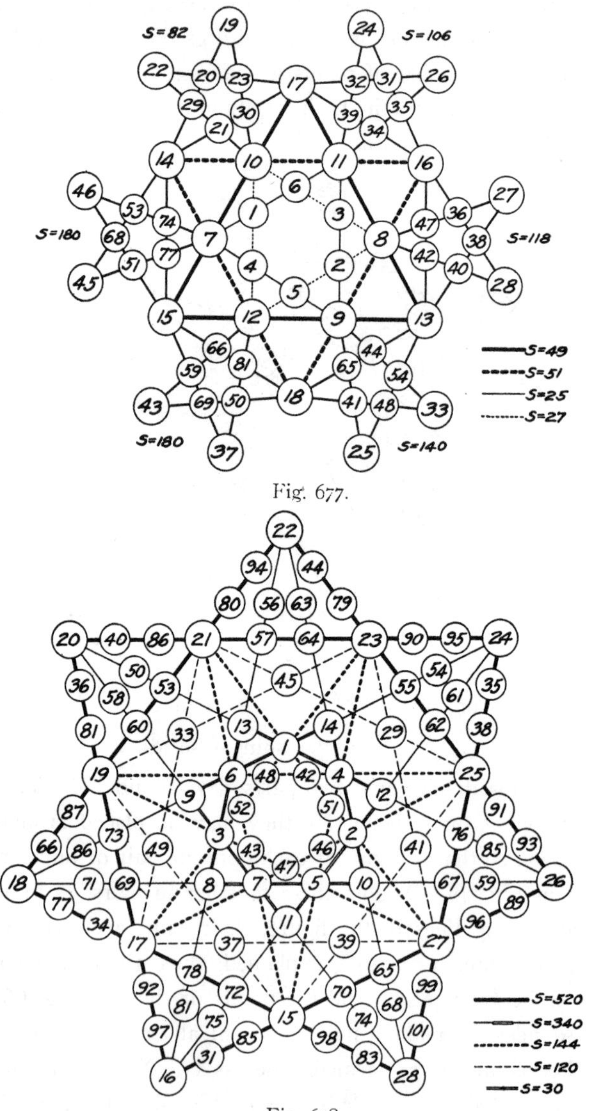

Fig. 677.

Fig. 678.

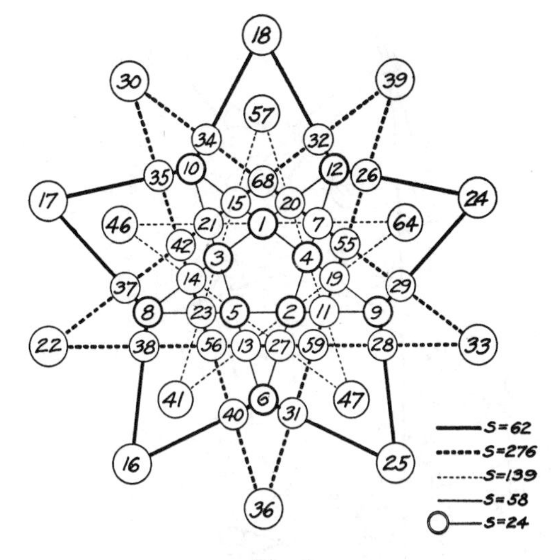

Fig. 679.

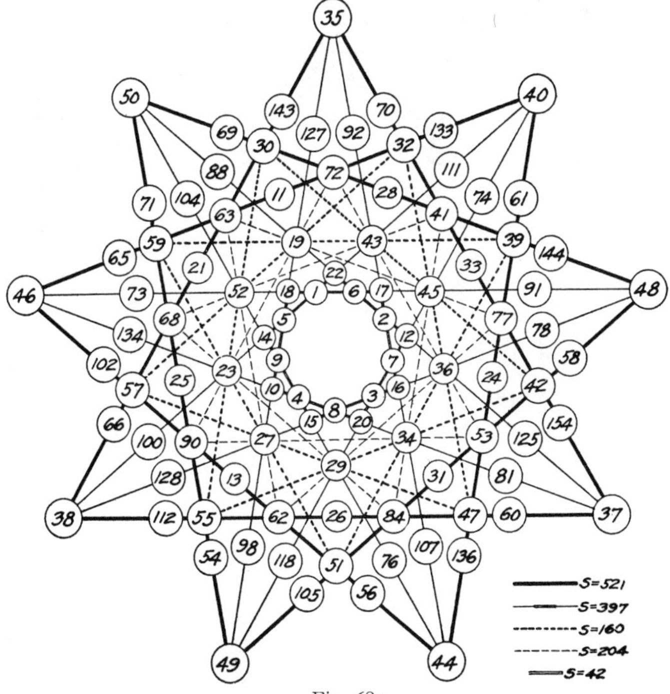

Fig. 680.

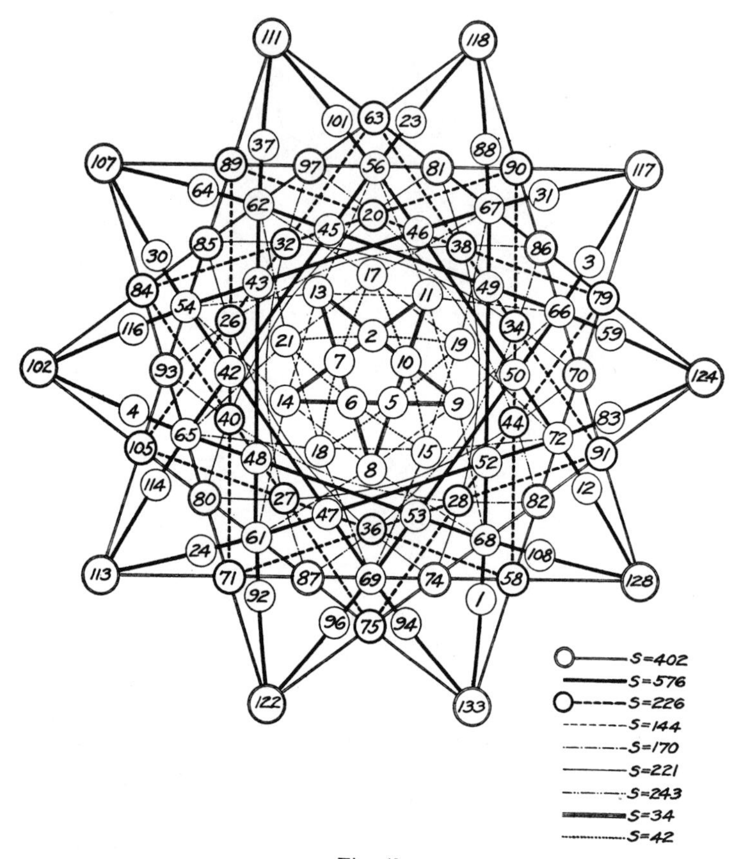

Fig. 681.

W. S. A.

CHAPTER XIV.

MAGIC OCTAHEDROIDS.

MAGIC IN THE FOURTH DIMENSION.

Definition of terms: *Row* is a general term; *rank* denotes a horizontal right-to-left row; *file* a row from front to back; and *column* a vertical row in a cube—not used of any horizontal dimension.

If n^2 numbers of a given series can be grouped so as to form a magic square and n such squares be so placed as to constitute a magic cube, why may we not go a step further and group n cubes in relations of the fourth dimension? In a magic square containing the natural series $1 \ldots n^2$ the summation is $\frac{n(n^2+1)}{2}$; in a magic cube with the series $1 \ldots n^3$ it is $\frac{n(n^3+1)}{2}$; and in an analogous fourth-dimension construction it naturally will be $\frac{n(n^4+1)}{2}$.

With this idea in mind I have made some experiments, and the results are interesting. The analogy with squares and cubes is not perfect, for rows of numbers can be arranged side by side to represent a visible square, squares can be piled one upon another to make a visible cube, but cubes cannot be so combined in drawing as to picture to the eye their higher relations. My expectation *a priori* was that some connection or relation, probably through some form of diagonal-of-diagonal, would be found to exist between the cubes containing the n^4 terms of a series. This particular feature did appear in the cases where n was odd. Here is how it worked out:

I. *When n is odd.*

1. Let $n=3$, then $S=123$.—The natural series $1 \ldots 81$ was di-

vided into three sub-series such that the sum of each would be one-third the sum of the whole. In dealing with any such series when n is odd there will be n sub-series, each starting with one of the first n numbers, and the difference between successive terms will be $n+1$, except after a multiple of n, when the difference is 1. In the present case the three sub-series begin respectively with 1, 2, 3, and the first is 1 5 9 10 14 18 19 23 27 28 32 36 37 41 45 46 50 54 55 59 63 64 68 72 73 77 81. These numbers were arranged in three squares constituting a magic cube, and the row of squares so formed was flanked on right and left by similar rows formed from the other two sub-series (see Fig. 682).

	I			II			III	
25	38	60	28	77	18	67	8	48
33	79	11	72	1	50	21	40	62
65	6	52	23	45	55	35	75	13
29	78	16	68	9	46	26	39	58
70	2	51	19	41	63	31	80	12
24	43	56	36	73	14	66	4	53
69	7	47	27	37	59	30	76	17
20	42	61	32	81	10	71	3	49
34	74	15	64	5	54	22	44	57

Fig. 682. (3^4)

It is not easy—perhaps it is not possible—to make an absolutely perfect cube of 3. These are not perfect, yet they have many striking features. Taking the three cubes separately we find that in each all the "straight" dimensions—rank, file and column—have the proper footing, 123. In the middle cube there are two plane diagonals having the same summation, and in cubes I and III one each. In cube II four cubic diagonals and four diagonals of vertical squares are correct; I and III each have one cubic diagonal and one vertical-square diagonal.

So much for the original cubes; now for some combinations. The three squares on the diagonal running down from left to right will make a magic cube with rank, file, column, cubic diagonals, two plane diagonals and four vertical-square diagonals (37 in all) correct. Two other cubes can be formed by starting with the top squares of II and III respectively and following the "broken diagonals" running downward to the right. In each of these S occurs at least 28 times (in 9 ranks, 9 files, 9 columns and one cubic diag-

I					II					II					IV					V				
317	473	604	10	161	192	348	479	510	36	67	223	354	385	536	567	98	229	260	411	442	598	104	135	286
110	136	292	448	579	610	11	167	323	454	485	511	42	198	329	360	386	542	73	204	235	261	417	573	79
423	554	85	236	267	298	429	585	111	142	173	304	460	611	17	48	179	335	486	517	548	54	210	361	392
211	367	398	529	60	86	242	273	404	560	586	117	148	279	435	461	617	23	154	310	336	492	523	29	185
504	35	186	342	498	379	535	61	217	373	254	410	561	92	248	129	285	436	592	123	4	160	311	467	623
606	12	168	324	455	481	512	43	199	330	356	387	543	74	205	231	262	418	574	80	106	137	293	449	580
299	430	581	112	143	174	305	456	612	18	49	180	331	487	518	549	55	206	362	393	424	555	81	237	268
87	243	274	405	556	587	118	149	280	431	462	618	24	155	306	337	493	524	30	181	212	368	399	530	56
380	531	62	218	374	255	406	562	93	249	130	281	437	593	124	5	156	312	468	624	505	31	187	343	499
193	349	480	506	37	68	224	355	381	537	568	99	230	256	412	443	599	105	131	287	318	474	605	6	162
175	301	457	613	19	50	176	332	488	519	550	51	207	363	394	425	551	82	238	269	300	426	582	113	144
588	119	150	276	432	463	619	25	151	307	338	494	525	26	182	213	369	400	526	57	88	244	275	401	557
251	407	563	94	250	126	282	438	594	125	1	157	313	469	625	501	32	188	344	500	376	532	63	219	375
69	225	351	382	538	569	100	226	257	413	444	600	101	132	288	319	475	601	7	163	194	350	476	507	38
482	513	44	200	326	357	388	544	75	201	232	263	419	575	76	107	138	294	450	576	607	13	169	325	451
464	620	21	152	308	339	495	521	27	183	214	370	396	527	58	89	245	271	402	558	589	120	146	277	433
127	283	439	595	121	2	158	314	470	621	502	33	189	345	496	377	533	64	220	371	252	408	564	95	246
570	96	227	258	414	445	596	102	133	289	320	471	602	8	164	195	346	477	508	39	70	221	352	383	539
358	389	545	71	202	233	264	420	571	77	108	139	295	446	577	608	14	170	321	452	483	514	45	196	327
46	177	333	489	520	546	52	208	364	395	421	552	83	239	270	296	427	583	114	145	171	302	458	614	20
3	159	315	466	622	503	34	190	341	497	378	534	65	216	372	253	409	565	91	247	128	284	440	591	122
441	597	103	134	290	316	472	603	9	165	191	347	478	509	40	66	222	353	384	540	566	97	288	259	415
234	265	416	572	78	109	140	291	447	578	609	15	166	322	453	484	515	41	197	328	359	390	541	72	203
547	53	209	365	391	422	553	84	240	266	297	428	584	115	141	172	303	459	615	16	47	178	334	490	516
340	491	522	28	184	215	366	397	528	59	90	241	272	403	559	590	116	147	278	434	465	616	22	153	309

Fig. 683. (5^4)

onal). Various other combinations may be found by taking the squares together in horizontal rows and noting how some columns and assorted diagonals have the proper summation, but the most important and significant are those already pointed out. In all the sum 123 occurs over 200 times in this small figure.

One most interesting fact remains to be noticed. While the

three cubes were constructed separately and independently the figure formed by combining them is an absolutely perfect square of 9, with a summation of 369 in rank, file and corner diagonal (besides all "broken" diagonals running downward to the right), and a perfect balancing of complementary numbers about the center. Any such pair, taken with the central number 41, gives us the familiar sum 123, and this serves to bind the whole together in a remarkable manner.

I				II				III				IV			
1	255	254	4	248	10	11	245	240	18	19	237	25	231	230	28
252	6	7	249	13	243	242	16	21	235	234	24	228	30	31	225
8	250	251	5	241	15	14	244	233	23	22	236	32	226	227	29
253	3	2	256	12	246	247	9	20	238	239	17	229	27	26	232
224	34	35	221	41	215	214	44	49	207	206	52	200	58	59	197
37	219	218	40	212	46	47	209	204	54	55	201	61	195	194	64
217	39	38	220	48	210	211	45	56	202	203	53	193	63	62	196
36	222	223	33	213	43	42	216	205	51	50	208	60	198	199	57
192	66	67	189	73	183	182	76	81	175	174	84	168	90	91	165
69	187	186	72	180	78	79	177	172	86	87	169	93	163	162	96
185	71	70	188	80	178	179	77	88	170	171	85	161	95	94	164
68	190	191	65	181	75	74	184	173	83	82	176	92	166	167	89
97	159	158	100	152	106	107	149	144	114	115	141	121	135	134	124
156	102	103	153	109	147	146	112	117	139	138	120	132	126	127	129
104	154	155	101	145	111	110	148	137	119	118	140	128	130	131	125
157	99	98	160	108	150	151	105	116	142	143	113	133	123	122	136

Fig. 684. (4^4)

2. Let $n = 5$, then $S = 1565$.—In Fig. 683 is represented a group of 5-cubes each made up of the numbers in a sub-series of the natural series 1...625. In accordance with the principle stated in a previous paragraph the central sub-series is 1 7 13 19 25 26 32 ... 625, and the other four can easily be discovered by inspection. Each of the twenty-five small squares has the summation 1565 in rank, file, corner diagonal and broken diagonals, twenty times altogether in each square, or 500 times for all.

Combining the five squares in col. I we have a cube in which all the 75 "straight" rows (rank, file and vertical column), all the horizontal diagonals and three of the four cubic diagonals foot up 1565. In cube III all the cubic diagonals are correct. Each cube also has seven vertical-square diagonals with the same summation. Taking together the squares in horizontal rows we find certain diagonals having the same sum, but the columns do not. The five squares in either diagonal of the large square, however, combine to produce almost perfect cubes, with rank, file, column and cubic diagonals all correct, and many diagonals of vertical squares.

A still more remarkable fact is that the squares in the broken diagonals running in either direction also combine to produce cubes as nearly perfect as those first considered. Indeed, the great square seems to be an enlarged copy of the small squares, and where the cells in the small ones unite to produce S the corresponding squares in the large figure unite to produce cubes more or less perfect. Many other combinations are discoverable, but these are sufficient to illustrate the principle, and show the interrelations of the cubes and their constituent squares. The summation 1565 occurs in this figure not less than 1400 times.

The plane figure containing the five cubes (or twenty-five squares) is itself a perfect square with a summation of 7825 for every rank, file, corner or broken diagonal. Furthermore all complementary pairs are balanced about the center, as in Fig. 682. Any square group of four, nine or sixteen of the small squares is magic, and if the group of nine is taken at the center it is "perfect." It is worthy of notice that all the powers of n above the first lie in the middle rank of squares, and that all other multiples of n are grouped in regular relations in the other ranks and have the same grouping in all the squares of any given rank. The same is true of the figure illustrating 7^4, which is to be considered next.

3. Let $n=7$, then $S=8407$.—This is so similar in all its properties to the 5-construction just discussed that it hardly needs separate description. It is more nearly perfect in all its parts than the 5^4, having a larger proportion of its vertical-square diagonals correct. Any square group of four, nine, sixteen, twenty-five or thirty

I						II						III					
1	1295	1294	3	1292	6	1278	20	21	1276	23	1273	37	1259	1258	39	1256	42
1290	8	1288	1287	11	7	25	1271	27	28	1268	1272	1254	44	1252	1251	47	43
1284	1283	15	16	14	1279	31	32	1264	1263	1265	36	1248	1247	51	52	50	1243
13	17	1281	1282	1280	18	1266	1262	34	33	35	1261	49	53	1245	1246	1244	54
12	1286	9	10	1289	1285	1267	29	1270	1269	26	30	48	1250	45	46	1253	1249
1291	2	4	1293	5	1296	24	1277	1275	22	1274	19	1255	38	40	1257	41	1260
1188	110	111	1186	113	1183	127	1169	1168	129	1166	132	1152	146	147	1150	149	1147
115	1181	117	118	1178	1182	1164	134	1162	1161	137	133	151	1145	153	154	1142	1146
121	122	1174	1173	1175	126	1158	1157	141	142	140	1153	157	158	1138	1137	1139	162
1176	1172	124	123	125	1171	139	143	1155	1156	1154	144	1140	1136	160	159	161	1135
1177	119	1180	1179	116	120	138	1160	135	136	1163	1159	1141	155	1144	1143	152	156
114	1187	1185	112	1184	109	1165	128	130	1167	131	1170	150	1151	1149	148	1148	145
217	1079	1078	219	1076	222	1062	236	237	1060	239	1057	253	1043	1042	255	1040	258
1074	224	1072	1071	227	223	241	1055	243	244	1052	1056	1038	260	1036	1035	263	259
1068	1067	231	232	230	1063	247	248	1048	1047	1049	252	1032	1031	267	268	266	1027
229	233	1065	1066	1064	234	1050	1046	250	249	251	1045	265	269	1029	1030	1028	270
228	1070	225	226	1073	1069	1051	245	1054	1053	242	246	264	1034	261	262	1037	1033
1075	218	220	1077	221	1080	240	1061	1059	238	1058	235	1039	254	256	1041	257	1044
865	431	430	867	428	870	414	884	885	412	887	409	901	395	394	903	392	906
426	872	424	423	875	871	889	407	891	892	404	408	390	908	388	387	911	907
420	419	879	880	878	415	895	896	400	399	401	900	384	383	915	916	914	379
877	881	417	418	416	882	402	398	898	897	899	397	913	917	381	382	380	918
876	422	873	874	425	421	403	893	406	405	890	894	912	386	909	910	389	385
427	866	868	429	869	432	888	413	411	886	410	883	391	902	904	393	905	396
864	434	435	862	437	859	451	845	844	453	842	456	828	470	471	826	473	823
439	857	441	442	854	858	840	458	838	837	461	457	475	821	477	478	818	822
445	446	850	849	851	450	834	833	465	466	464	829	481	482	814	813	815	486
852	848	448	447	449	847	463	467	831	832	830	468	816	812	484	483	485	811
853	443	856	855	440	444	462	836	459	460	839	835	817	479	820	819	476	480
438	863	861	436	860	433	841	452	454	843	455	846	474	827	825	472	824	469
756	542	543	754	545	751	559	737	736	561	734	564	720	578	579	718	581	715
547	749	549	550	746	750	732	566	730	729	569	565	583	713	585	586	710	714
553	554	742	741	743	558	726	725	573	574	572	721	589	590	706	705	707	594
744	740	556	555	557	739	571	575	723	724	722	576	708	704	592	591	593	703
745	551	748	747	548	552	570	728	567	568	731	727	709	587	712	711	584	588
546	755	753	544	752	541	733	560	562	735	563	738	582	719	717	580	716	577

Fig. 685, First Part. (6^4 : S = 3891)

IV						V						VI					
1225	71	70	1227	68	1230	1224	74	75	1222	77	1219	1206	92	93	1204	95	1201
66	1232	64	63	1235	1231	79	1217	81	82	1214	1218	97	1199	99	100	1196	1200
60	59	1239	1240	1238	55	85	86	1210	1209	1211	90	103	104	1192	1191	1193	108
1237	1241	57	58	56	1242	1212	1208	88	87	89	1207	1194	1190	106	105	107	1189
1236	62	1233	1234	65	61	1213	83	1216	1215	80	84	1195	101	1198	1197	98	102
67	1226	1228	69	1229	72	78	1223	1221	76	1220	73	96	1205	1203	94	1202	91
180	1118	1119	178	1121	175	181	1115	1114	183	1112	186	199	1097	1096	201	1094	204
1123	173	1125	1126	170	174	1110	188	1108	1107	191	187	1092	206	1090	1089	209	205
1129	1130	166	165	167	1134	1104	1103	195	196	194	1099	1086	1085	213	214	212	1081
168	164	1132	1131	1133	163	193	197	1101	1102	1100	198	211	215	1083	1084	1082	216
169	1127	172	171	1124	1128	192	1106	189	190	1109	1105	210	1088	207	208	1091	1087
1122	179	177	1120	176	1117	1111	182	184	1113	185	1116	1093	200	202	1095	203	1098
1009	287	286	1011	284	1014	1008	290	291	1006	293	1003	990	308	309	988	311	985
282	1016	280	279	1019	1015	295	1001	297	298	998	1002	313	983	315	316	980	984
276	275	1023	1024	1022	271	301	302	994	993	995	306	319	320	976	975	977	324
1021	1025	273	274	272	1026	996	992	304	303	305	991	978	974	322	321	323	973
1020	278	1017	1018	281	277	997	299	1000	999	296	300	979	317	982	981	314	318
283	1010	1012	285	1013	288	294	1007	1005	292	1004	289	312	989	987	310	986	307
361	935	934	363	932	366	360	938	939	358	941	355	342	956	957	340	959	337
930	368	928	927	371	367	943	353	945	946	350	354	961	335	963	964	332	336
924	923	375	376	374	919	949	950	346	345	347	954	967	968	328	327	329	972
373	377	921	922	920	378	348	344	952	951	953	343	330	326	970	969	971	325
372	926	369	370	929	925	349	947	352	351	944	948	331	965	334	333	962	966
931	362	364	933	365	936	942	359	357	940	356	937	960	341	339	958	338	955
504	794	795	502	797	499	505	791	790	507	788	510	523	773	772	525	770	528
799	497	801	802	494	498	786	512	784	783	515	511	768	530	766	765	533	529
805	806	490	489	491	810	780	779	519	520	518	775	762	761	537	538	536	757
492	488	808	807	809	487	517	521	777	778	776	522	535	539	759	760	758	540
493	803	496	495	800	804	516	782	513	514	785	781	534	764	531	532	767	763
798	503	501	796	500	793	787	506	508	789	509	792	769	524	526	771	527	774
612	686	687	610	689	607	613	683	682	615	680	618	631	665	664	633	662	636
691	605	693	694	602	606	678	620	676	675	623	619	660	638	658	657	641	637
697	698	598	597	599	702	672	671	627	628	626	667	654	653	645	646	644	649
600	596	700	699	701	595	625	629	669	670	668	630	643	647	651	652	650	648
601	695	604	603	692	696	624	674	621	622	677	673	642	656	639	640	659	655
690	611	609	688	608	685	679	614	616	681	617	684	661	632	634	663	635	666

Fig. 685, Second Part. ($6^4 : S = 3891$)

six small squares is magic, and if the group of nine or twenty-five be taken at the center of the figure it is "perfect." The grouping of multiples and powers of n is very similar to that already described for 5^4.

II. *When n is even.*

I. Let $n=4$, then $S=514$.—The numbers may be arranged in

I III

1	4095	4094	4	5	4091	4090	8	4032	66	67	4029	4028	70	71	4025
4088	10	11	4085	4084	14	15	4081	73	4023	4022	76	77	4019	4018	80
4080	18	19	4077	4076	22	23	4073	81	4015	4014	84	85	4011	4010	88
25	4071	4070	28	29	4067	4066	32	4008	90	91	4005	4004	94	95	4001
4065	31	30	4068	4069	27	26	4072	96	4002	4003	93	92	4006	4007	89
24	4074	4075	21	20	4078	4079	17	4009	87	86	4012	4013	83	82	4016
16	4082	4083	13	12	4086	4087	9	4017	79	78	4020	4021	75	74	4024
4089	7	6	4092	4093	3	2	4096	72	4026	4027	69	68	4030	4031	65
4064	34	35	4061	4060	38	39	4057	97	3999	3998	100	101	3995	3994	104
41	4055	4054	44	45	4051	4050	48	3992	106	107	3989	3988	110	111	3985
49	4047	4046	52	53	4043	4042	56	3984	114	115	3981	3980	118	119	3977
4040	58	59	4037	4036	62	63	4033	121	3975	3974	124	125	3971	3970	128
64	4034	4035	61	60	4038	4039	57	3969	127	126	3972	3973	123	122	3976
4041	55	54	4044	4045	51	50	4048	120	3978	3979	117	116	3982	3983	113
4049	47	46	4052	4053	43	42	4056	112	3986	3987	109	108	3990	3991	105
40	4058	4059	37	36	4062	4063	33	3993	103	102	3996	3997	99	98	4000

II IV

Fig. 686, 8^4, First Part (One cube written).

either of two ways. If we take the diagram for the 4-cube as given in Chapter II, page 78, and simply extend it to cover the larger numbers involved we shall have a group of four cubes in which all the "straight" dimensions have $S=514$, but no diagonals except the four cubic diagonals. Each horizontal row of squares will produce a cube having exactly the same properties as those in the four vertical rows. If the four squares in either diagonal of the figure be piled together neither vertical columns nor cubic diagonals will have the correct summation, but all the diagonals of vertical squares in either direction will. Regarding the whole group of sixteen squares as a plane square we find it magic, having

V VII

3968	130	131	3965	3964	134	135	3961	193	3903	3902	196	197	3899	3898	200
137	3959	3958	140	141	3955	3954	144	3896	202	203	3893	3892	206	207	3889
145	3951	3950	148	149	3947	3946	152	3888	210	211	3885	3884	214	215	3881
3944	154	155	3941	3940	158	159	3937	217	3879	3878	220	221	3875	3874	224
160	3938	3939	157	156	3942	3943	153	3873	223	222	3876	3877	219	218	3880
3945	151	150	3948	3949	147	146	3952	216	3882	3883	213	212	3886	3887	209
3953	143	142	3956	3957	139	138	3960	208	3890	3891	205	204	3894	3895	201
136	3962	3963	133	132	3966	3967	129	3897	199	198	3900	3901	195	194	3904
161	3935	3934	164	165	3931	3930	168	3872	226	227	3869	3868	230	231	3865
3928	170	171	3925	3924	174	175	3921	233	3863	3862	236	237	3859	3858	240
3920	178	179	3917	3916	182	183	3913	241	3855	3854	244	245	3851	3850	248
185	3911	3910	188	189	3907	3906	192	3848	250	251	3845	3844	254	255	3841
3905	191	190	3908	3909	187	186	3912	256	3842	3843	253	252	3846	3847	249
184	3914	3915	181	180	3918	3919	177	3849	247	246	3852	3853	243	242	3856
176	3922	3923	173	172	3926	3927	169	3857	239	238	3860	3861	235	234	3864
3929	167	166	3932	3933	163	162	3936	232	3866	3867	229	228	3870	3871	225

VI VIII

Fig. 686, 8^4, Second Part (One cube writen).

the summation 2056 in every rank, file and corner diagonal, 1028 in each half-rank or half-file, and 514 in each quarter-rank or quarter-file. Furthermore all complementary pairs are balanced about the center.

The alternative arrangement shown in Fig. 684 makes each of the small squares perfect in itself, with every rank, file and corner diagonal footing up 514 and complementary pairs balanced about the center. As in the other arrangement the squares in each vertical or horizontal row combine to make cubes whose "straight" dimensions all have the right summation. In addition the new form has the two plane diagonals of each original square (eight for each cube), but sacrifices the four cubic diagonals in each cube. In lieu of these we find a complete set of "bent diagonals" ("Franklin") like those described for the magic cube of six in Chapter IX.

If the four squares in either diagonal of the large figure be piled up it will be found that neither cubic diagonal nor vertical column is correct, but that all diagonals of vertical squares facing toward front or back are. Taken as a plane figure the whole group makes up a magic square of 16 with the summation 2056 in every rank, file or corner diagonal, half that summation in half of each of those dimensions, and one-fourth of it in each quarter dimension.

2. Let $n=6$, then S=3891.—With the natural series 1...1296 squares were constructed which combined to produce the six magic cubes of six indicated by the Roman numerals in Figure 685. These have all the characteristics of the 6-cube described in Chapter IX—108 "straight" rows, 12 plane diagonals and 25 "bent" diagonals in each cube, with the addition of 32 vertical-square diagonals if the squares are piled in a certain order. A seventh cube with the same features is made by combining the squares in the lowest horizontal row—i. e., the bottom squares of the numbered cubes. The feature of the cubic bent diagonals is found on combining any three of the small squares, no matter in what order they are taken. In view of the recent discussion of this cube it seems unnecessary to give any further account of it now.

The whole figure, made up as it is of thirty-six magic squares, is itself a magic square of 36 with the proper summation (23346) for every rank, file and corner diagonal, and the corresponding fractional part of that for each half, third or sixth of those dimensions. Any square group of four, nine, sixteen or twenty-five of the small squares will be magic in all its dimensions.

3. Let $n=8$, then S=16388.—The numbers 1...4096 may be arranged in several different ways. If the diagrams in Chapter II be adopted we have a group of eight cubes in which rank, file, column and cubic diagonal are correct (and in which the halves of these dimensions have the half summation), but all plane diagonals are irregular. If the plan be adopted of constructing the small squares of complementary couplets, as in the 6-cube, the plane diagonals are equalized at the cost of certain other features. I have used therefore a plan which combines to some extent the advantages of both the others.

It will be noticed that each of the small squares in Fig. 686 is perfect in that it has the summation 16388 for rank, file and corner diagonal (also for broken diagonals if each of the separated parts contain two, four or six—not an odd number of cells), and in balancing complementary couplets. When the eight squares are piled one upon the other a cube results in which rank, file, column, the plane diagonals of each horizontal square, the four ordinary cubic diagonals and 32 cubic bent diagonals all have S=16388. What is still more remarkable, the *half* of each of the "straight" dimensions and of each cubic diagonal has half that sum. Indeed this cube of eight can be sliced into eight cubes of 4 in each of which every rank, file, column and cubic diagonal has the footing 8194; and each of these 4-cubes can be subdivided into eight tiny 2-cubes in each of which the eight numbers foot up 16388.

So much for the features of the single cube here presented. As a matter of fact only the one cube has actually been written out. The plan of its construction, however, is so simple and the relations of numbers so uniform in the powers of 8 that it was easy to investigate the properties of the whole 8^4 scheme without having the squares actually before me. I give here the initial number of each of the eight squares in each of the eight cubes, leaving it for some one possessed of more leisure to write them all out and verify my statements as to the intercubical features. It should be remembered that in each square the number diagonally opposite the one here given is its complement, i. e., the number which added to it will give the sum 4097.

I	II	III	IV	V	VI	VII	VIII
1	3840	3584	769	3072	1281	1537	2304
4064	289	545	3296	1057	2784	2528	1825
4032	321	577	3264	1089	2752	2496	1857
97	3744	3488	865	2976	1377	1633	2208
3968	385	641	3200	1153	2688	2432	1921
161	3680	3424	929	2912	1441	1697	2144
193	3648	3392	961	2880	1473	1729	2112
3872	481	737	3104	1249	2592	2336	2017
16388	16388	16388	16388	16388	16388	16388	16388

Each of the sixty-four numbers given above will be at the upper left-hand corner of a square and its complement at the lower right-hand corner. The footings given are for these initial numbers, but the arrangement of numbers in the squares is such that the footing will be the same for every one of the sixty-four columns in each cube. If the numbers in each horizontal line of the table above be added they will be found to have the same sum: consequently the squares headed by them must make a cube as nearly perfect as the example given in Fig. 686, which is cube I of the table above. But the sum of half the numbers in each line is half of 16388, and hence each of the eight cubes formed by taking the squares in the horizontal rows is capable of subdivision into 4-cubes and 2-cubes, like our original cube. We thus have sixteen cubes, each with the characteristics described for the one presented in Fig. 686.

If we pile the squares lying in the diagonal of our great square (starting with 1, 289, etc., or 2304, 2528, etc.) we find that its columns and cubic diagonals are not correct; but all the diagonals of its vertical squares are so, and even here the remarkable feature of the half-dimension persists.

Of course there is nothing to prevent one's going still further and examining constructions involving the fifth or even higher powers, but the utility of such research may well be doubted. The purpose of this article is to suggest in sketch rather than to discuss exhaustively an interesting field of study for some one who may have time to develop it.

H. M. K.

FOUR-FOLD MAGICS.*

A magic square has two magic directions parallel to its sides through any cell—a row and a column; a magic cube has three magic directions parallel to its edges, a row, a column and a "line," the latter being measured at right angles to the paper-plane. By analogy, if for no other reason, a magic 4-fold should have four magic directions parallel to its linear edges, a row, a column, a line, and an "*i*." [The *i* is a convenient abbreviation for the imaginary direction, after the symbol $i=\sqrt{-1}$.] It is quite easy to determine by analogy how the imaginary direction is to be taken. If we look at a cube, set out as so many square sections on a plane, we see that the directions we have chosen to call rows and columns are shown in the square sections, and the third direction along a line is found by taking any cell in the first square plate, the similarly situated cell in the second plate, then that in the third and so on. In an octrahedroid the rows, columns and lines are given by the several cubical sections, viewed as solids, while the fourth or imaginary direction is found by starting at any cell in the first cube, passing to the corresponding cell of the second cube, then to that of the third, and so on.

If we denote each of the nine subsidiaries of order 3 in Fig. 687 by the number in its central cell, and take the three squares 45, 1, 77, in that order, to form the plates of a first cube; 73, 41, 9 to form a second cube, and 5, 81, 37 for a third cube, we get an associated octahedroid, which is magic along the four directions parallel to its edges and on its 8 central hyperdiagonals. We find the magic sum

* The subject has been treated before in:

Frost (A. H.), "The Properties of Nasik Cubes," *Quarterly Journal of Mathematics,* London, 1878, p. 93.

"C. P." (C. Planck), "Magic Squares, Cubes, etc.," *The English Mechanic,* London, March 16, 1888.

Arnoux (Gabriel), *Arithmétique graphique,* Paris, 1894, Gauthier-Villars et Fils.

Planck (C.), *The Theory of Path Nasiks,* 1905. Printed for private circulation. There are copies at the British Museum, the Bodleian, Oxford, and the University Library, Cambridge.

on 9 rows, 9 columns and 18 diagonals, the nine subsidiaries equally weighted and magic in rows and columns, and further the square is 9-ply, that is, the nine numbers in any square section of order 3 give the magic sum of the great square.

It will be convenient here to turn aside and examine the evolution of the Nasik idea and the general analogy between the figures of various dimensions in order that we may determine how the Nasik concept ought to be expanded when we apply it in the higher dimensions. This method of treatment is suggested by Professor Kingery's remark, p. 352, "It is not easy—perhaps it is not possible—to make an absolutely perfect cube of 3." If we insist on magic central

65	6	52	29	78	16	20	42	61
36	73	14	27	37	59	72	1	50
22	44	57	67	8	48	31	80	12
69	7	47	33	79	11	24	43	56
28	77	18	19	41	63	64	5	54
26	39	58	71	3	49	35	75	13
70	2	51	34	74	15	25	38	60
32	81	10	23	45	55	68	9	46
21	40	62	66	4	53	30	76	17

Fig. 687.

diagonals we know that, in the restricted sense, there is only one magic square of order 3, but if we reckon reflections and reversions as different there are 8. If we insist on magic central great diagonals in the cube, as by analogy we ought to do, then, in the restricted sense, there are just 4 magic cubes of order 3. But each of these can be placed on any one of six bases and then viewed from any one of four sides, and each view thus obtained can be duplicated by reflection. In the extended sense, therefore, there are 192 magic cubes of order 3. None of these, however, has the least claim to be considered "perfect." This last term has been used with several different meanings by various writers on the subject. From the present writer's point of view the Nasik idea, as presently to be developed, ought to stand pre-eminent; next in importance comes the ply property, then the adornment of magic subsidiaries, with the properties of association, bent diagonals of Franklin, etc., etc., taking subordinate places.

The lattice idea certainly goes back to prehistoric time, and what we now call the rows and columns of a rectangular lattice first appealed to man because they disclose contiguous rectilinear series of cells, that is sets of cells, whose centers are in a straight line, and each of which has linear contact with the next. It must soon have been noticed that two other series exist in every square, which fulfil the same conditions, only now the contact is punctate instead of linear. They are what we call the central diagonals. It was not until the congruent nature of the problem was realized that it became apparent that a square lattice has as many diagonals as rows and columns together. Yet the ancient Hindus certainly recognized this congruent feature. The eccentric diagonals have been called "broken diagonals," but they are really not broken if we remember that we tacitly assume all space of the dimensions under consideration saturated with contiguous replicas of the figure before us, cells similarly situated in the several replicas being considered identical. A. H. Frost* nearly 50 years ago invented the term "Nasik" to embrace that species of square which shows magic summations on all its contiguous rectilinear series of cells, and later extended the idea by analogy to cubes,† and with less success to a figure in four dimensions. If the Nasik criterion be applied to 3-dimensional magics what does it require? We must have 3 magic directions through any cell parallel to the edges, (planar contact), 6 such directions in the diagonals of square sections parallel to the faces (linear contact), and 4 directions parallel to the great diagonals of the cube (point contact), a total of 13 magic directions through every cell. It has long been known that the smallest square which can be nasik is of order 4, or if the square is to be associated, (that

* *Quarterly Journal of Mathematics,* London, 1865, and 1878, pp. 34 and 93.

† The idea of the crude magic cube is, of course, much older: Fermat gives a 4^3 in his letter to Mersenne of the 1st of April, 1640. *Œuvres de Fermat,* Vol. II, p. 191.

is with every pair of complementary numbers occupying cells which are equally displaced from the center of the figure in opposite directions), then the smallest Nasik order is 5. Frost stated definitely* that in the case of a cube the smallest Nasik order is 9; Arnoux† was of opinion that it would be 8, though he failed to construct such a magic. It is only quite recently‡ that the present writer has shown that the smallest Nasik order in k dimensions is always 2^k, (or 2^k+1 if we require association).

It is not difficult to perceive that if we push the Nasik analogy to higher dimensions the number of magic directions through any cell of a k-fold must be $\frac{1}{2}(3^k-1)$, for we require magic directions from every cell through each cell of the surrounding little k-fold of order 3. In a 4-fold Nasik, therefore, there are 40 contiguous rectilinear summations through any cell. But how are we to determine these 40 directions and what names are we to assign to the magic figures in the 4th and higher dimensions? By far the best nomenclature for the latter purpose is that invented by Stringham,§ who called the regular m-dimensional figure, which has n $(m-1)$-dimensional boundaries, an m-fold n-hedroid. Thus the square is a 2-fold tetrahedroid (tetragon), the cube a 3-fold hexahedroid (hexahedron); then come the 4-fold octahedroid, the 5-fold decahedroid, and so on. Of course the 2-fold octahedroid is the plane octagon, the 3-fold tetrahedroid the solid tetrahedron; but since the regular figure in k dimensions which is analogous to the square and cube has always $2k$ $(k-1)$-dimensional boundaries—is in fact a k-fold $2k$-hedroid—the terms octahedroid, decahedroid, etc., as applied to magics, are without ambiguity, and may be appropriately used for magics in 4, 5, etc. dimensions, while retaining the familiar "square," "cube," for the lower dimensions.

To obtain a complete knowledge of these figures, requires a study of analytical geometry of the 4th and higher dimensions, but, by analogy, on first principles, we can obtain sufficient for our purpose. If we had only a linear one-dimensional space at command

* *Quarterly Journal,* Vol. XV, p. 110.

† *Arithmétique graphique,* Paris, 1894, p. 140.

‡ *Theory of Path Nasiks,* 1905.

§ *American Journal of Mathematics,* Vol. III, 1880.

we could represent a square of order n in two ways, ("aspects"), either by laying the n rows, in order, along our linear dimension, or by dealing similarly with the n columns. In the first aspect, by rows, the cells which form any column cannot appear as contiguous, though they actually are so when we represent the figure as a square

34	74	15	65	6	52	24	43	56
23	45	55	36	73	14	64	5	54
66	4	53	22	44	57	35	75	13
20	42	61	33	79	11	70	2	51
72	1	50	19	41	63	32	81	10
31	80	12	71	3	49	21	40	62
69	7	47	25	38	60	29	78	16
28	77	18	68	9	46	27	37	59
26	39	58	30	76	17	67	8	48

Fig. 688. P_1- and P_2-aspects.

69	20	34	7	42	74	47	61	15
28	72	23	77	1	45	18	50	55
26	31	66	39	80	4	58	12	53
25	33	65	38	79	6	60	11	52
68	19	36	9	41	73	46	63	14
30	71	22	76	3	44	17	49	57
29	70	24	78	2	43	16	51	56
27	32	64	37	81	5	59	10	54
67	21	35	8	40	75	48	62	13

Fig. 689. V-aspect.

69	7	47	28	77	18	26	39	58
20	42	61	72	1	50	31	80	12
34	74	15	23	45	55	66	4	53
25	38	60	68	9	46	30	76	17
33	79	11	19	41	63	71	3	49
65	6	52	36	73	14	22	44	57
29	78	16	27	37	59	67	8	48
70	2	51	32	81	10	21	40	62
24	43	56	64	5	54	35	75	13

Fig. 690. H-aspect.

on a plane. Similarly we can represent a cube on a plane in three aspects. Suppose the paper-plane is placed vertically before us and the cube is represented by n squares on that plane (P-plane aspect). We get a second aspect by taking, in order, the first column of each square to form the first square of the new aspect, all the second columns, in order, to form the second square of the new aspect, and so on (V-plane aspect). We obtain a third aspect by dealing similarly with the rows (H-plane aspect). Here the "lines," which appear as contiguous cells in the V- or H-plane aspects do not so appear in the P-plane aspect, though they actually are contiguous when we examine the cube as a solid in three dimensions. Now consider an octahedroid represented by n cubes in a space of three dimensions. We get a second aspect by taking the n anterior, vertical square plates of each cube, in order, to form a first new cube; the n plates immediately behind the anterior plate in each cube to form a second new cube, and so on. Evidently we obtain a third aspect, in like manner, by slicing each cube into vertical, antero-posterior plates, and a fourth aspect by using the horizontal plates. Carrying on the same reasoning, it becomes clear that we can represent a k-fold of order n, in $k—1$ dimensions, by n $(k—1)$-folds, in k different aspects. Thus we can represent a 5-fold decahedroid of order n, in 4-dimensional space, by n 4-fold octahedroids, and this in 5 different ways or aspects.

Return now to Fig. 687 and the rule which follows it, for forming from it the magic octahedroid of order 3. If we decide to represent the three cubic sections of the octahedroid by successive columns of squares we get Fig. 688.

If we obtain a second aspect by using the square plates of the paper-plane, as explained above, we find that this is equivalent to taking the successive rows of squares from Fig. 688 to form our three cubes, instead of taking the columns of squares. Thus the presentation plane shows two different aspects of an octahedroid: this is due to the fact that the fourth dimension is the square of the second. We may call these aspects P_1- and P_2-aspects. The aspect obtained by using antero-posterior vertical planes is shown in Fig. 689, that from horizontal planes in Fig. 690. We may call these the V- and H-aspects. If we use the rows of squares in Figs. 689 or 690 we get correct representations of the octahedroid, but these are not new aspects, they are merely repetitions of P_1, for they give new views of the same three cubes as shown in P_1. In the same way, if we turned all the P-plane plates of a cube upside down we should not call that a new aspect of the cube. The aspects P_2, V, H can be obtained from P_1 by turning the octahedroid as a whole in 4-dimensional space, just as the V-plane and H-plane aspects of a cube can be obtained from the P-plane aspect by turning the cube in 3-dimensional space. Fig. 690, above, is Fig. 688 turned through a right angle about the plane of xy; we can turn about a plane in 4 dimensions just as we turn about a straight line in 3 dimensions or about a point in 2 dimensions. It will be noticed that in the four aspects each of the 4 directions parallel to an edge becomes in turn imaginary, so that it cannot be made to appear as a series of contiguous cells in 3-dimensional space; yet if we had a 4-dimensional space at command, these four directions could all be made to appear as series of contiguous cells. There is one point, however, which must not be overlooked. When we represent a cube as so many squares, the rows and columns appear as little squares having linear contact, but actually, in the cube, the cells are all cubelets having planar contact. Similarly, in an octahedroid represented as so many cubes the rows and columns appear as cubelets having planar contact, but in the octahedroid the cells are really little octahedroids having solid, 3-dimensional contact.

When we examine the above octahedroid (Figs. 688-690) in all its aspects we see that there are through every cell 4 different directions parallel to the edges, 12 directions parallel to the diagonals of the square faces, and 16 directions parallel to the great diagonals of the several cubical sections. There remain for consideration the hyperdiagonals, which bear to the octahedroid the same relation that the great diagonals bear to a cube. If we represent a cube by squares on a plane we can obtain the great diagonals by starting at any corner cell of an outside plate, then passing to the next cell of the corresponding diagonal of the succeeding plate, and so on. Similarly we obtain the hyperdiagonals of the octahedroid by starting from any corner cell of an outside cube, passing to the next cell on the corresponding great diagonal of the succeeding cube, and so on. Evidently there are 8 central hyperdiagonals, for we can start at any one of the 8 corners of one outside cube and end at the opposite corner of the other outside cube. There are therefore, through any cell, 8 different directions parallel to the central hyperdiagonals. With the directions already enumerated this makes a total of 40 directions through each cell and agrees with the result already stated.* Evidently the number of k-dimensional diagonals of a k-fold is 2^{k-1}, and if the analogy with the magic square is to be carried through then all the central k-dimensional diagonals of a k-fold ought always to be magic.

The smallest octahedroid which can have all these 40 directions magic is 16^4, and the writer has given one of the 256 square plates of this magic and a general formula by which the number occupying any specified cell can be determined. But it will be interesting to determine how nearly we can approach this ideal in the lower orders. The octahedroid of order 3 can be but crude, and practically Fig. 688 cannot be improved upon. All rows, columns, lines, and "*i*"s are magic, and likewise the 8 central hyper-diagonals. Of course, since the figure is associated, all central rectilinear paths are magic, but this is of little account and other asymmetrical magic diagonal summations are purely accidental and therefore negligible.

Turning to the next odd order, 5: Professor Kingery's Fig. 683 is not a magic octahedroid as it stands, but a magic can be obtained from it by taking the diagonals of subsidiary squares to form the 5 cubes. Denoting each subsidiary by the number in its central cell, we may use 602, 41, 210, etc. for the first cube; 291, 460 etc. for the second cube; 85, 149, etc. for the middle cube, etc., etc. But few of the plane diagonals through any cell of this octahedroid are magic. In fact no octahedroid of lower order than 8 can have all its plain diagonals magic; but by sacrificing this property we can obtain a 5^4 with many more magic properties than the above.

In Fig. 691 the great square is magic, Nasik and 25-ply: the 25 subsidiaries are purposely not Nasik, but they are all magic in rows

* If we call the diagonals in square sections parallel to faces 2-dimensional, those parallel to the great diagonals of cubical sections 3-dimensional, etc., etc., then the number of m-dimensional diagonals of a k-fold is $2^{m-1}k!/m!(k-m)!$ In fact the number required is the $(m+1)$th term of the expansion of $\frac{1}{2}(1+2)^k$. It will be noticed that this reckons rows, columns etc. as "diagonals of one dimension."

and columns. If we take up the subsidiaries in the way just described, viz., 513, 221, etc., for the first cube; 205, 413, etc., for the second cube, and so on, we get a 5^4, which has 20 contiguous rectilinear summations through any cell, viz., the 4 directions parallel to the edges and the whole of the 16 three-dimensional diagonals parallel to the great diagonals of any cubical section. If the reader

495	58	271	589	152	478	66	259	597	165	486	54	267	585	173	499	62	255	593	156	482	75	263	576	169
178	391	84	297	615	186	379	92	285	623	199	387	80	293	606	182	400	88	276	619	195	383	96	289	602
511	204	417	110	323	524	212	405	118	306	507	225	413	101	319	520	208	421	114	302	503	216	409	122	315
349	537	230	443	6	332	550	238	426	19	345	533	246	439	2	328	541	234	447	15	336	529	242	435	23
32	375	563	126	469	45	358	571	139	452	28	366	559	147	465	36	354	567	135	473	49	362	555	143	456
70	258	596	164	477	53	266	584	172	490	61	254	592	160	498	74	262	580	168	481	57	275	588	151	494
378	91	284	622	190	386	79	292	610	198	399	87	280	618	181	382	100	288	601	194	395	83	296	614	177
211	404	117	310	523	224	412	105	318	506	207	425	113	301	519	220	408	121	314	502	203	416	109	322	515
549	237	430	18	331	532	250	438	1	344	545	233	446	14	327	528	241	434	22	340	536	229	442	10	348
357	575	138	451	44	370	558	146	464	27	353	566	134	472	40	361	554	142	460	48	374	562	130	468	31
270	583	171	489	52	253	591	159	497	65	261	579	167	485	73	274	587	155	493	56	257	600	163	476	69
78	291	609	197	390	86	279	617	185	398	99	287	605	193	381	82	300	613	176	394	95	283	621	189	377
411	104	317	510	223	424	112	305	518	206	407	125	313	501	219	420	108	321	514	202	403	116	309	522	215
249	437	5	343	531	232	450	13	326	544	245	433	21	339	527	228	441	9	347	540	236	429	17	335	548
557	150	463	26	369	570	133	471	39	352	553	141	459	47	365	561	129	467	35	373	574	137	455	43	356
595	158	496	64	252	578	166	484	72	265	586	154	492	60	273	599	162	480	68	256	582	175	488	51	269
278	616	184	397	90	286	604	192	385	98	299	612	180	393	81	282	625	188	376	94	295	608	196	389	77
111	304	517	210	423	124	312	505	218	406	107	325	513	201	419	120	308	521	214	402	103	316	509	222	415
449	12	330	543	231	432	25	338	526	244	445	8	346	539	227	428	16	334	547	240	436	4	342	535	248
132	475	38	351	569	145	458	46	364	552	128	466	34	372	565	136	454	42	360	573	149	462	30	368	556
170	483	71	264	577	153	491	59	272	590	161	479	67	260	598	174	487	55	268	581	157	500	63	251	594
603	191	384	97	290	611	179	392	85	298	624	187	380	93	281	607	200	388	76	294	620	183	396	89	277
311	504	217	410	123	324	512	205	418	106	307	525	213	401	119	320	508	221	414	102	303	516	209	422	115
24	337	530	243	431	7	350	538	226	444	20	333	546	239	427	3	341	534	247	440	11	329	542	235	448
457	50	363	551	144	470	33	371	564	127	453	41	359	572	140	461	29	367	560	148	474	37	355	568	131

Fig. 691.

will write out the four aspects of the octahedroid, in the way already explained, he will be able to verify this statement. As an example, the 20 summations through the cell containing the number 325, which lies in the first plate of the first cube of the P_1-aspect, are here shown:

ROW	COLUMN	LINE	"z"	CUBICAL DIAGONALS															
				P_1-ASPECT				P_2-ASPECT				V-ASPECT				H ASPECT			
325	325	325	325	325	325	325	325	325	325	325	325	325	325	325	325	325	325	325	325
513	8	508	512	534	388	607	3	538	392	611	7	533	387	608	4	413	103	507	509
201	466	216	204	143	576	169	456	126	589	152	469	141	579	166	458	501	406	219	218
419	154	404	416	477	44	451	164	494	31	468	151	479	41	454	162	119	214	401	402
107	612	112	108	86	232	13	617	82	228	9	613	87	233	12	616	207	517	113	111

Since there are 20 magic summations through each of the 625 cells and each summation involves 5 cells, the total number of different symmetrical magic summations in this octahedroid is 2500. This does not include the 8 central hyperdiagonals, which are also magic, for this is not a symmetrical property since *all* the hyperdiagonals are not magic.

The next odd order, 7, was the one which Frost attacked. Glass models of his 7 cubes were for many years to be seen at the South Kensington Museum, London, and possibly are still there. He does not appear to have completely grasped the analogy between magics in 3 and 4 dimensions, and from the account he gives in *The Quarterly Journal,* he evidently assumed that the figure was magic on all its plane diagonals. Actually it is magic on all plane diagonals only in the P-aspect; in the other 3 aspects it is Nasik in one set of planes but only semi-Nasik in the other two sets of planes, therefore of the 12 plane diagonals through any cell of the octahedroid only 9 are magic.* Frost obtained his figure by direct application of the method of paths; the present writer using the method of formative square has obtained an example with one additional plane magic diagonal. It is shown as a great square of order 49, magic on its 49 rows, 49 columns and 98 diagonals, and 49-ply, that is *any* square bunch of 49 numbers gives the same sum as a row or column. The 49 subsidiaries are equally weighted Nasiks, magic on their 7 rows, 7 columns and 14 diagonals. If the subsidiaries be taken up along the Indian paths, as in the previous examples, we get 7 cubes forming an octahedroid of order 7. This is magic on the 4 directions parallel to the edges, is completely plane Nasik in

* Probably the reader will have alrealy noticed that although there are 4 aspects, and 6 plane diagonals appear in each aspect, yet there are only 12 plane diagonals in all, since, with this method of enumeration, each diagonal occurs twice.

the P_1 and P_2-aspects, and in the other two aspects it is Nasik in two sets of planes and crude in the third set. Therefore of the 12 plane diagonals through any cell 10 are magic. It is practically certain that we can go no further in this direction with this order, but by giving up the magic plane diagonals we can, as with 5^4 above, obtain a larger number of magic summations on the higher diagonals.

When we consider the even orders we find those $2 \equiv$ (mod 4) of little interest. The powerful methods used for the other orders are now useless if we insist on using consecutive numbers: we must employ other methods. The best methods here are either to use an extension of Thompson's method of pseudo-cubes, as employed by Mr. Worthington in his construction of 6^3 (pp. 201-206),* or, best of all, to use the method of reversions.

With orders $\equiv 0 (\text{mod } 4)$ we can give a greater number of ornate features than with any other orders. We quote one example below (Fig. 692).

The columns of Fig. 692 give the 4 cubes of an octahedroid of order 4, which is crude in plane diagonals, but is magic on every other contiguous rectilinear path, it has therefore 28 such paths through each cell. The 28 magic paths through the cell containing the number 155 are displayed below.

				CUBICAL DIAGONALS															
ROW	COLUMN	LINE	"z"	P_1-ASPECT				P_2-ASPECT				V-ASPECT				H-ASPECT			
155	155	155	155	155	155	155	155	155	155	155	155	155	155	155	155	155	155	155	155
38	70	98	101	2	50	242	194	5	53	245	197	77	125	113	65	36	33	225	228
91	171	151	154	103	103	103	103	106	106	106	106	166	166	166	166	86	86	86	86
230	118	110	104	254	206	14	62	248	200	8	56	116	68	80	128	237	240	48	45

HYPERDIAGONALS							
155	155	155	155	155	155	155	155
256	208	16	64	253	205	13	61
102	102	102	102	102	102	102	102
1	49	241	193	4	52	244	196

But this does not exhaust the magic properties, for this figure is 4-ply in every plane section parallel to any face of the octahedroid.

* It was by this method that Firth in the 80's constructed what was, almost certainly, the first correct magic cube of order 6. *Mr. Worthington's introduction of magic central diagonals on all the faces is new. Though, of course, not a symmetrical summation, this is a very pleasing feature.*

If the reader will examine the figure in its four aspects he will find that 6 such planes can be drawn through any cell, and since a given number is a member of four different 4-ply bundles in each plane, it follows that each number is a member of 24 different bundles. If we add the 28 rectilinear summations through any cell we see

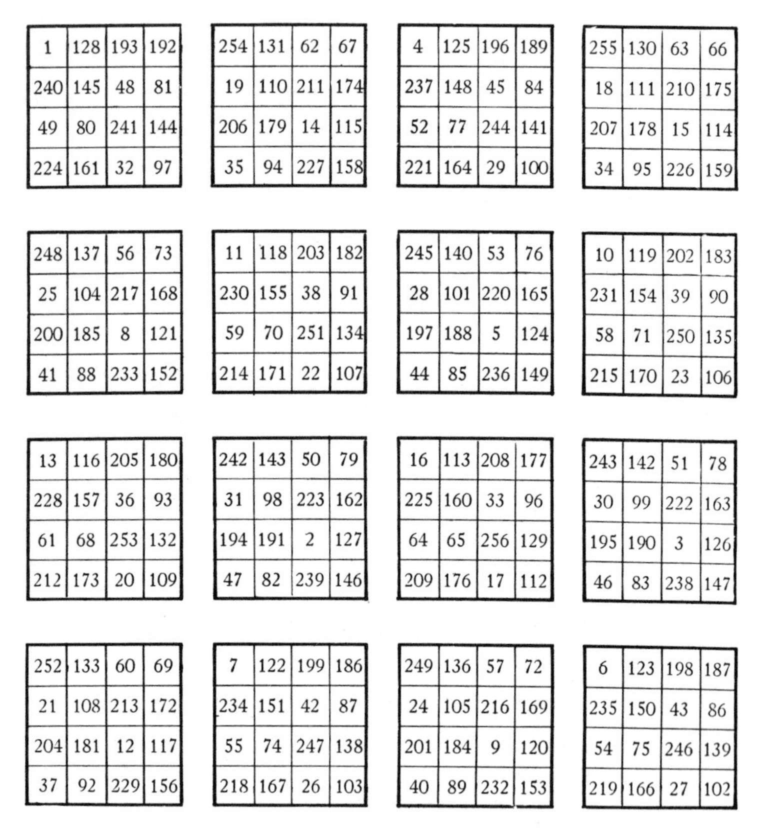

1	128	193	192
240	145	48	81
49	80	241	144
224	161	32	97

254	131	62	67
19	110	211	174
206	179	14	115
35	94	227	158

4	125	196	189
237	148	45	84
52	77	244	141
221	164	29	100

255	130	63	66
18	111	210	175
207	178	15	114
34	95	226	159

248	137	56	73
25	104	217	168
200	185	8	121
41	88	233	152

11	118	203	182
230	155	38	91
59	70	251	134
214	171	22	107

245	140	53	76
28	101	220	165
197	188	5	124
44	85	236	149

10	119	202	183
231	154	39	90
58	71	250	135
215	170	23	106

13	116	205	180
228	157	36	93
61	68	253	132
212	173	20	109

242	143	50	79
31	98	223	162
194	191	2	127
47	82	239	146

16	113	208	177
225	160	33	96
64	65	256	129
209	176	17	112

243	142	51	78
30	99	222	163
195	190	3	126
46	83	238	147

252	133	60	69
21	108	213	172
204	181	12	117
37	92	229	156

7	122	199	186
234	151	42	87
55	74	247	138
218	167	26	103

249	136	57	72
24	105	216	169
201	184	9	120
40	89	232	153

6	123	198	187
235	150	43	86
54	75	246	139
219	166	27	102

Fig. 692.

that each of the 256 numbers takes part in 52 different summations. The total number of different magic summations in the octahedroid is therefore $\frac{256\times 52}{4}=3328$. The six planes parallel to the faces through 155 are shown in Fig. 693, and from them the 24 different bundles in which 155 is involved can be at once determined.

The reader might object that the border cells of a square section cannot be involved in 4 bundles of that section; but this would be to overlook the congruent property. The number 107, which

11	118	203	182
230	155	38	91
59	70	251	134
214	171	22	107

19	110	211	174
230	155	38	91
31	98	223	162
234	151	42	87

131	118	143	122
110	155	98	151
179	70	191	74
94	171	82	167

25	104	217	168
230	155	38	91
28	101	220	165
231	154	39	90

137	118	140	119
104	155	101	154
185	70	188	71
88	171	85	170

145	110	148	111
104	155	101	154
157	98	160	99
108	151	105	150

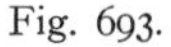

Fig. 693.

occupies a corner cell of the first section given above is contained in the following bundles:

251	134
22	107

134	59
107	214

22	107
203	182

107	214
182	11

It is noticeable that the four corner cells of a square form one of its 4-ply bundles.

It would have been desirable to indicate the methods by which the above examples have been constructed, but exigencies of space forbid. The four orders dealt with, 3, 5, 7, 4, were all obtained in different ways. Fig. 692 was constructed by direct application, in four dimensions, of the method of paths; in fact, it is the octahedroid

$$\left|\begin{array}{cccc} 2 & 2 & 2 & 1 \\ 2 & 2 & 1 & 2 \\ 2 & 1 & 2 & 2 \\ 1 & 2 & 2 & 2 \end{array}\right| 4.$$

The whole of its magic properties may be deduced by examination of the determinant and its adjoint, without any reference to the constructed figure. There is therefore nothing empirical about this method.

C. P.

CHAPTER XV.

ORNATE MAGIC SQUARES.

GENERAL RULE FOR CONSTRUCTING ORNATE MAGIC SQUARES OF ORDERS $\equiv 0 \pmod 4$.

TAKE a square lattice of order $4m$ and draw heavy lines at every fourth vertical bar and also at every fourth horizontal bar, thus dividing the lattice into m^2 subsquares of order 4. The "period" consists of the $4m$ natural numbers 1, 2, 3.... $4m$. Choose from these *any* two pairs of complementary numbers, that is, pairs whose sum is $4m+1$ and arrange these four numbers, four times repeated, as in a Jaina square (first type) in the left-hand square of the top row of subsquares in the large lattice. It is essential that the Jaina pattern shall contain only one complementary couplet in each of its four columns, i. e., if the two pairs are a_1 a_2 and b_1 b_2, every column must consist entirely of a's, or entirely of b's. The first Jaina type can be obtained by using the paths (1, 2) (2, 1) and the order a_1 b_1 a_2 b_2 four times repeated. This gives the square shown in Fig. 694, which fulfils the conditions. Proceed in the same way with each of the m subsquares in the top row, *using a different pair of complementaries in each subsquare.* Since the period 1, 2, 3.... $4m$ contains $2m$ complementary pairs and two pairs are used for each subsquare, it follows that when the top row of subsquares is filled up, all the $4m$ numbers will have been used.

Now fill all the remaining rows of subsquares in the large lattice with replicas of the top row. The *outline* so constructed can always be turned over either of its central diagonals without repetition. The resulting square will therefore contain the first $(4m)^2$

numbers without repetition or omission, and it will always have the following magic properties.

A. *The Great Square.....*

1. is magic on its $4m$ rows and $4m$ columns;

a_1	b_1	a_2	b_2
a_2	b_2	a_1	b_1
a_1	b_1	a_2	b_2
a_2	b_2	a_1	b_1

Fig. 694.

10	51	15	54	12	49	13	56
23	46	18	43	21	48	20	41
50	11	55	14	52	9	53	16
47	22	42	19	45	24	44	17
26	35	31	38	28	33	29	40
7	62	2	59	5	64	4	57
34	27	39	30	36	25	37	32
63	6	58	3	61	8	60	1

Fig. 695.

2. is pandiagonal, i. e., magic on its $8m$ diagonals;

3. has Franklin's property of *bent diagonals* in an extended sense; i. e., we can start at any cell in the top row, and proceeding downward bend the diagonal at *any* heavy horizontal bar. It

2	3	7	6	4	1	5	8
7	6	2	3	5	8	4	1
2	3	7	6	4	1	5	8
7	6	2	3	5	8	4	1
2	3	7	6	4	1	5	8
7	6	2	3	5	8	4	1
2	3	7	6	4	1	5	8
7	6	2	3	5	8	4	1

Fig. 696.

8	48	8	48	8	48	8	48
16	40	16	40	16	40	16	40
48	8	48	8	48	8	48	8
40	16	40	16	40	16	40	16
24	32	24	32	24	32	24	32
0	56	0	56	0	56	0	56
32	24	32	24	32	24	32	24
56	0	56	0	56	0	56	0

Fig. 697.

matters not how many times we bend, or at which of the heavy bars, providing only that when the traverse is completed, the number of cells passed over in the one direction (downward to the right) shall be exactly equal to the number passed over in the other direc-

27	46	111	106	3	58	135	94	63	22	75	130
112	105	28	45	136	93	4	57	76	129	64	21
34	39	118	99	10	51	142	87	70	15	82	123
117	100	33	40	141	88	9	52	81	124	69	16
25	48	109	108	1	60	133	96	61	24	73	132
113	104	29	44	137	92	5	56	77	128	65	20
36	37	120	97	12	49	144	85	72	13	84	121
116	101	32	41	140	89	8	53	80	125	68	17
30	43	114	103	6	55	138	91	66	19	78	127
110	107	26	47	134	95	2	59	74	131	62	23
31	42	115	102	7	54	139	90	67	18	79	126
119	98	35	38	143	86	11	50	83	122	71	14

Fig. 698. S = 870

115	110	131	158	3	78	243	190	51	94	195	174	19	46	227	222
130	159	114	111	242	191	2	79	194	175	50	95	226	223	18	47
126	99	142	147	14	67	254	179	62	83	206	163	30	35	238	211
143	146	127	98	255	178	15	66	207	162	63	82	239	210	31	34
118	107	134	155	6	75	246	187	54	91	198	171	22	43	230	219
132	157	116	109	244	189	4	77	196	173	52	93	228	221	20	45
123	102	139	150	11	70	251	182	59	86	203	166	27	38	235	214
141	148	125	100	253	180	13	68	205	164	61	84	237	212	29	36
117	108	133	156	5	76	245	188	53	92	197	172	21	44	229	220
129	160	113	112	241	192	1	80	193	176	49	96	225	224	17	48
124	101	140	149	12	69	252	181	60	85	204	165	28	37	236	213
144	145	128	97	256	177	16	65	208	161	64	81	240	209	32	33
119	106	135	154	7	74	247	186	55	90	199	170	23	42	231	218
136	153	120	105	248	185	8	73	200	169	56	89	232	217	24	41
122	103	138	151	10	71	250	183	58	87	202	167	26	39	234	215
137	152	121	104	249	184	9	72	201	168	57	88	233	216	25	40

Fig. 699. S = 2056

tion (downward to the left). Similarly we may start at any cell in the left-hand column and, proceeding diagonally to the right, bend the diagonal at *any* heavy vertical bar under the same limitations.

It will be noticed that when the order of the square is $\equiv 4$ (mod 8), i. e., when m is odd, the central bars are not *heavy bars,*

1	382	20	399	3	384	18	397	5	386	16	395	7	388	14	393	9	390	12	391
40	379	21	362	38	377	23	364	36	375	25	366	34	373	27	368	32	371	29	370
381	2	400	19	383	4	398	17	385	6	396	15	387	8	394	13	389	10	392	11
380	39	361	22	378	37	363	24	376	35	365	26	374	33	367	28	372	31	369	30
41	342	60	359	43	344	58	357	45	346	56	355	47	348	54	353	49	350	52	351
80	339	61	322	78	337	63	324	76	335	65	326	74	333	67	328	72	331	69	330
341	42	360	59	343	44	358	57	345	46	356	55	347	48	354	53	340	50	352	51
340	79	321	62	338	77	323	64	336	75	325	66	334	73	327	68	332	71	329	70
81	302	100	319	83	304	98	317	85	306	96	315	87	308	94	313	89	400	92	311
120	299	101	282	118	297	103	284	116	295	105	286	114	293	107	288	112	291	109	290
301	82	320	99	303	84	318	97	305	86	316	95	307	88	314	93	309	90	312	91
300	119	281	102	298	117	283	104	296	115	285	106	294	113	287	108	292	111	289	110
121	262	140	279	123	264	138	277	125	266	136	275	127	268	134	273	129	270	132	271
160	259	141	242	158	257	143	244	156	255	145	246	154	253	147	248	152	251	149	250
261	122	280	139	263	124	278	137	265	126	276	135	267	128	274	133	269	130	272	131
260	159	241	142	258	157	243	144	256	155	245	146	254	153	247	148	252	151	249	150
161	222	180	239	163	224	178	237	165	226	176	235	167	228	174	233	169	230	172	231
200	219	181	202	198	217	183	204	196	215	185	206	194	213	187	208	192	211	189	210
221	162	240	179	223	164	238	177	225	166	236	175	227	168	234	173	229	170	232	171
220	199	201	182	218	197	203	184	216	195	205	186	214	193	207	188	212	191	209	190

Fig. 700.

and also the number of rows of subsquares is odd. We cannot therefore in these cases get a magic bent diagonal traverse from top to bottom of the square, but we may stop at the last heavy bar before reaching the bottom of the square, when we shall have a sum $4(m-1)$ times the mean, or we may carry the diagonal beyond the bottom of the square and traverse the top row of subsquares a second time, when the sum will be $4(m+1)$ times the mean. We can get in these cases a diagonal traverse $4m$ times the mean by

1	552	553	48	49	504	505	96	97	456	457	144	145	408	409	192	193	360	361	240	241	312	313	288
575	26	23	530	527	74	71	482	479	122	119	434	431	170	167	386	383	218	215	338	335	266	263	290
24	529	576	25	72	481	528	73	120	433	480	121	168	385	432	169	216	337	384	217	264	289	336	265
554	47	2	551	506	95	50	503	458	143	98	455	410	191	146	407	362	239	194	359	314	287	242	311
3	550	555	46	51	502	507	94	99	454	459	142	147	406	411	190	195	358	363	238	243	310	315	286
573	28	21	532	525	76	69	484	477	124	117	436	429	172	165	388	381	220	213	340	333	268	261	292
22	531	574	27	70	483	526	75	118	435	478	123	166	387	430	171	214	339	382	219	262	291	334	267
556	45	4	549	508	93	52	501	460	141	100	453	412	189	148	405	364	237	196	357	316	285	244	309
5	548	557	44	53	500	509	92	101	452	461	140	149	404	413	188	197	356	365	236	245	308	317	284
571	30	19	534	523	78	67	486	475	126	115	438	427	174	163	390	379	222	211	342	331	270	259	294
20	533	572	29	68	485	524	77	116	437	476	125	164	389	428	173	212	341	380	221	260	293	332	269
558	43	6	547	510	91	54	499	462	139	102	451	414	187	150	403	366	235	198	355	318	283	246	307
7	546	559	42	55	498	511	90	103	450	463	138	151	402	415	186	199	354	367	234	247	306	319	282
569	32	17	536	521	80	65	488	473	128	113	440	425	176	161	392	377	224	209	344	329	272	257	296
18	535	570	31	66	487	522	79	114	439	474	127	162	391	426	175	210	343	378	223	258	295	330	271
560	41	8	545	512	89	56	497	464	137	104	449	416	185	152	401	368	233	200	353	320	281	248	305
9	544	561	40	57	496	513	88	105	448	465	136	153	400	417	184	201	352	369	232	249	304	321	280
567	34	15	538	519	82	63	490	471	130	111	442	423	178	159	394	375	226	207	346	327	274	255	298
16	537	568	33	64	489	520	81	112	441	472	129	160	393	424	177	208	345	376	225	256	297	328	273
562	39	10	543	514	87	58	495	466	135	106	447	418	183	154	399	370	231	202	351	322	279	250	303
11	542	563	38	59	494	515	86	107	446	467	134	155	398	419	182	203	350	371	230	251	302	323	278
565	36	13	540	517	84	61	492	469	132	109	444	421	180	157	396	373	228	205	348	325	276	253	300
14	539	566	35	62	491	578	83	110	443	470	131	158	395	422	179	206	347	374	227	254	299	326	275
564	37	13	541	516	85	60	493	468	133	108	445	420	181	156	397	372	229	204	349	324	277	252	301

Fig. 701.

inserting at any point one vertical series of four cells between any two heavy bars and then continuing diagonally.

4. The great square is 4-ply, and therefore 4-symmetrical, i. e., we may choose any vertical and any horizontal bar (not necessarily heavy bars) and we shall find that any four cells, symmetrically situated with regard to these two bars as axes, will contain numbers whose sum is four times the mean. It follows that any $4m$ cells which form a symmetrical figure with regard to any such axes will contain numbers whose sum is the magic sum of the great square.

B. *The Subsquares.....*

5. are balanced Jaina squares, i. e., each of them has the 36 summations of a Jaina and in each case the magic sum is four times the mean number of the great square.

6. They have the property of subsidiary minors, i. e., if we

1.16 — 2.15				13.4 — 14.3				12.5 — 11.6				8.9 — 7.10			
1	2	16	15	13	14	4	3	12	11	5	6	8	7	9	10
16	15	1	2	4	3	13	14	5	6	12	11	9	10	8	7
1	2	16	15	13	14	4	3	12	11	5	6	8	7	9	10
16	15	1	2	4	3	13	14	5	6	12	11	9	10	8	7

Fig. 702.

erase any p rows of subsquares, and any p columns of the same and draw the remaining rows and columns together, we have a square with *all* the properties of the original great square.

EXAMPLES

In every case the Jaina pattern quoted above is used. Fig. 695 is an example of order 8 and the complementaries have been paired thus: 2,7 with 3,6; and 4,5 with 1,8. The La Hireian primaries of Fig. 695 are shown in Figs. 696 and 697.

* * *

Fig. 698 is an example of an order 12 square in which the pairing of the complementaries is 3, 10 with 4, 9; 1, 12 with 5, 8; and 6, 7 with 2, 11.

A square of order 16 is shown in Fig. 699. The couplets in this square are taken thus:

8 and 9 with 7 and 10; 1 and 16 with 5 and 12;
4 and 13 with 6 and 11; 2 and 15 with 3 and 14.

Figs. 700 and 701 show respectively squares of orders 20 and 24 in which the couplets are taken in numerical order, i. e., for order 20, 1 and 20 with 2 and 19; 3 and 18 with 4 and 17, etc.

In Fig. 701 there are 1008 magic diagonal summations. Since we

1	32	241	240	193	224	49	48	177	176	65	96	113	112	129	160
242	239	2	31	50	47	194	223	66	95	178	175	130	159	114	111
16	17	256	225	208	209	64	33	192	161	80	81	128	97	144	145
255	226	15	18	63	34	207	210	79	82	191	162	143	146	127	98
13	20	253	228	205	212	61	36	189	164	77	84	125	100	141	148
254	227	14	19	62	35	206	211	78	83	190	163	142	147	126	99
4	29	244	237	196	221	52	45	180	173	68	93	116	109	132	157
243	238	3	30	51	46	195	222	67	94	179	174	131	158	115	110
12	21	252	229	204	213	60	37	188	165	76	85	124	101	140	149
251	230	11	22	59	38	203	214	75	86	187	166	139	150	123	102
5	28	245	236	197	220	53	44	181	172	69	92	117	108	133	156
246	235	6	27	54	43	198	219	70	91	182	171	134	155	118	107
8	25	248	233	200	217	56	41	184	169	72	89	120	105	136	153
247	234	7	26	55	42	199	218	71	90	183	170	135	154	119	106
9	24	249	232	201	216	57	40	185	168	73	88	121	104	137	152
250	231	10	23	58	39	202	215	74	87	186	167	138	151	122	108

Fig. 703.

can bend at any heavy bar, the number of bent diagonals from top to bottom, starting at a given cell in the top row, is the same as the number of combinations of 6 things 3 at a time, viz., 20. Therefore there are $20 \times 24 = 480$ bent diagonals from top to bottom and 480 more from side to side. Adding the 48 continuous diagonals we get 1008.

In the foregoing pages the question of magic knight paths has not been considered. It is, however, easy for all orders > 8 and $\equiv 0$

(mod 8) to add the knight Nasik property *without sacrificing any of the other features,* by a proper choice of the complementary couplets for the subsquare outlines. The example shown in Fig. 702 will explain. It shows the top row of subsquares in a scheme for order 16. The numbers above the squares indicate the couplets used, the Jaina pattern, Fig. 694, being used throughout. The rule is simple: the leading numbers, 1, 13, 12, 8 must sum four times the mean of the period, i. e., 34, while of course no one of them may be a complement of any other. Their complementaries 16, 4, 5, 9, will then have the same sum, and the second members in each square will be similarly related. The square is completed by filling the remaining rows with replicas and turning over a central diagonal. Fig. 703 is a square of order 16 constructed from the outline Fig. 702. It has all the properties of the 16^2 shown in Fig. 699 and is also magic on its 64 knight paths.

The following is an arrangement of the couplets for a square of order 24:

1.24–4.21	8.17–5.20	10.15–13.12	11.14–16.9	22.3–18.7	23.2–19.6

C. P.

ORNATE MAGIC SQUARES OF COMPOSITE ODD ORDERS.

When we consider these orders in the light of the general rule used for orders ≡ 0 (mod 4) it appears at first sight that they cannot be made to fulfil all the conditions; but it is not essential to the *ply* property, nor to the balanced magic subsquares that the numbers be taken in complementary pairs for the subsquares of the outline. All that is necessary is that the groups of numbers chosen shall all have the same sum.

Suppose, as an illustration, we are dealing with order 15. If we can arrange the first 15 natural numbers in five balanced columns, three in a column, and form five magic outlines of order 3, using a different column thrice repeated for each outline, we shall have five balanced magic outlines like Fig. 704. These can be arranged in the first row of subsquares with replicas in the following rows. If we can turn this outline upon itself in some way to avoid repetitions, we shall have a magic square which will be 9-ply and with magic subsquares. But will it be pandiagonal?

2	7	15
7	15	2
15	2	7

Fig. 704.

2	6	12	11	9
15	13	8	3	1
7	5	4	10	14

Fig. 705.

In the small outlines of 9 cells made from Fig. 704 as a pattern, it will be noticed that like numbers must always occur in parallel diagonals; therefore if we arrange the five small squares so that like numbers always lie along / diagonals, the great outline will

2	7	15	6	5	13	12	4	8	11	10	3	9	14	1
7	15	2	5	13	6	4	8	12	10	3	11	14	1	9
15	2	7	13	6	5	8	12	4	3	11	10	1	9	14

Fig. 706.

be "boxed" and therefore magic in \ diagonals, but in the / diagonals we shall have in every case only five different numbers each occurring thrice. The problem is thus reduced to finding a

2	12	9	6	11	15	8	1	13	3	7	4	14	5	10
9	6	11	2	12	1	13	3	15	8	14	5	10	7	4
11	2	12	9	6	3	15	8	1	13	10	7	4	14	5
12	9	6	11	2	8	1	13	3	15	4	14	5	10	7
6	11	2	12	9	13	3	15	8	1	5	10	7	4	14

Fig. 707.

magic rectangle 3×5. We therefore construct such a rectangle by the method of "Complementary Differences"* as shown in Fig. 705.

In Fig. 706 we have the five magic outlines constructed from the five columns of the rectangle, and placed side by side with like

* See "The Construction of Magic Squares and Rectangles by the Method of Complementary Differences," by W. S. Andrews, pp. 257 ff.

numbers always in the / diagonals, and so disposed that *the number* in any / diagonal is always succeeded (when the diagonal passes across into a neighboring square) by the number which succeeds it in its row in the rectangle.

If an associated square is required the magic rectangle must be associated and the large rectangle of subsquares must also be associated as a whole. It will be noticed that all these conditions will be fulfilled in practice if we write the successive columns of the

155	28	171	125	88	156	20	178	126	80	163	21	170	133	81
44	211	114	14	181	39	224	106	9	194	31	219	119	1	189
139	98	57	199	68	147	94	53	207	64	143	102	49	203	72
157	30	167	127	90	152	22	180	122	82	165	17	172	135	77
40	213	116	10	183	41	220	108	11	190	33	221	115	3	191
140	103	51	200	73	141	95	58	201	65	148	96	50	208	66
164	16	174	134	76	159	29	166	129	89	151	24	179	121	84
34	218	117	4	188	42	214	113	12	184	38	222	109	8	192
142	105	47	202	75	137	97	60	197	67	150	92	52	210	62
160	18	176	130	78	161	25	168	131	85	153	26	175	123	86
35	223	111	5	193	36	215	118	6	185	43	216	110	13	186
149	91	54	209	61	144	104	46	204	74	136	99	59	196	69
154	23	177	124	83	162	19	173	132	79	158	27	169	128	87
37	225	107	7	195	32	217	120	2	187	45	212	112	15	182
145	93	56	205	63	146	100	48	206	70	138	101	55	198	71

Fig. 708. S = 1695

magic rectangle Fig. 705 along the \ central diagonals of the successive square outlines in the larger rectangle Fig. 706 and fill in all the / diagonals with replicas. If now all the remaining rows of subsquares be filled with replicas of the top row it will be found that the whole outline *cannot* be turned over either of its central diagonals without repetitions in the magic, but it *can* be turned successfully *in its own plane,* about its central point through one right angle, without repetitions. (This will bring the top row in coincidence with the left-hand column, so that the right-hand square in Fig. 706 is turned on its side and lies over the left-hand square.) The resulting magic is shown in Fig. 709. It is magic on its 15 rows, 15 columns, 30 diagonals and 60 knight paths, also 9-ply and associated. The 25 subsquares of order 3 all sum 339 on their 3 rows and 3 columns. (It is easy to see that only one of them can have magic central diagonals, for a magic of order 3 can only have this property when it is associated, and in this case the mean number must occupy the central cell, but there is here only one mean num-

2	127	210	6	125	208	12	124	203	11	130	198	9	134	196
202	15	122	200	13	126	199	8	132	205	3	131	209	1	129
135	197	7	133	201	5	128	207	4	123	206	10	121	204	14
32	157	150	36	155	148	42	154	143	41	160	138	39	164	136
142	45	152	140	43	156	139	38	162	145	33	161	149	31	159
165	137	37	163	141	35	158	147	34	153	146	40	151	144	44
107	172	60	111	170	58	117	169	53	116	175	48	114	179	46
52	120	167	50	118	171	49	113	177	55	108	176	59	106	174
180	47	112	178	51	110	173	57	109	168	56	115	166	54	119
182	82	75	186	80	73	192	79	68	191	85	63	189	89	61
67	195	77	65	193	81	64	188	87	70	183	86	74	181	84
90	62	187	88	66	185	83	72	184	78	71	190	76	69	194
212	22	105	216	20	103	222	19	98	221	25	93	219	29	91
97	225	17	95	223	21	94	218	27	100	213	26	104	211	24
30	92	217	28	96	215	23	102	214	18	101	220	16	99	224

Fig. 709. $S = 1695$

ber, viz., 113, therefore only the central subsquare can have magic diagonals.)

In exactly the same manner as above described, by using the long rows of the magic rectangle, Fig. 705, instead of the short columns, we can construct another ornate magic of order 15.

Fig. 707 shows the first row of 25-celled subsquares constructed from the *rows* of the rectangle, and using a magic square of order 5 as pattern. If we fill the two remaining rows of subsquares with replicas the outline can be turned over either of its central diagonals. The resulting square is shown in Fig. 710. It is magic on 15 rows,

15 columns, 30 diagonals and 60 knight paths, also 25-ply and associated. Also the nine subsquares of order 5 are balanced nasiks, summing 565 on their 5 rows, 5 columns and 10 diagonals.

The above method can of course be used when the order is the square of an odd number, e. g., orders 9, 25, etc. These have previously been dealt with by a simpler method which is not applicable when the order is the product of different odd numbers.

17	132	159	171	86	30	128	151	178	78	22	124	164	170	85
174	81	26	122	162	166	88	18	135	158	179	80	25	127	154
131	152	177	84	21	123	165	173	76	28	130	157	169	89	20
87	24	126	161	167	83	16	133	153	180	79	29	125	160	172
156	176	77	27	129	163	168	90	23	121	155	175	82	19	134
212	12	39	111	191	225	8	31	118	183	217	4	44	110	190
114	186	221	2	42	106	193	213	15	38	119	185	220	7	34
11	32	117	189	216	3	45	113	181	223	10	37	109	194	215
192	219	6	41	107	188	211	13	33	120	184	224	5	40	112
36	116	182	222	9	43	108	195	218	1	35	115	187	214	14
92	207	144	51	71	105	203	136	58	63	97	199	149	50	70
54	66	101	197	147	46	73	93	210	143	59	65	100	202	139
206	137	57	69	96	198	150	53	61	103	205	142	49	74	95
72	99	201	146	47	68	91	208	138	60	64	104	200	145	52
141	56	62	102	204	148	48	75	98	196	140	55	67	94	209

Fig. 710. S = 1695

A similar distinction arises in the case of orders $\equiv 0 \pmod 4$ previously considered. These were first constructed by a rule which applied only to orders of form 2^m, e. g., 4, 8, 16, 32, etc., but the general rule is effective in every case.

There are two other ornate squares of order 15, shown in Figs. 708 and 711, these four forms of ornate squares being numbered in ascending order of difficulty in construction. Fig. 708 is constructed by using the paths $\begin{vmatrix} 3, 5 \\ 5, 3 \end{vmatrix}$ and taking the period from the *continuous* diagonal of the magic rectangle Fig. 705.

Fig. 708 is magic on 15 rows, 15 columns, 30 diagonals, 60 knight paths, and is 9-ply, 25-ply and associated.

The square shown in Fig. 711 has been only recently obtained; for many years the conditions therein fulfilled were believed to be impossible. It is magic on 15 rows, 15 columns and 30 diagonals, and is 3×5 rectangular ply, i. e., *any* rectangle 3×5 with long axis horizontal contains numbers whose sum is the magic sum of the square. Also the 15 subrectangles are balanced magics, summing

37	93	191	81	163	32	99	185	89	160	45	102	188	79	151
167	219	5	59	115	180	222	8	49	106	172	213	11	51	118
135	27	143	199	61	127	18	146	201	73	122	24	140	209	70
97	183	86	156	43	92	189	80	164	40	105	192	83	154	31
212	9	50	119	175	225	12	53	109	166	217	3	56	111	178
30	147	203	64	121	22	138	206	66	133	17	144	200	74	130
187	78	161	36	103	182	84	155	44	100	195	87	158	34	91
2	54	110	179	220	15	57	113	169	211	7	48	116	171	223
150	207	68	124	16	142	198	71	126	28	137	204	65	134	25
82	153	41	96	193	77	159	35	104	190	90	162	38	94	181
47	114	170	224	10	60	117	173	214	1	52	108	176	216	13
210	72	128	19	136	202	63	131	21	148	197	69	125	29	145
157	33	101	186	88	152	39	95	194	85	165	42	98	184	76
107	174	215	14	55	120	177	218	4	46	112	168	221	6	58
75	132	23	139	196	67	123	26	141	208	62	129	20	149	205

Fig. 711. $S = 1695$

565 in their three long rows and 339 in their five short columns. This square is not associated, and only half of its knight paths are magic.

The three squares of order 15, shown in Figs. 708, 709, and 710 are described as magic on their 60 knight paths, but actually they are higher Nasiks of Class II, as defined at the end of my pamphlet on *The Theory of Path Nasiks.** Further, the squares in Figs. 709 and 710 have the following additional properties.

* *The Theory of Path Nasiks,* by C. Planck, M.A., M.R.C.S., printed by A. J. Lawrence, Rugby, Eng.

Referring to the square in Fig. 710 showing subsquares of order 5; if we superpose the diagonals of these subsquares in the manner described in my paper on "Fourfold Magics" (above, page 363, last paragraph), we obtain three magic parallelopipeds $5\times5\times3$. Denoting each subsquare by the number in its central cell, the three parallelopipeds will be:

I.	53,	169,	117.
II.	177,	113,	49.
III.	109, .	57,	173.

These three together form an octahedroid $5\times5\times3\times3$ which is associated and magic in each of the four directions parallel to its edges.

If we deal in like manner with Fig. 709 which has subsquares of order 3 we obtain five magic parallelopipeds of order $3\times3\times5$ together forming an associated magic octahedroid of order $3\times3\times5\times5$. Since the lengths of the edges are the same as those of the octahedroid formed from Fig. 710 square, these two four-dimensional figures are identical but the distribution of the numbers in their cells is not the same. They can however be made completely identical both in form and distribution of numbers by a slight change in our method of dealing with the square Fig. 709, i. e., by taking the square plates to form the parallelopipeds from the knight paths instead of the diagonals. Using the path $(-1, 2)$ we get 225, 106, 3, 188, 43 for the first plates of each parallelopiped, and then using $(2, -1)$ for the successive plates of each, we obtain the parallelopipeds:

I.	225,	8,	31,	118,	183
II.	106,	193,	213,	15,	38
III.	3,	45,	113,	181,	223
IV.	188,	211,	13,	33,	120
V.	43,	108,	195,	218,	1

This octahedroid is completely identical with that previously obtained from Fig. 710, as can be easily verified by taking any number at random and writing down the four series of numbers through its containing cell parallel to the edges, first in one octahedroid and then in the other. The sets so obtained will be found identical.

C. P.

THE CONSTRUCTION OF ORNATE MAGIC SQUARES OF ORDERS 8, 12 AND 16 BY TABLES.

The following simple method for constructing ornate magic squares of the above orders is presented in the belief that it is new and original. All squares of orders $4m$ can be made by this method, so it will suffice to explain in detail only the rules for constructing squares of order 8.

1	7	6	4
8	2	3	5

Fig. 712.

1	6	4	7
8	3	5	2

Fig. 713.

I. Make a magic rectangle with the first eight digits as shown in Fig. 712. This is the only *form* in which this rectangle can be

1	6	8	3
7	4	2	5

Fig. 714.

	1		6		8		3		
1	1	56	41	32	57	16	17	40	(8)
2	2	55	42	31	58	15	18	39	(7)
3	3	54	43	30	59	14	19	38	(6)
4	4	53	44	29	60	13	20	37	(5)
5	5	52	45	28	61	12	21	36	(4)
6	6	51	46	27	62	11	22	35	(3)
7	7	50	47	26	63	10	23	34	(2)
8	8	49	48	25	64	9	24	33	(1)
	7		4		2		5		

Fig. 715.

made, i. e., no complementary couplet therein can be inverted without destroying the magic feature, but the relative positions of the couplets can naturally be shifted without affecting it.

II. Draw a *table* diagram such as Fig. 714, and write the row numbers of the magic rectangle Fig. 712, alternately at the top and bottom of the eight columns as shown by dotted lines.

III. Following the arithmetical order of the numbered columns, write in the numbers 1 to 64 downward and upward, thus making the table, Fig. 715.

1	1	56	41	32	25	48	49	8	8
(2)	63	10	23	34	39	18	15	58	(7)
6	6	51	46	27	30	43	54	3	3
(5)	60	13	20	37	36	21	12	61	(4)
4	4	53	44	29	28	45	52	5	5
(3)	62	11	22	35	38	19	14	59	(6)
7	7	50	47	26	31	42	55	2	2
(8)	57	16	17	40	33	24	9	64	(1)

Fig. 716.

	1		6		8		3	
1	1	48	21	60	29	52	9	40
5	33	16	53	28	61	20	41	8
3	2	47	22	59	30	51	10	39
7	34	15	54	27	62	19	42	7
2	3	46	23	58	31	50	11	38
6	35	14	55	26	63	18	43	6
4	4	45	24	57	32	49	12	37
8	36	13	56	25	64	17	44	5
		4		7		5		2

Fig. 717.

1	4	6	7
8	5	3	2

Fig. 718.

1	6	7	4
8	3	2	5

Fig. 719.

1	1	48	21	60	25	56	13	36	8
(5)	32	49	12	37	8	41	20	61	(4)
6	35	14	55	26	59	22	47	2	3
(2)	62	19	42	7	38	11	50	31	(7)
7	34	15	54	27	58	23	46	3	2
(3)	63	18	43	6	39	10	51	30	(6)
4	4	45	24	57	28	53	16	33	5
(8)	29	52	9	40	5	44	17	64	(1)

Fig. 720.

NOTE. A variety of different tables may be made on the above principle by changing the progression, and each table will produce a different magic square. Any number that will divide n^2 (which in this case is 64) without remainder may be used as an increment. Thus in the present case 2, 4, 8, 16 and 32 are available. When the

1	1	106	74	143	75	36	3	108	110	107	73	34	12
(6)	129	54	56	17	55	124	127	52	20	53	57	126	(7)
11	31	76	104	113	105	6	33	78	140	77	103	4	2
(5)	132	51	59	14	58	121	130	49	23	50	60	123	(8)
3	7	100	80	137	81	30	9	102	116	101	79	28	10
(4)	135	48	62	11	61	118	133	46	26	47	63	120	(9)
9	25	82	98	119	99	12	27	84	134	83	97	10	4
(10)	117	66	44	29	43	136	115	64	8	65	45	138	(3)
8	22	85	95	122	96	15	24	87	131	86	94	13	5
(2)	141	42	68	5	67	112	139	40	32	41	69	114	(11)
7	19	88	92	125	93	18	21	90	128	89	91	16	6
(12)	111	72	38	35	37	142	109	70	2	71	39	144	(1)

Fig. 721.

	1		8		9		12		5		4		
1	1	106	74	143	75	36	111	72	38	35	37	142	(12)
2	4	103	77	140	78	33	114	69	41	32	40	139	(11)
3	7	100	80	137	81	30	117	66	44	29	43	136	(10)
4	10	97	83	134	84	27	120	63	47	26	46	133	(9)
5	13	94	86	131	87	24	123	60	50	23	49	130	(8)
6	16	91	89	128	90	21	126	57	53	20	52	127	(7)
7	19	88	92	125	93	18	129	54	56	17	55	124	(6)
8	22	85	95	122	96	15	132	51	59	14	58	121	(5)
9	25	82	98	119	99	12	135	48	62	11	61	118	(4)
10	28	79	101	116	102	9	138	45	65	8	64	115	(3)
11	31	76	104	113	105	6	141	42	68	5	67	112	(2)
12	34	73	107	110	108	3	144	39	71	2	70	109	(1)
		7		11		3		6		2		10	

Fig. 722.

1	7	8	11	9	3
12	6	5	2	4	10

Fig. 723.

1	11	3	9	8	7
12	2	10	4	5	6

Fig. 724.

addition produces a number larger than 64, the lowest *unused* number of the series is substituted. For example, if 32 is made the increment, the numbers in the columns of the table will run thus:

1, 33, 2, 34, 3, 35 etc.

because

$1+32=33,\qquad 33+32=65$ substitute 2
$2+32=34,\qquad 34+32=66$ " 3 etc.

IV. The table must now be indexed with some arrangement of the numbers 1 to 8 under the following conditions: The first

		1	2	3	4	5	6	7	8	9	10	11	12	13	14	15	16	
1	1	1	192	97	224	41	152	73	248	113	208	17	176	89	232	57	136	16
2	(9)	128	193	32	161	88	233	56	137	16	177	112	209	40	153	72	249	(8)
3	13	133	60	229	92	173	20	205	116	245	76	149	44	221	100	189	4	4
4	(5)	252	69	156	37	212	109	180	13	140	53	236	85	164	29	196	125	(12)
5	6	6	187	102	219	46	147	78	243	118	203	22	171	94	227	62	131	11
6	(14)	123	198	27	166	83	238	51	142	11	182	107	214	35	158	67	254	(3)
7	10	130	63	226	95	170	23	202	119	242	79	146	47	218	103	186	7	7
8	(2)	255	66	159	34	215	106	183	10	143	50	239	82	167	26	199	122	(15)
9	15	135	58	231	90	175	18	207	114	247	74	151	42	223	98	191	2	2
10	(7)	250	71	154	39	210	111	178	15	138	55	234	87	162	31	194	127	(10)
11	3	3	190	99	222	43	150	75	246	115	206	19	174	91	230	59	134	14
12	(11)	126	195	30	163	86	235	54	139	14	179	110	211	38	155	70	251	(6)
13	12	132	61	228	93	172	21	204	117	244	77	148	45	220	101	188	5	5
14	(4)	253	68	157	36	213	108	181	12	141	52	237	84	165	28	197	124	(13)
15	8	8	185	104	217	48	145	80	241	120	201	24	169	96	225	64	129	9
16	(16)	121	200	25	168	81	240	49	144	9	184	105	216	33	160	65	256	(1)

Fig. 725.

four digits used must include no complementary couplet, and the last four digits must be selected so as to balance each of the first four with its complementary. The straight arithmetical series is used in Fig. 715 as it fulfils the above conditions, but any series, such as shown in the subjoined examples, will produce magic results, and each arrangement will make a different magic square.

1	2	3	4	5	6
5	3	7	2	6	8
3	5	1	6	2	5
7	8	5	1	8	2
2	1	4	8	1	7
6	4	8	3	7	4
4	6	2	7	3	1
8	7	6	5	4	3 etc.

	1		13		6		10		16		4		11		7		
1	1	192	97	224	41	152	73	248	121	200	25	168	81	240	49	144	16
2	2	191	98	223	42	151	74	247	122	199	26	167	82	239	50	143	15
3	3	190	99	222	43	150	75	246	123	198	27	166	83	238	51	142	14
4	4	189	100	221	44	149	76	245	124	197	28	165	84	237	52	141	13
5	5	188	101	220	45	148	77	244	125	196	29	164	85	236	53	140	12
6	6	187	102	219	46	147	78	243	126	195	30	163	86	235	54	139	11
7	7	186	103	218	47	146	79	242	127	194	31	162	87	234	55	138	10
8	8	185	104	217	48	145	80	241	128	193	32	161	88	233	56	137	9
9	129	64	225	96	169	24	201	120	249	72	153	40	209	112	177	16	8
10	130	63	226	95	170	23	202	119	250	71	154	39	210	111	178	15	7
11	131	62	227	94	171	22	203	118	251	70	155	38	211	110	179	14	6
12	132	61	228	93	172	21	204	117	252	69	156	37	212	109	180	13	5
13	133	60	229	92	173	20	205	116	253	68	157	36	213	108	181	12	4
14	134	59	230	91	174	19	206	115	254	67	158	35	214	107	182	11	3
15	135	58	231	90	175	18	207	114	255	66	159	34	215	106	183	10	2
16	136	57	232	89	176	17	208	113	256	65	160	33	216	105	184	9	1
		8		12		3		15		9		5		14		2	

Fig. 726.

The index numbers are written in columns on each side of the table, those on one side being in reverse order to those on the other side. One set of these numbers may be conveniently written in circles for identification, or any other way of distinguishing the similar numbers may be used.

V. Make another 2×4 magic rectangle with a re-arrangement

1	13	6	10	15	3	12	8
16	4	11	7	2	14	5	9

Fig. 727.

1	8	13	12	6	3	10	15
16	9	4	5	11	14	7	2

Fig. 728.

1	13	5	9	7	11	3	15
16	4	12	8	10	6	14	2

Fig. 729.

	1	16	13	4	5	12	9	8	7	10	11	6	3	14	15	2
1	1	136	89	224	41	176	113	248	73	208	17	152	97	232	57	192
16	121	256	33	168	81	216	9	144	49	184	105	240	25	160	65	200
13	132	5	220	93	172	45	244	117	204	77	148	21	228	101	188	61
4	252	125	164	37	212	85	140	13	180	53	236	109	156	29	196	69
5	6	131	94	219	46	171	118	243	78	203	22	147	102	227	62	187
12	126	251	38	163	86	211	14	139	54	179	110	235	30	155	70	195
9	135	2	223	90	175	42	247	114	207	74	151	18	231	98	191	58
8	255	122	167	34	215	82	143	10	183	50	239	106	159	26	199	66
7	130	7	218	95	170	47	242	119	202	79	146	23	226	103	186	63
10	250	127	162	39	210	87	138	15	178	55	234	111	154	31	194	71
11	3	134	91	222	43	174	115	246	75	206	19	150	99	230	59	190
6	123	254	35	166	83	214	11	142	51	182	107	238	27	158	67	198
3	133	4	221	92	173	44	245	116	205	76	149	20	229	100	189	60
14	253	124	165	36	213	84	141	12	181	52	237	108	157	28	197	68
15	8	129	96	217	48	169	120	241	80	201	24	145	104	225	64	185
2	128	249	40	161	88	209	16	137	56	177	112	233	32	153	72	193

Fig. 730.

of couplets, such as shown in Fig. 713. Any other arrangement that differs from Fig. 712 would, however, answer equally well.

VI. Draw an 8×8 lattice (Fig. 716) and write opposite the *alternate cells* of the two outside columns the eight numbers in Fig. 713 in their linear order, from the top of the lattice downward, and the same numbers (*in circles*) opposite the remaining alternate cells from the bottom of the lattice upward.

Inspection of Figs. 715 and 716 will assist a clear understanding of the above directions.

The magic square is now made by filling the cells of the lattice with the numbers from the table in linear groups of four, according

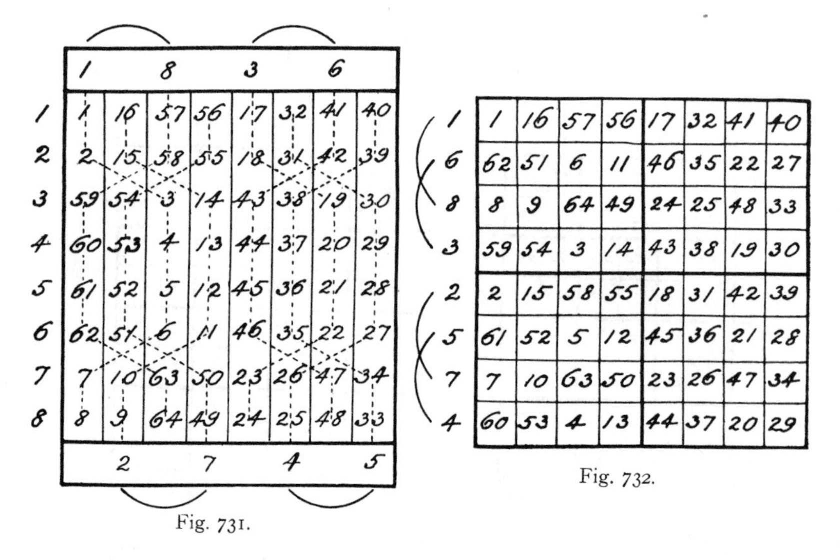

Fig. 731.

Fig. 732.

to their *index numbers*. The linear groups of four numbers in the left-hand half of square are written from left to right and those in the right-hand half of square from right to left.

Another example of an order 8 magic square, including rectangles and table, is shown in Figs. 717, 718, 719 and 720. The progressive increment in the table, Fig. 717, is 32, as referred to in a previous paragraph, and the index numbers are written in the order shown in the first column of numbers on page 392.

The magic squares, Figs. 716 and 720, are 4-ply, associated and pandiagonal.

In using the above rules there are at least three different ways for producing variations.

a. By changing the progression in the table.

b. By making divisions in the table (as in Fig. 726).

c. By using different arrangements of couplets in rectangles.

d. By using different arrangements of index numbers.

It is therefore evident that the possible number of variants is very large, and each of them will possess the same ornate qualities as those above described.

A magic square of order 12 is given in Fig. 721, and the table used in its construction with two 2×6 magic rectangles in Figs. 722, 723 and 724. This square is 4-ply, associated and pandiagonal.

1 3 8 6

1	1	40	17	56	57	32	41	16
2	2	39	18	55	58	31	42	15
3	59	30	43	14	3	38	19	54
4	60	29	44	13	4	37	20	53
5	61	28	45	12	5	36	21	52
6	62	27	46	11	6	35	22	51
7	7	34	23	50	63	26	47	10
8	8	33	24	49	64	25	48	9

5 7 4 2

Fig. 733.

1	1	40	17	56	57	32	41	16
4	60	29	44	13	4	37	20	53
2	2	39	18	55	58	31	42	15
3	59	30	43	14	3	38	19	54
8	8	33	24	49	64	25	48	9
5	61	28	45	12	5	36	21	52
7	7	34	23	50	63	26	47	10
6	62	27	46	11	6	35	22	51

Fig. 734.

A magic square of order 16 with its table and rectangles are shown in Figs. 725, 726, 727 and 728. In addition to the ornate features common to the squares shown in Figs. 716, 720 and 721, this square is also knight Nasik. Fig. 725 can readily be changed into a balanced, quartered, 4-ply, pandiagonal Franklin magic square by one transposition, as shown in Fig. 730, which is indexed by the rectangle Fig. 729. By this change it ceases to be associated and knight Nasik, but acquires other ornate features besides becoming a Franklin square. It contains nine magic subsquares of order 8, each of which is pandiagonal; also, the numbers in the corner

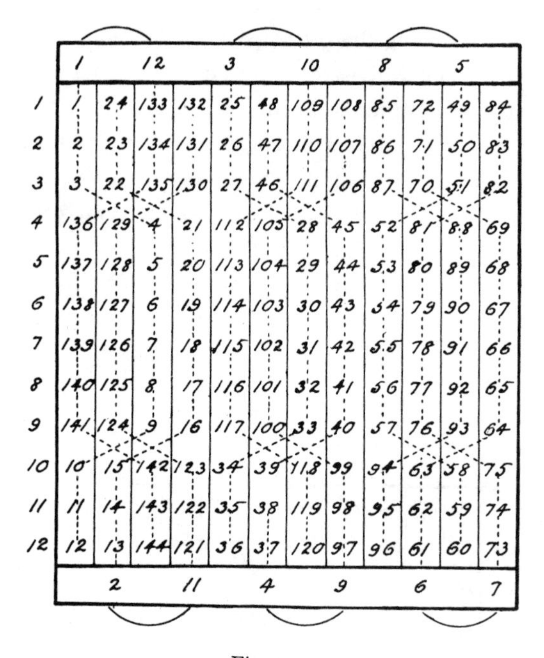

	1		12		3		10		8		5	
1	1	24	133	132	25	48	109	108	85	72	49	84
2	2	23	134	131	26	47	110	107	86	71	50	83
3	3	22	135	130	27	46	111	106	87	70	51	82
4	136	129	4	21	112	105	28	45	52	81	88	69
5	137	128	5	20	113	104	29	44	53	80	89	68
6	138	127	6	19	114	103	30	43	54	79	90	67
7	139	126	7	18	115	102	31	42	55	78	91	66
8	140	125	8	17	116	101	32	41	56	77	92	65
9	141	124	9	16	117	100	33	40	57	76	93	64
10	10	15	142	123	34	39	118	99	94	63	58	75
11	11	14	143	122	35	38	119	98	95	62	59	74
12	12	13	144	121	36	37	120	97	96	61	60	73
		2		11		4		9		6		7

Fig. 735.

1	1	24	133	132	25	48	109	108	85	72	49	84
6	138	127	6	19	114	103	30	43	54	79	90	67
12	12	13	144	121	36	37	120	97	96	61	60	73
7	139	126	7	18	115	102	31	42	55	78	91	66
2	2	23	134	131	26	47	110	107	86	71	50	83
5	137	128	5	20	113	104	29	44	53	80	89	68
11	11	14	143	122	35	38	119	98	95	62	59	74
8	140	125	8	17	116	101	32	41	56	77	92	65
3	3	22	135	130	27	46	111	106	87	70	57	82
4	136	129	4	21	112	105	28	45	52	81	88	69
10	10	15	142	123	34	39	118	99	94	63	58	75
9	141	124	9	16	117	100	33	40	57	76	93	64

Fig. 736.

cells of any 4×4, 8×8, 12×12 square and the corner cells of the great square sum $S/4=514$, as do also the corner numbers in any 2×4, 2×6, 2×8 rectangle etc.

The "table" method for constructing ornate magics is not limited to the foregoing rules. For a long time the writer endeavored in vain to make tables that would be competent to produce Franklin

	1		16		3		14		7		10		5		12	
1	1	32	241	240	33	64	209	208	97	128	145	144	65	96	177	176
2	2	31	242	239	34	63	210	207	98	127	146	143	66	95	178	175
3	3	30	243	238	35	62	211	206	99	126	147	142	67	94	179	174
4	4	29	244	237	36	61	212	205	100	125	148	141	68	93	180	173
5	245	236	5	28	213	204	37	60	149	140	101	124	181	172	69	92
6	246	235	6	27	214	203	38	59	150	139	102	123	182	171	70	91
7	247	234	7	26	215	202	39	58	151	138	103	122	183	170	71	90
8	248	233	8	25	216	201	40	57	152	137	104	121	184	169	72	89
9	249	232	9	24	217	200	41	56	153	136	105	120	185	168	73	88
10	250	231	10	23	218	199	42	55	154	135	106	119	186	167	74	87
11	251	230	11	22	219	198	43	54	155	134	107	118	187	166	75	86
12	252	229	12	21	220	197	44	53	156	133	108	117	188	165	76	85
13	13	20	253	228	45	52	221	196	109	116	157	132	77	84	189	164
14	14	19	254	227	46	51	222	195	110	115	158	131	78	83	190	163
15	15	18	255	226	47	50	223	194	111	114	159	130	79	82	191	162
16	16	17	256	225	48	49	224	193	112	113	160	129	80	81	192	161
		2		15		4		13		8		9		6		11

Fig. 737.

squares directly without any transpositions, until it occurred to him that this might be accomplished by *bending the columns of the table*. This simple device worked out with perfect success, thus adding another link to the scheme for making *all kinds* of the $4m$ squares by this method. The bending of the table columns also leads to the construction of a number of other ornate variants, as will be shown in examples to follow.

Fig. 731 is a table constructed with the straight series 1 to 64. the bending of the columns being shown by the dotted lines. As in tables previously explained, each column of numbers is started and finished following the arithmetical sequence of the numerals at the top and bottom of the table, but the four middle numbers of each column are bent three spaces out of line either to right or left. It will be seen that the column numerals are written in couplets

1	1	32	241	240	33	64	209	208	97	128	145	144	65	96	177	176
5	245	236	5	28	213	204	37	60	149	140	101	124	181	172	69	92
16	16	17	256	225	48	49	224	193	112	113	160	129	80	81	192	161
12	252	229	12	21	220	197	44	53	156	133	108	117	188	165	76	85
2	2	31	242	239	34	63	210	207	98	127	146	143	66	95	178	175
6	246	235	6	27	214	203	38	59	150	139	102	123	182	171	70	91
15	15	18	255	226	47	50	223	194	111	114	159	130	79	82	191	162
11	251	230	11	22	219	198	43	54	155	134	107	118	187	166	75	86
3	3	30	243	238	35	62	211	206	99	126	147	142	67	94	179	174
7	247	234	7	26	215	202	39	58	151	138	103	122	183	170	71	90
14	14	19	254	227	46	51	222	195	110	115	158	131	78	83	190	163
10	250	231	10	23	218	199	42	55	154	135	106	119	186	167	74	87
4	4	29	244	237	36	61	212	205	100	125	148	141	68	93	180	173
8	248	233	8	25	216	201	40	57	152	137	104	121	184	169	72	89
13	13	20	253	228	45	52	221	196	109	116	157	132	77	84	189	164
9	249	232	9	24	217	200	41	56	153	136	105	120	185	168	73	88

Fig. 738.

$= n + 1$, as marked by brackets. The relative positions of these couplets may, however, be varied.

The horizontal lines of the table are indexed with the first eight digits in straight series, but either of the series shown on page 3, or an equivalent, may be used.

This form of table differs essentially in one feature from those previously described, there being no vertical central division, and each complete line of eight numbers is copied into the magic square as written in the table. A table made in this way with bent columns is in fact a square that is magic in its lines and columns but not in its diagonals. The re-arrangement of its lines by the index numbers corrects its diagonals and imparts its ornate features.

An 8×8 lattice is now drawn and indexed on one side with the

	1		3		16		14		2		4		15		13	
1	1	248	5	244	31	234	27	238	3	246	7	242	29	236	25	240
2	63	202	59	206	33	216	37	212	61	204	57	208	35	214	39	210
3	95	170	91	174	65	184	69	180	93	172	89	176	67	182	71	178
4	97	152	101	148	127	138	123	142	99	150	103	146	125	140	121	144
5	129	120	133	116	159	106	155	110	131	118	135	114	157	108	153	112
6	191	74	187	78	161	88	165	84	189	76	185	80	163	86	167	82
7	223	42	219	46	193	56	197	52	221	44	217	48	195	54	199	50
8	225	24	229	20	255	10	251	14	227	22	231	18	253	12	249	16
9	2	247	6	243	32	233	28	237	4	245	8	241	30	235	26	239
10	64	201	60	205	34	215	38	211	62	203	58	207	36	213	40	209
11	96	169	92	173	66	183	70	179	94	171	90	175	68	181	72	177
12	98	151	102	147	128	137	124	141	100	149	104	145	126	139	122	143
13	130	119	134	115	160	105	156	109	132	117	136	113	158	107	154	111
14	192	73	188	77	162	87	166	83	190	75	186	79	164	85	168	81
15	224	41	220	45	194	55	198	51	222	43	218	47	196	53	200	49
16	226	23	230	19	256	9	252	13	228	21	232	17	254	11	250	15
		12		10		5		7		11		9		6		8

Fig. 739.

first eight digits, so selected that *alternate* numbers form couplets $= n+1$ in each subdivision of the square.

Finally, the lines from the table (Fig. 731) are transferred to the lattice in accordance with the index numbers, and the square thus made (Fig. 732) is 4-ply, pandiagonal, and Franklin; also each corner subsquare of order 4 is a magic pandiagonal.

NOTE. In some cases the numbers of the indexing couplets are more widely separated, as in Fig. 734; while in other cases they may be written adjoining each other. In all cases, however, a symmetrical arrangement of couplets is observed, but their positions, as shown in these examples, is an essential feature only in connection with the particular squares illustrated.

Fig. 733 shows another table in which the columns are bent

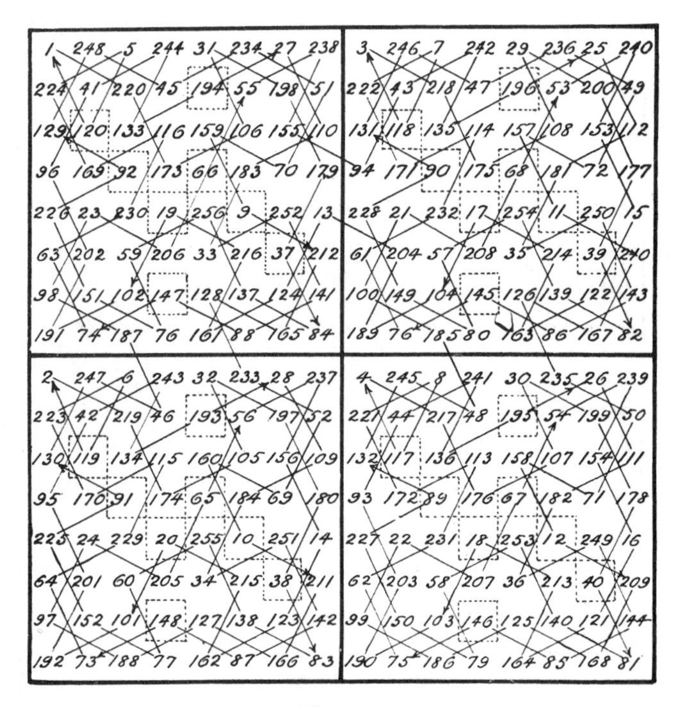

Fig. 740.

through a space of four columns, which produces the magic square, Fig. 734. This square is 4-ply, pandiagonal and knight Nasik.

Fig. 735 is a table with bent columns from which the square of order 12, shown in Fig. 736, is constructed. This square is 4-ply and pandiagonal, and it contains nine pandiagonal subsquares of order 4, as shown by the heavy bars in the lattice.

A table and square of order 16 are shown in Figs. 737 and 738. The square is 4-ply, pandiagonal and Franklin, and it also possesses many other interesting features. It is composed of 16 subsquares of order 4, as shown by the heavy bars, and each subsquare is magic and pandiagonal.

Fig. 739 is a table from which our final example of magic square, shown in Figs. 740 and 741, is constructed. The table series is made with increments of 32 and the columns are bent as marked by the dotted lines. The square is 4-ply (and therefore 4 symmetrical) quartered, pandiagonal, knight Nasik, Franklin and magic in its reflected diagonals. Also, any 9×9 square has its

1	248	5	244	31	234	27	238	3	246	7	242	29	236	25	240
224	41	220	45	194	55	198	51	222	43	218	47	196	53	200	49
129	120	133	116	159	106	155	110	131	118	135	114	157	108	153	112
96	169	92	173	66	183	70	179	94	171	90	175	68	181	72	177
226	23	230	19	256	9	252	13	228	21	232	17	254	11	250	15
63	202	59	206	33	216	37	212	61	204	57	208	35	214	39	210
98	151	102	147	128	137	124	141	100	149	104	145	126	139	122	143
191	74	187	76	161	88	165	84	189	76	185	80	163	86	167	82
2	247	6	243	32	233	28	237	4	245	8	241	30	235	26	239
223	42	219	46	193	56	197	52	221	44	217	48	195	54	199	50
130	119	134	115	160	105	156	109	132	117	136	113	158	107	154	111
95	170	91	174	65	184	69	180	93	172	89	176	67	182	71	178
225	24	229	20	255	10	251	14	227	22	231	18	253	12	249	16
64	201	60	205	34	215	38	211	62	203	58	207	36	213	40	209
97	152	101	148	127	138	123	142	99	150	103	146	125	140	121	144
192	73	188	77	162	87	166	83	190	75	186	79	164	85	168	81

Fig. 741.

corner numbers in arithmetical sequence. Fig. 740 shows it laid out in one continuous re-entrant knight's tour. The first number of each of the 32 periods of 8 numbers is enclosed in a dotted cell and an arrowhead points the direction of progression. The numbers in each of these periods sum $S/2 = 1028$, also, the numbers in each half period sum $S/4 = 514$. Although this feature exists in many other squares, it may not be commonly known. Fig. 741 is the same square written in the usual way to facilitate the checking up of its several ornate qualities.

F. A. W.

THE CONSTRUCTION OF ORNATE MAGIC SQUARES OF ORDER 16 BY MAGIC RECTANGLES.

In the preceding paper Mr. Woodruff presents a remarkable magic of order 16 which is 4-ply, pandiagonal, associated and knight Nasik, a combination of ornate properties which has probably never been accomplished before in this order of square, and it is constructed moreover by a unique method of his own devising. (See Fig. 725.)

An analysis of Mr. Woodruff's magic by the La Hireian plan shows its primary to be composed of sundry 2×8 rectangles having no particular numerical arrangement that indicates intentional de-

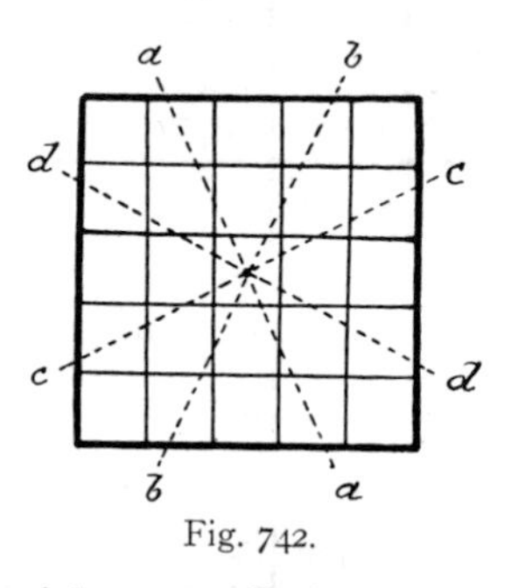

Fig. 742.

sign. This feature might naturally be expected in a square made by a new method, but it suggested to the writer that squares similar to Mr. Woodruff's in their ornate qualifications might be formed by applying the well-known method of magic rectangles on the La Hireian principle, as described in the present paper.

In using 2×8 magic rectangles for making ornate squares of order 16 by the La Hireian method, it is found that certain rectangles will produce knight Nasik squares while others will not. By inspection of the arrangement of the numbers in any 2×8 magic rectangle, guided by a simple rule, it may easily be determined if a knight Nasik square will result from its use, and if not, how the numbers may be re-arranged to produce Nasik results.

There are four knight paths through each cell of a square, as shown by dotted lines in Fig. 742, and the numbers included in each of these paths must obviously sum the magic constant of the square to be constructed if the latter is to be knight Nasik.

The La Hireian primary of order 16, shown in Fig. 743, is made up of sixteen 2×8 magic rectangles, as indicated by the heavy

1	8	15	10	3	6	13	12	12	13	6	3	10	15	8	1
16	9	2	7	14	11	4	5	5	4	11	14	7	2	9	16
1	8	15	10	3	6	13	12	12	13	6	3	10	15	8	1
16	9	2	7	14	11	4	5	5	4	11	14	7	2	9	16
1	8	15	10	3	6	13	12	12	13	6	3	10	15	8	1
16	9	2	7	14	11	4	5	5	4	11	14	7	2	9	16
1	8	15	10	3	6	13	12	12	13	6	3	10	15	8	1
16	9	2	7	14	11	4	5	5	4	11	14	7	2	9	16
1	8	15	10	3	6	13	12	12	13	6	3	10	15	8	1
16	9	2	7	14	11	4	5	5	4	11	14	7	2	9	16
1	8	15	10	3	6	13	12	12	13	6	3	10	15	8	1
16	9	2	7	14	11	4	5	5	4	11	14	7	2	9	16
1	8	15	10	3	6	13	12	12	13	6	3	10	15	8	1
16	9	2	7	14	11	4	5	5	4	11	14	7	2	9	16
1	8	15	10	3	6	13	12	12	13	6	3	10	15	8	1
16	9	2	7	14	11	4	5	5	4	11	14	7	2	9	16

Fig. 743.

bars. Starting from any cell in Fig. 743, the sum of the numbers included in the complete knight paths, indicated by *aa* and *bb* in Fig. 742, will sum $136 = S$, but the paths *cc* and *dd* will sum either 104 or 168, and therefore this primary is incompetent to produce a knight Nasik magic square.

The knight paths *aa* and *bb* are necessarily Nasik, as they include the numbers in one or other of the long rows of numbers in the magic rectangles which sum 68. The other two knight paths, *cc* and *dd*, fail to be Nasik because they include the numbers enclosed in circles in Fig. 743, or their complementaries, and these numbers do not sum 68. It therefore follows that in order to produce a knight Nasik primary, the magic rectangle from which it is formed must show a summation of 68 for the numbers enclosed in circles in Fig. 743 and their complementaries. A re-arrangement of the couplets in the 2×8 magic rectangle, without inverting any couplet, is shown in the La Hireian primary square, Fig. 744. By

1	15	3	13	12	6	10	8	8	10	6	12	13	3	15	1
16	2	14	4	5	11	7	9	9	7	11	5	4	14	2	16
1	15	3	13	12	6	10	8	8	10	6	12	13	3	15	1
16	2	14	4	5	11	7	9	9	7	11	5	4	14	2	16
1	15	3	13	12	6	10	8	8	10	6	12	13	3	15	1
16	2	14	4	5	11	7	9	9	7	11	5	4	14	2	16
1	15	3	13	12	6	10	8	8	10	6	12	13	3	15	1
16	2	14	4	5	11	7	9	9	7	11	5	4	14	2	16
1	15	3	13	12	6	10	8	8	10	6	12	13	3	15	1
16	2	14	4	5	11	7	9	9	7	11	5	4	14	2	16
1	15	3	13	12	6	10	8	8	10	6	12	13	3	15	1
16	2	14	4	5	11	7	9	9	7	11	5	4	14	2	16
1	15	3	13	12	6	10	8	8	10	6	12	13	3	15	1
16	2	14	4	5	11	7	9	9	7	11	5	4	14	2	16
1	15	3	13	12	6	10	8	8	10	6	12	13	3	15	1
16	2	14	4	5	11	7	9	9	7	11	5	4	14	2	16

Fig. 744.

this re-arrangement, the numbers in circles are made to sum 68, and the rectangle is therefore competent to produce a knight Nasik square. A second La Hireian primary (Fig. 745) is made by changing the numbers in Fig. 744 to their root numbers and then turning this primary around its central point 90° to the right, thus changing the horizontal lines in Fig. 744 into the vertical columns in Fig. 745. The final magic square, Fig. 746, is constructed in the usual way by adding together the numbers in these two primaries, cell by cell. Like its two primaries, this square is 4-ply, associated, pandiagonal and knight Nasik.

If the magic square shown in Fig. 746 is divided into 2×8 rectangles in the same way as Fig. 744 or Fig. 745, these rectangles will show the same features in summations as the rectangles of the primary squares, i. e., each summation will be S/2.

Using the natural series 1 to 16 inclusive, it is only possible to construct four distinct forms of 2×8 magic rectangles, as shown in Figs. 747 and 748. The four columns of numbers in Fig. 747

240	0	240	0	240	0	240	0	240	0	240	0	240	0	240	0
16	224	16	224	16	224	16	224	16	224	16	224	16	224	16	224
208	32	208	32	208	32	208	32	208	32	208	32	208	32	208	32
48	192	48	192	48	192	48	192	48	192	48	192	48	192	48	192
64	176	64	176	64	176	64	176	64	176	64	176	64	176	64	176
160	80	160	80	160	80	160	80	160	80	160	80	160	80	160	80
96	144	96	144	96	144	96	144	96	144	96	144	96	144	96	144
128	112	128	112	128	112	128	112	128	112	128	112	128	112	128	112
128	112	128	112	128	112	128	112	128	112	128	112	128	112	128	112
96	144	96	144	96	144	96	144	96	144	96	144	96	144	96	144
160	80	160	80	160	80	160	80	160	80	160	80	160	80	160	80
64	176	64	176	64	176	64	176	64	176	64	176	64	176	64	176
48	192	48	192	48	192	48	192	48	192	48	192	48	192	48	192
208	32	208	32	208	32	208	32	208	32	208	32	208	32	208	32
16	224	16	224	16	224	16	224	16	224	16	224	16	224	16	224
240	0	240	0	240	0	240	0	240	0	240	0	240	0	240	0

Fig. 745.

show the selection of numbers in the upper and lower rows of the four forms of 2×8 rectangles, the numbers in circles being those used in the upper rows of the respective rectangles.

The designs below the rectangles in Fig. 748, Forms I, II, III and IV, show the geometric arrangement of the numbers as written in the upper and lower lines of same. In the upper row of Form III rectangle there is a departure from the column sequence of numbers in order to make it suitable for constructing Nasik magic squares, and it is rather curious that this change is required only in this one rectangle out of the four. The relative positions of the couplets in each form of 2×8 rectangle may naturally be re-arranged in a great many different ways without disturbing their general magic qualities, although in some cases such re-arrangement will upset the magic summation of the numbers in a *zig-zag* line of cells, which, as previously noted, is of vital importance when the square is to be knight Nasik.

241	15	243	13	252	6	250	8	248	10	246	12	253	3	255	1
32	226	30	228	21	235	23	233	25	231	27	229	20	238	18	240
209	47	211	45	220	38	218	40	216	42	214	44	221	35	223	33
64	194	62	196	53	203	55	201	57	199	59	197	52	206	50	208
65	191	67	189	76	182	74	184	72	186	70	188	77	179	79	177
176	82	174	84	165	91	167	89	169	87	171	85	164	94	162	96
97	159	99	157	108	150	106	152	104	154	102	156	109	147	111	145
144	114	142	116	133	123	135	121	137	119	139	117	132	126	130	128
129	127	131	125	140	118	138	120	136	122	134	124	141	115	143	113
112	146	110	148	101	155	103	153	105	151	107	149	100	153	98	160
161	95	163	93	172	86	170	88	168	90	166	92	173	83	175	81
80	178	78	180	69	187	71	185	73	183	75	181	68	190	66	192
49	207	51	205	60	198	58	200	56	202	54	204	61	195	63	193
224	34	222	36	213	43	215	41	217	39	219	37	212	46	210	48
17	239	19	237	28	230	26	232	24	234	22	236	29	227	31	225
256	2	254	4	245	11	247	9	249	7	251	5	244	14	242	16

Fig. 746.

Inspection of these examples will show that the couplet 1—16 is common to all four forms, but in every other case there is a difference. Thus the couplet 2—15 is only found in Form I, and it is inverted in the other three forms. The couplet 3—14 exists only in Form II, being elsewhere inverted. The couplet 4—13 is seen in Forms III and IV, and is inverted in Forms I and II—and so forth.

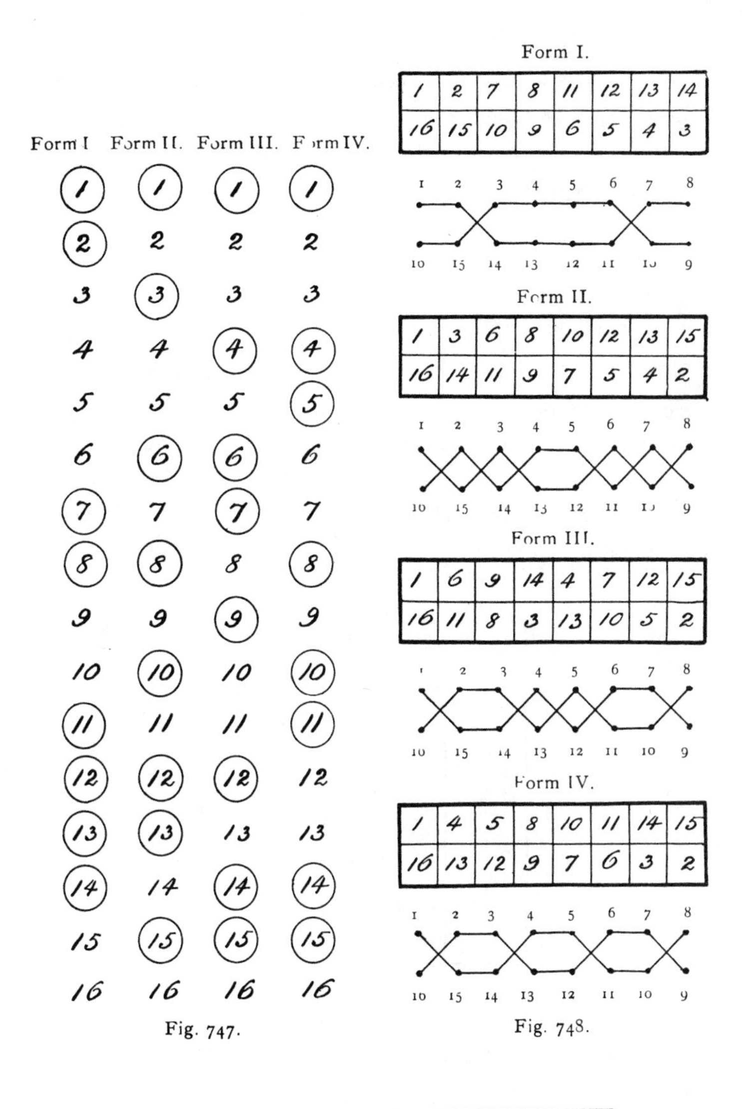

Fig. 747.

Fig. 748.

1	2	3	21	20	19	18	10	11	17	16	12
24	23	22	4	5	6	7	15	14	8	9	13

Fig. 749.

The above described method will produce knight Nasik squares of all orders ≡ 0 (mod 8) excepting order 8, but it will not apply in this respect to orders ≡ 4 (mod 8).

Fig. 749 shows a 2×12 magic rectangle that may be used for a magic square of order 24 covering the knight Nasik property.

W. S. A.

PANDIAGONAL-CONCENTRIC MAGIC SQUARES OF ORDERS 4*m*.

These squares are composed of a central pandiagonal square surrounded by one or more bands of numbers, each band, together with its enclosed numbers, forming a pandiagonal magic square.

The squares described here are of orders 4*m* and the bands or borders are composed of double strings of numbers. The central square and bands are constructed simultaneously instead of by the

45	28	35	22	47	26	33	24
49	8	63	10	51	6	61	12
31	42	17	40	29	44	19	38
3	54	13	60	1	56	15	58
46	27	36	21	48	25	34	23
50	7	64	9	52	5	62	11
32	41	18	39	30	43	20	37
4	53	14	59	2	55	16	57

Fig. 750.

usual method of first forming the nucleus square and arranging the bands successively around it.

A square of the 8th order is shown in Fig. 750, both the central 4^2 and 8^2 being pandiagonal. It is 4^2 ply, i. e., any square group of 16 numbers gives a constant total of $8(n^2+1)$, where n = the number of cells on the edge of the magic. It is also magic in all of its Franklin diagonals; i. e., each diagonal string of numbers bending at right angles on either of the horizontal or vertical center lines of the square, as is shown by dotted lines, gives constant totals. In any size concentric square of the type here described, all of its concentric squares of orders $8m$ will be found to possess the Franklin bent diagonals.

The analysis of these pandiagonal-concentric squares is best illustrated by their La Hireian method of construction, which is

1	9	7	3	4	6	4	6	7	3	1	9
1	9	7	3	4	6	4	6	7	3	1	9
7	3	1	9	4	6	4	6	1	9	7	3
7	3	1	9	4	6	4	6	1	9	7	3
2	8	8	2	5	5	5	5	8	2	2	8
2	8	8	2	5	5	5	5	8	2	2	8
8	2	2	8	5	5	5	5	2	8	8	2
8	2	2	8	5	5	5	5	2	8	8	2
3	7	9	1	6	4	6	4	9	1	3	7
3	7	9	1	6	4	6	4	9	1	3	7
9	1	3	7	6	4	6	4	3	7	9	1
9	1	3	7	6	4	6	4	3	7	9	1

Fig. 751.

here explained in connection with the 12th order square. The square lattice of the subsidiary square, Fig. 751, is, for convenience of construction, divided into square sections of 16 cells each. In each of the corner sections (regardless of the size of the square to be formed) are placed four 1's, their position to be as shown in Fig. 751. Each of these 1's is the initial number of the series 1, 2, 3,. . . . . $(n/4)^2$, which must be written in the lattice in natural order, each number falling in the same respective cell of a 16-cell section as the initial number. Two of these series are indicated in Fig. 751 by circles enclosing the numbers, and inspection will show that each of the remaining series of numbers is written in the lattice in the same manner, though they are in a reversed or reflected order. Any size subsidiary square thus filled possesses all the magic features of the final square.

99	54	72	45
108	9	135	18
63	90	36	81
0	117	27	126

Fig. 752.

A second subsidiary square of the 4th order is constructed with the series 0, $(n/4)^2$, $2(n/4)^2$, $3(n/4)^2$,......$15(n/4)^2$, which must be so arranged as to produce a pandiagonal magic such as is shown

100	63	79	48	103	60	76	51	106	57	73	54
109	18	142	21	112	15	139	24	115	12	136	27
70	93	37	90	67	96	40	87	64	99	43	84
7	120	28	135	4	123	31	132	1	126	34	129
101	62	80	47	104	59	77	50	107	56	74	53
110	17	143	20	113	14	140	23	116	11	137	26
71	92	38	89	68	95	41	86	65	98	44	83
8	119	29	134	5	122	32	131	2	125	35	128
102	61	81	46	105	58	78	49	108	55	75	52
111	16	144	19	114	13	141	22	117	10	138	25
72	91	39	88	69	94	42	85	66	97	45	82
9	118	30	133	6	121	33	130	3	124	36	127

Fig. 753.

in Fig. 752. It is obvious that if this square is pandiagonal, several of these squares may be contiguously arranged to form a larger square that is pandiagonal and 4^2-ply, and also has the concentric features previously mentioned.

Fig. 752 is now added to each section of Fig. 751, cell to cell, which will produce the final magic square in Fig. 753.

With a little practice, any size square of order $4m$ may be constructed without the use of subsidiary squares, by writing the numbers directly into the square and following the same order of numeral procession as shown in Fig. 754. Other processes of direct con-

1	224	61	228	5	220	57	232	9	216	53	236	13	212	49	240
113	176	77	148	117	172	73	152	121	168	69	156	125	164	65	160
205	20	241	48	201	24	245	44	197	28	249	40	193	32	253	36
189	100	129	96	185	104	133	92	181	108	137	88	177	112	141	84
2	223	62	227	6	219	58	231	10	215	54	235	14	211	50	239
114	175	78	147	118	171	74	151	122	167	70	155	126	163	66	159
206	19	242	47	202	23	246	43	198	27	250	39	194	31	254	35
190	99	130	95	186	103	134	91	182	107	138	87	178	111	142	83
3	222	63	226	7	218	59	230	11	214	55	234	15	210	51	238
115	174	79	146	119	170	75	150	123	166	71	154	127	162	67	158
207	18	243	46	203	22	247	42	199	26	251	38	195	30	255	34
191	98	131	94	187	102	135	90	183	106	139	86	179	110	143	82
4	221	64	225	8	217	60	229	12	213	56	233	16	209	52	237
116	173	80	145	120	169	76	149	124	165	72	153	128	161	68	157
208	17	244	45	204	21	248	41	200	25	252	37	196	29	256	33
192	97	132	93	188	101	136	89	184	105	140	85	180	109	144	81

Fig. 754.

struction may be discovered by numerous arrangements and combinations of the subsidiary squares.

Fig. 754 contains pandiagonal squares of the 4th, 8th, 12th and 16th orders and is 4^2-ply. The 8th and 16th order squares are also magic in their Franklin bent diagonals.

These concentric squares involve another magic feature in respect to zig-zag strings of numbers. These strings pass from side to side, or from top to bottom, and bend at right angles after every fourth cell as indicated by the dotted line in Fig. 754. It should be noted, however, that in squares of orders $8m+4$ the central four numbers of a zig-zag string must run parallel to the side of the square, and the string must be symmetrical in respect to the center line of the square which divides the string in halves. For example in a square of the 20th order, the zig-zag string should be of this form

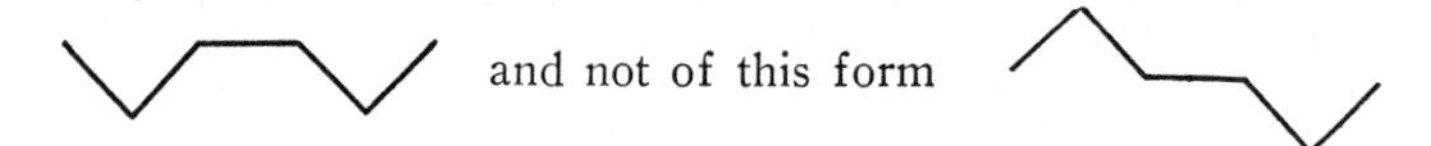

In fact any group or string of numbers in these squares, that is symmetrical to the horizontal or vertical center line of the magic and is selected in accordance with the magic properties of the 16-cell subsidiary square, will give the sum $[r(n^2+1)]/2$, where r = the number of cells in the group or string, and n = the number of cells in the edge of the magic. One of these strings is exemplified in Fig. 754 by the numbers enclosed in circles.

To explain what is meant above in reference to selecting the numbers in accordance with the magic properties of the 16-cell subsidiary square, note that the numbers, 27, 107, 214, 166, in the exemplified string, form a magic row in the small subsidiary square, 70, 235, 179, 30 and 251, 86, 14, 163 form magic diagonals, and 66, 159, 255, 34 and 141, 239, 82, 52 form ply groups.

H. A. S.

INDEX.

Abraham ben Ezra, 124.
Acoustic figures, Chladni, 117-120, 153.
Alternation, Squares constructed by, 102-112.
American Journal of Mathematics, 366n.
Andrews, W. S., 113, 114, 125, 126, 177, 189, 194, 195, 196, 229, 257n, 277, 350, 384n, 410.
Archimedes, 156.
Archytus, 148.
Aristotle, 151, 153, 156.
Arithmetical series, 291, 393.
Arithmétique graphique (Arnoux), 363n, 366.
Arnoux, *Arithmétique graphique,* 363n, 366.
Arrow heads indicating sequence, 10.
Associated or regular magic rectangles, 258-262.
Associated or regular squares, 229, 233, 236, 238, 243, 253, 396; Definition of, 1, 255, 256, 268, 270, 287, 385; of odd numbers, 2ff; of even numbers, 18ff.
Astronomer Poet (The) Omar, 157.

Babylonia, Magic square in, 123, 125.
Bachet de Mezeriac's method of constructing odd magic squares, 17-18.
Ball, Rouse, 314.
Batteux's series of the Pythagorean harmonic scale, 154-155.
Beverly, Mr., 175.
Binate transposition, Alternation by, 106, 111.
Black, E., 186.
Break move, 283; in odd magic squares, 7, 144f; Examples of, 8.
Break-step, 276.
Browne, C. A., 158, 159; Magic square of, 150, 158.
Burge, 154, 155.

Caïssan squares, 165.
Cartesian geometry, 315.
Carus, Dr. Paul, viii, 1n, 79, 84, 87, 88, 112, 128, 153, 162, 167.
Cayley, 315.
Chessboard, Magic Squares and other Problems on a, 187.
China, Magic square in, 1, 122, 125, 187.
Chinese Philosophy (Dr. Paul Carus) 1n.
Chinese Reader's Manual (Mayers), 123.
Chladni acoustic figures, 117-120, 153.
Cicero, 151.
Collinson, Peter, 89.
Complementary differences, 257ff, 277, 384.
Composite magic squares, 44ff, 260n, 383.
Concentric magic squares, 47ff, 215, 410, 413; Modifications of, 172.
Concentric spheres, 332.
Confucius, 123.
Constructive plans, Notes on various, 178ff.
Continuous squares, 236, 256. See also "Nasik."
Crantor, 154.

De la Hire's method, 225, 231, 248; of constructing odd magic squares, 14, 17; even magic squares, 34ff.
See also "La Hirean."
Donaldson, 151n.
Dudeney, Henry E., 183.
Dürer's picture, Melancholy, 146, 147.

Egypt, Magic diagrams in, 146.
English Mechanic, 363n.
Equilibrium, Figure of, 252, 257.
Euler, 315.
Even magic cubes, 76ff; squares, 18ff; squares by De la Hire's method, 34ff.
Exponential method, 284.
Exponential La Hireian method, 285-290, 293.

Factorial method, 290-292, 294.
Falkener, Edward, *Games Ancient and Modern,* 163.
Fermat, 314, 315, 365n.
Figures, Order of (*o, ro, i, ri*), 79, 113ff.
Firth, W., 189, 298, 304, 314, 373n.
Franklin, Benjamin, 89, 93, 94, 112, 146; *Letters and Papers on Philosophical Subjects* by, 89; Parton's *Life and Times of,* 96.
Franklin cube, 189.
Franklin squares, 88, 89ff, 94, 95, 105, 111, 167, 168, 178, 180, 193, 401, 402, 403; Properties of, 96, 98.
Franklin's property of bent diagonals, 377.
Frenicle, 89.
Frierson, L. S., 62, 145, 167, 168, 169, 170, 172, 178, 210, 257, 277, 287; Magic cross of, 171; Pentagram of, 172; Squares of, 167-172.
Frost, Rev. A. H., 164, 362f, 363n, 372.
Fuh-Hi, 122.
Fujisawa, Prof., 124.

Games Ancient and Modern (Edward Falkener), 163.
Geometric magic cubes, 283, 293f.
Geometric magic squares, 283ff.
Golden Verses, 149.
Great Britain, Proceedings of the Royal Institution of, 187.
Gwalior, India, 165.

Harmonic figures, 147.
Harmonic scale, Pythagorean, 153, 154.
Harmony of the Spheres, viii.
Ho, Map of, 122.

Index cubes, 306, 308, 314.
Index squares, 300, 307, 314.
India, 123, 125, 165, 187.
Indian magic squares, 165, 167, 168.
Inlaid squares, 214.

Jaina, inscription, 124; square, 87, 94, 125ff, 165, 166, 287, 331, 376, 381, 383; square modified by Dr. Carus, 127, 167, 181.
Jones, Sir William, 165.
Jowett, 148, 151.

Kensington Museum, South, 164, 372.
Kesson, Mr., 165.
Khajuraho (India), Jaina inscription in, 125.
Kielhorn, Prof., 124, 125.
Kingery, H. M., 189, 196, 362, 364, 370, 382.
Knight's move in magic squares, The, 4, 5-7, 12, 144f, 175, 405.

La Hireian method, 189, 198, 244, 268, 270, 273, 315-317, 331, 381, 411; Exponential, 285-290, 293.
See also "De la Hire."
La Hireian, Non-, 263, 305.
La Loubère, 165.
Latin squares, 315.
Letters and Papers on Philosophical Subjects (By Benjamin Franklin), 89.
Life and Times of Benjamin Franklin (By James Parton), 96.
Logan, Mr., 89, 91.

Loh, The Scroll of, 122.
Lozenge magic squares, 244ff.
Lusus numerorum, vii, 158, 161.

MacMahon, Major P. A., 187.
Magic circles, 321ff.
Magic cubes, Characteristics of, 64; Even, 76ff; General notes on, 84ff; Geometric, 283ff, 293f; Odd, 64ff; of the sixth order, 189ff.
Magic octahedroids, 317, 320, 351ff.
Magic rectangles, 170, 268, 270f, 291f, 384; Associated or regular, 258-262.
Magic series, Law of, 231
Magic spheres, 331ff; Concentric, 332.
Magic squares, and combinations, 163ff; and Other Problems on the Chessboard, 187; and Pythagorean numbers, 146ff; Associated or regular, 2ff, 18ff, 229ff, 233, 236, 238, 243, 253, 255, 256, 268, 270, 287, 385, 396; by alternation, 102ff; Composite, 44ff, 260n, 383; Concentric, 47ff, 215, 410, 413; Concentric, modified, 172; Continuous, 236, 256; Construction of, 14, 54ff, 178ff; Definition of, 1; Earliest record of, 1; Even, 18ff, 34ff; Franklin, 88, 89ff, 94, 95, 111, 167, 168, 178, 180, 193; Franklin, analyzed by Dr. Carus, 96ff; Frierson, 166; Frierson's analysis of, 129; Geometric, 283ff; Indian (La Loubère), 165; in symbols, 120f; Inlaid, 214; Jaina, 87, 94, 125ff, 165, 166, 376, 383; Knight's move in, 4, 5-7, 12, 144f, 175, 405; Lozenge, 244ff; Mathematical study of, 129ff; Nasik, 234, 236, 237f, 242, 255f, 287, 290, 291, 370, 383, 402, 403, 405, 408; Number series in, 137ff; Odd, 1ff, 248ff, 260n, 383; Oddly-even, 191, 217ff, 225ff; of form $4m$, 296; of form $4p+2$, 267ff, 290n; of form $8p+2$, 277ff; Ornate, 260n, 376ff; Overlapping, 207; Pan-diagonal, 229ff, 233, 235, 268, 269, 291, 292, 377, 396, 401, 402, 410ff; Pure, 232, 236; Serrated, 241ff; with predetermined summations, 54.
Magic stars, 5-pointed, 339-342; 6-pointed, 342-344; 7-pointed, 344; 8-pointed, 345-347.
Map of Ho, The, 122.
Mathematical Recreations (Rouse Ball), 314.
Mathematical study of magic squares, 129ff; value of magic squares, 187.
Mathematics, Quarterly Journal of, 363.
Mayers, 123.
Melancholy, Dürer's picture of, 146, 147.
Mersenne, 314, 365n.
Method of De la Hire, 225; of reversions, 298, 318; Scheffler's, 14; Thompson's, 304, 315, 373.
Méziriac's (Bachet de) method of constructing odd magic squares, 17.
Morton, Frederic A., 339, 348.
Moscopulus, 188.

Nasik Cubes, The Properties of (Frost), 363n.
Nasik idea, Evolution of the, 364.
Nasik squares, 234, 236, 237f, 242, 255, 256, 287, 290, 291, 370, 383, 402, 403, 405, 408; defined, 365; and cubes, 164; Non-, 370.
Nasiks, The Theory of Path (C. Planck), 273n, 363n, 388.
Natural squares, 295; Deformed, 315.
Number series, 137ff.

Odd magic cubes, 64ff.
Odd magic squares, 1ff, 248ff, 260n, 383; Bachet de Mezeriac's method of constructing, 17; Breakmoves in, 7; Examples of breakmoves in, 8; General principles of, 7.
Oddly-even magic squares, 196, 217ff, 225ff.
Omar, the astronomer poet, 157.
Orders of figures, (*o, ro, i, ri*),79, 113ff.
Ornate magics, 260n, 376ff.

Pan-diagonal magic squares, 227, 229ff, 233, 235, 268, 269f, 291, 292, 377, 396, 401, 402, 410ff.

Parton, James, 96, 100.
Path method, 273.
Pentagram, Magic, 172.
"Phaedrus" of Plato, 148.
Philolaus, 148, 157.
Philosophical Magazine, 175.
Philosophical Subjects, Letters and Papers on (Benjamin Franklin), 89.
Philosophy, Chinese (Dr.Paul Carus), 1n; Pythagorean, 148.
Planck, C., 189, 240, 257, 258, 260n, 267, 268, 277, 290, 291, 292, 320, 375, 390; "Magic Squares, Cubes, etc.", 363n; *The Theory of Path Nasiks,* 273n, 363n, 388n.
Plato, 148ff, 159.
Platonic school, 157.
Plutarch, 146, 149, 150, 154, 156n.
"Politics" of Aristotle, 153.
Predetermined summations, Magic squares with, 54.
Primary squares, 256, 285, 290, 292; Construction by, 13-18, 224, 232.
Proceedings of the Royal Institution of Great Britain, 187.
"Procreation of the Soul" (Plutarch), 149, 154, 156n.
Pseudo-cubes, 306; Method of, 304.
Pure magic square, 232, 236.
Pythagoras, vii, 123, 124, 147, 156; Harmonic scale, of, 153, 154; Philosophy of, 148; School of, 147.
Pythagorean numbers, 146ff.

Quarterly Journal of Mathematics, 365n, 366n, 372.
Quaternate transposition, Alternation by, 109.
Queen, The, 165.

Rectangles, Magic, 170.
Reflections on Magic Squares, 79, 87, 113ff, 153.
"Republic" of Plato, 148, 153, 156, 157, 158.
Reversions, Method of, 298, 318; Theory of, 295ff.
Royal Institution of Great Britain, Proceedings of, 187.

Savage, D. F., 216, 225.
Sayles, Harry A., 176, 189, 201, 244, 247, 283, 294, 331, 339.
Scheffler, Prof., 14.
Schilling, Prof., 124.
Schleiermacher, 151n.
Schneider, 151n.
Schubert, Prof. Hermann, 151n.
Scroll of Loh, The, 122.
Series, Arithmetical, 291, 393; Magic, 231; Number, 137ff.
Shuldham, Chas. D., 173.
Siamese twins, 209.
Smith, David Eugene, 124, 127.
"Soul of the World and Nature" (Timaeus), 154.
South Kensington Museum (London), 164, 372.
Spheres, Harmony of the, vi.
State, Number of the, 153.
Stifelius, 92.
Stringham, 366.
Symbols, Magic Squares in, 120f.

Tetractys, 149, 151.
Theory of Path Nasiks (C. Planck), 273n, 366n.
Thompson's method, 304, 315, 373.
"Timaeus" of Plato, 148, 149, 154, 156n.
Timaeus the Locrian, 154.
Transposition, Alternation by, 106-112.

Ventres, D. B., 86.
Verses, Golden, 149.
Virgil, 124.

Worthington, John, 189, 206, 373.

Yang and yin, 122, 123.
Yih King, 122, 123.

DIAGRAMS OF COMPLETED MAGICS.

MAGIC SQUARES:

Order 3: 2, 54, 55, 58, 59, 60, 62, 128, 159, 164, 284, 304;
" 4: 18, 19, 61, 62, 91, 94, 116, 125, 127, 136, 140, 141, 163, 166, 173, 179, 180, 181, 182, 183, 211, 224, 265, 291, 297, 343, 412;
" 5: 2, 4, 11, 12, 15, 16, 17, 46, 47, 57, 113, 141, 173, 210, 235, 244, 246, 250, 253, 263, 289, 291;
" 6: 19, 20, 24, 36, 40, 50, 51, 57, 118, 163, 172, 184, 185, 186, 187, 215, 219, 220, 226, 228, 238, 264, 265, 269, 270, 292, 297, 300;
" 7: 4, 48, 245, 251;
" 8: 25, 27, 28, 43, 52, 90, 97, 101, 116, 126, 165, 167, 169, 170, 175, 180, 243, 256, 377, 391, 396, 410;
" 9: 13, 44, 45, 49, 134, 144, 172, 173, 208, 212, 215, 247;
" 10: 30, 53, 221, 222, 228, 272, 275, 281, 282, 298;
Order 12: 31, 45, 116, 301, 392, 398, 412;
" 13: 240;
" 14: 33, 213, 302;
" 15: 214, 386, 387, 388;
" 16: 91, 97, 108, 110, 382, 393, 395, 400, 403, 408, 413;
" 20: 379;
" 24: 380;
" 25: 370;
" 27: 150.

MAGIC CUBES:

Order 3: 65, 66, 69, 85, 86, 203, 293, 352, 364.
" 4: 78, 86, 293, 305, 308;
" 5: 73, 76, 87;
" 6: 191, 197, 202, 205, 305, 312, 313;
" 8: 81, 82.
" 10: 310, 311;

MAGIC OCTAHEDROIDS:

Order 3: 352, 367;
" 4: 315, 316, 317, 318, 354, 374;
" 5: 353, 371;
" 6: 356-357;
" 8: 358-359.

A CATALOGUE OF SELECTED DOVER BOOKS IN ALL FIELDS OF INTEREST

AMERICA'S OLD MASTERS, James T. Flexner. Four men emerged unexpectedly from provincial 18th century America to leadership in European art: Benjamin West, J. S. Copley, C. R. Peale, Gilbert Stuart. Brilliant coverage of lives and contributions. Revised, 1967 edition. 69 plates. 365pp. of text.
21806-6 Paperbound $3.00

FIRST FLOWERS OF OUR WILDERNESS: AMERICAN PAINTING, THE COLONIAL PERIOD, James T. Flexner. Painters, and regional painting traditions from earliest Colonial times up to the emergence of Copley, West and Peale Sr., Foster, Gustavus Hesselius, Feke, John Smibert and many anonymous painters in the primitive manner. Engaging presentation, with 162 illustrations. xxii + 368pp.
22180-6 Paperbound $3.50

THE LIGHT OF DISTANT SKIES: AMERICAN PAINTING, 1760-1835, James T. Flexner. The great generation of early American painters goes to Europe to learn and to teach: West, Copley, Gilbert Stuart and others. Allston, Trumbull, Morse; also contemporary American painters—primitives, derivatives, academics—who remained in America. 102 illustrations. xiii + 306pp. 22179-2 Paperbound $3.50

A HISTORY OF THE RISE AND PROGRESS OF THE ARTS OF DESIGN IN THE UNITED STATES, William Dunlap. Much the richest mine of information on early American painters, sculptors, architects, engravers, miniaturists, etc. The only source of information for scores of artists, the major primary source for many others. Unabridged reprint of rare original 1834 edition, with new introduction by James T. Flexner, and 394 new illustrations. Edited by Rita Weiss. 6⅝ x 9⅝.
21695-0, 21696-9, 21697-7 Three volumes, Paperbound $15.00

EPOCHS OF CHINESE AND JAPANESE ART, Ernest F. Fenollosa. From primitive Chinese art to the 20th century, thorough history, explanation of every important art period and form, including Japanese woodcuts; main stress on China and Japan, but Tibet, Korea also included. Still unexcelled for its detailed, rich coverage of cultural background, aesthetic elements, diffusion studies, particularly of the historical period. 2nd, 1913 edition. 242 illustrations. lii + 439pp. of text.
20364-6, 20365-4 Two volumes, Paperbound $6.00

THE GENTLE ART OF MAKING ENEMIES, James A. M. Whistler. Greatest wit of his day deflates Oscar Wilde, Ruskin, Swinburne; strikes back at inane critics, exhibitions, art journalism; aesthetics of impressionist revolution in most striking form. Highly readable classic by great painter. Reproduction of edition designed by Whistler. Introduction by Alfred Werner. xxxvi + 334pp.
21875-9 Paperbound $3.00

ALPHABETS AND ORNAMENTS, Ernst Lehner. Well-known pictorial source for decorative alphabets, script examples, cartouches, frames, decorative title pages, calligraphic initials, borders, similar material. 14th to 19th century, mostly European. Useful in almost any graphic arts designing, varied styles. 750 illustrations. 256pp. 7 x 10. 21905-4 Paperbound $4.00

PAINTING: A CREATIVE APPROACH, Norman Colquhoun. For the beginner simple guide provides an instructive approach to painting: major stumbling blocks for beginner; overcoming them, technical points; paints and pigments; oil painting; watercolor and other media and color. New section on "plastic" paints. Glossary. Formerly *Paint Your Own Pictures*. 221pp. 22000-1 Paperbound $1.75

THE ENJOYMENT AND USE OF COLOR, Walter Sargent. Explanation of the relations between colors themselves and between colors in nature and art, including hundreds of little-known facts about color values, intensities, effects of high and low illumination, complementary colors. Many practical hints for painters, references to great masters. 7 color plates, 29 illustrations. x + 274pp.
20944-X Paperbound $3.00

THE NOTEBOOKS OF LEONARDO DA VINCI, compiled and edited by Jean Paul Richter. 1566 extracts from original manuscripts reveal the full range of Leonardo's versatile genius: all his writings on painting, sculpture, architecture, anatomy, astronomy, geography, topography, physiology, mining, music, etc., in both Italian and English, with 186 plates of manuscript pages and more than 500 additional drawings. Includes studies for the Last Supper, the lost Sforza monument, and other works. Total of xlvii + 866pp. 7⅞ x 10¾.
22572-0, 22573-9 Two volumes, Paperbound $12.00

MONTGOMERY WARD CATALOGUE OF 1895. Tea gowns, yards of flannel and pillow-case lace, stereoscopes, books of gospel hymns, the New Improved Singer Sewing Machine, side saddles, milk skimmers, straight-edged razors, high-button shoes, spittoons, and on and on . . . listing some 25,000 items, practically all illustrated. Essential to the shoppers of the 1890's, it is our truest record of the spirit of the period. Unaltered reprint of Issue No. 57, Spring and Summer 1895. Introduction by Boris Emmet. Innumerable illustrations. xiii + 624pp. 8½ x 11⅝.
22377-9 Paperbound $8.50

THE CRYSTAL PALACE EXHIBITION ILLUSTRATED CATALOGUE (LONDON, 1851). One of the wonders of the modern world—the Crystal Palace Exhibition in which all the nations of the civilized world exhibited their achievements in the arts and sciences—presented in an equally important illustrated catalogue. More than 1700 items pictured with accompanying text—ceramics, textiles, cast-iron work, carpets, pianos, sleds, razors, wall-papers, billiard tables, beehives, silverware and hundreds of other artifacts—represent the focal point of Victorian culture in the Western World. Probably the largest collection of Victorian decorative art ever assembled—indispensable for antiquarians and designers. Unabridged republication of the Art-Journal Catalogue of the Great Exhibition of 1851, with all terminal essays. New introduction by John Gloag, F.S.A. xxxiv + 426pp. 9 x 12.
22503-8 Paperbound $5.00

"ESSENTIAL GRAMMAR" SERIES

All you really need to know about modern, colloquial grammar. Many educational shortcuts help you learn faster, understand better. Detailed cognate lists teach you to recognize similarities between English and foreign words and roots—make learning vocabulary easy and interesting. Excellent for independent study or as a supplement to record courses.

ESSENTIAL FRENCH GRAMMAR, Seymour Resnick. 2500-item cognate list. 159pp.
(EBE) 20419-7 Paperbound $1.50

ESSENTIAL GERMAN GRAMMAR, Guy Stern and Everett F. Bleiler. Unusual shortcuts on noun declension, word order, compound verbs. 124pp.
(EBE) 20422-7 Paperbound $1.25

ESSENTIAL ITALIAN GRAMMAR, Olga Ragusa. 111pp.
(EBE) 20779-X Paperbound $1.25

ESSENTIAL JAPANESE GRAMMAR, Everett F. Bleiler. In Romaji transcription; no characters needed. Japanese grammar is regular and simple. 156pp.
21027-8 Paperbound $1.50

ESSENTIAL PORTUGUESE GRAMMAR, Alexander da R. Prista. vi + 114pp.
21650-0 Paperbound $1.35

ESSENTIAL SPANISH GRAMMAR, Seymour Resnick. 2500 word cognate list. 115pp.
(EBE) 20780-3 Paperbound $1.25

ESSENTIAL ENGLISH GRAMMAR, Philip Gucker. Combines best features of modern, functional and traditional approaches. For refresher, class use, home study. x + 177pp.
21649-7 Paperbound $1.75

A PHRASE AND SENTENCE DICTIONARY OF SPOKEN SPANISH. Prepared for U. S. War Department by U. S. linguists. As above, unit is idiom, phrase or sentence rather than word. English-Spanish and Spanish-English sections contain modern equivalents of over 18,000 sentences. Introduction and appendix as above. iv + 513pp.
20495-2 Paperbound $3.50

A PHRASE AND SENTENCE DICTIONARY OF SPOKEN RUSSIAN. Dictionary prepared for U. S. War Department by U. S. linguists. Basic unit is not the word, but the idiom, phrase or sentence. English-Russian and Russian-English sections contain modern equivalents for over 30,000 phrases. Grammatical introduction covers phonetics, writing, syntax. Appendix of word lists for food, numbers, geographical names, etc. vi + 573 pp. 6⅛ x 9¼.
20496-0 Paperbound $5.50

CONVERSATIONAL CHINESE FOR BEGINNERS, Morris Swadesh. Phonetic system, beginner's course in Pai Hua Mandarin Chinese covering most important, most useful speech patterns. Emphasis on modern colloquial usage. Formerly *Chinese in Your Pocket.* xvi + 158pp.
21123-1 Paperbound $1.75

DESIGN BY ACCIDENT; A BOOK OF "ACCIDENTAL EFFECTS" FOR ARTISTS AND DESIGNERS, James F. O'Brien. Create your own unique, striking, imaginative effects by "controlled accident" interaction of materials: paints and lacquers, oil and water based paints, splatter, crackling materials, shatter, similar items. Everything you do will be different; first book on this limitless art, so useful to both fine artist and commercial artist. Full instructions. 192 plates showing "accidents," 8 in color. viii + 215pp. 8⅜ x 11¼. 21942-9 Paperbound $3.75

THE BOOK OF SIGNS, Rudolf Koch. Famed German type designer draws 493 beautiful symbols: religious, mystical, alchemical, imperial, property marks, runes, etc. Remarkable fusion of traditional and modern. Good for suggestions of timelessness, smartness, modernity. Text. vi + 104pp. 6⅛ x 9¼.
20162-7 Paperbound $1.50

HISTORY OF INDIAN AND INDONESIAN ART, Ananda K. Coomaraswamy. An unabridged republication of one of the finest books by a great scholar in Eastern art. Rich in descriptive material, history, social backgrounds; Sunga reliefs, Rajput paintings, Gupta temples, Burmese frescoes, textiles, jewelry, sculpture, etc. 400 photos. viii + 423pp. 6⅜ x 9¾. 21436-2 Paperbound $5.00

PRIMITIVE ART, Franz Boas. America's foremost anthropologist surveys textiles, ceramics, woodcarving, basketry, metalwork, etc.; patterns, technology, creation of symbols, style origins. All areas of world, but very full on Northwest Coast Indians. More than 350 illustrations of baskets, boxes, totem poles, weapons, etc. 378 pp.
20025-6 Paperbound $3.00

THE GENTLEMAN AND CABINET MAKER'S DIRECTOR, Thomas Chippendale. Full reprint (third edition, 1762) of most influential furniture book of all time, by master cabinetmaker. 200 plates, illustrating chairs, sofas, mirrors, tables, cabinets, plus 24 photographs of surviving pieces. Biographical introduction by N. Bienenstock. vi + 249pp. 9⅞ x 12¾. 21601-2 Paperbound $5.00

AMERICAN ANTIQUE FURNITURE, Edgar G. Miller, Jr. The basic coverage of all American furniture before 1840. Individual chapters cover type of furniture—clocks, tables, sideboards, etc.—chronologically, with inexhaustible wealth of data. More than 2100 photographs, all identified, commented on. Essential to all early American collectors. Introduction by H. E. Keyes. vi + 1106pp. 7⅞ x 10¾.
21599-7, 21600-4 Two volumes, Paperbound $11.00

PENNSYLVANIA DUTCH AMERICAN FOLK ART, Henry J. Kauffman. 279 photos, 28 drawings of tulipware, Fraktur script, painted tinware, toys, flowered furniture, quilts, samplers, hex signs, house interiors, etc. Full descriptive text. Excellent for tourist, rewarding for designer, collector. Map. 146pp. 7⅞ x 10¾.
21205-X Paperbound $3.00

EARLY NEW ENGLAND GRAVESTONE RUBBINGS, Edmund V. Gillon, Jr. 43 photographs, 226 carefully reproduced rubbings show heavily symbolic, sometimes macabre early gravestones, up to early 19th century. Remarkable early American primitive art, occasionally strikingly beautiful; always powerful. Text. xxvi + 207pp. 8⅜ x 11¼. 21380-3 Paperbound $4.00

POEMS OF ANNE BRADSTREET, edited with an introduction by Robert Hutchinson. A new selection of poems by America's first poet and perhaps the first significant woman poet in the English language. 48 poems display her development in works of considerable variety—love poems, domestic poems, religious meditations, formal elegies, "quaternions," etc. Notes, bibliography. viii + 222pp.
22160-1 Paperbound $2.50

THREE GOTHIC NOVELS: THE CASTLE OF OTRANTO BY HORACE WALPOLE; VATHEK BY WILLIAM BECKFORD; THE VAMPYRE BY JOHN POLIDORI, WITH FRAGMENT OF A NOVEL BY LORD BYRON, edited by E. F. Bleiler. The first Gothic novel, by Walpole; the finest Oriental tale in English, by Beckford; powerful Romantic supernatural story in versions by Polidori and Byron. All extremely important in history of literature; all still exciting, packed with supernatural thrills, ghosts, haunted castles, magic, etc. xl + 291pp.
21232-7 Paperbound $3.00

THE BEST TALES OF HOFFMANN, E. T. A. Hoffmann. 10 of Hoffmann's most important stories, in modern re-editings of standard translations: Nutcracker and the King of Mice, Signor Formica, Automata, The Sandman, Rath Krespel, The Golden Flowerpot, Master Martin the Cooper, The Mines of Falun, The King's Betrothed, A New Year's Eve Adventure. 7 illustrations by Hoffmann. Edited by E. F. Bleiler. xxxix + 419pp. 21793-0 Paperbound $3.00

GHOST AND HORROR STORIES OF AMBROSE BIERCE, Ambrose Bierce. 23 strikingly modern stories of the horrors latent in the human mind: The Eyes of the Panther, The Damned Thing, An Occurrence at Owl Creek Bridge, An Inhabitant of Carcosa, etc., plus the dream-essay, Visions of the Night. Edited by E. F. Bleiler. xxii + 199pp. 20767-6 Paperbound $2.00

BEST GHOST STORIES OF J. S. LEFANU, J. Sheridan LeFanu. Finest stories by Victorian master often considered greatest supernatural writer of all. Carmilla, Green Tea, The Haunted Baronet, The Familiar, and 12 others. Most never before available in the U. S. A. Edited by E. F. Bleiler. 8 illustrations from Victorian publications. xvii + 467pp. 20415-4 Paperbound $3.00

MATHEMATICAL FOUNDATIONS OF INFORMATION THEORY, A. I. Khinchin. Comprehensive introduction to work of Shannon, McMillan, Feinstein and Khinchin, placing these investigations on a rigorous mathematical basis. Covers entropy concept in probability theory, uniqueness theorem, Shannon's inequality, ergodic sources, the E property, martingale concept, noise, Feinstein's fundamental lemma, Shanon's first and second theorems. Translated by R. A. Silverman and M. D. Friedman. iii + 120pp. 60434-9 Paperbound $2.00

SEVEN SCIENCE FICTION NOVELS, H. G. Wells. The standard collection of the great novels. Complete, unabridged. *First Men in the Moon, Island of Dr. Moreau, War of the Worlds, Food of the Gods, Invisible Man, Time Machine, In the Days of the Comet.* Not only science fiction fans, but every educated person owes it to himself to read these novels. 1015pp. (USO) 20264-X Clothbound $6.00

A HISTORY OF COSTUME, Carl Köhler. Definitive history, based on surviving pieces of clothing primarily, and paintings, statues, etc. secondarily. Highly readable text, supplemented by 594 illustrations of costumes of the ancient Mediterranean peoples, Greece and Rome, the Teutonic prehistoric period; costumes of the Middle Ages, Renaissance, Baroque, 18th and 19th centuries. Clear, measured patterns are provided for many clothing articles. Approach is practical throughout. Enlarged by Emma von Sichart. 464pp. 21030-8 Paperbound $3.50

ORIENTAL RUGS, ANTIQUE AND MODERN, Walter A. Hawley. A complete and authoritative treatise on the Oriental rug—where they are made, by whom and how, designs and symbols, characteristics in detail of the six major groups, how to distinguish them and how to buy them. Detailed technical data is provided on periods, weaves, warps, wefts, textures, sides, ends and knots, although no technical background is required for an understanding. 11 color plates, 80 halftones, 4 maps. vi + 320pp. 6⅛ x 9⅛. 22366-3 Paperbound $5.00

TEN BOOKS ON ARCHITECTURE, Vitruvius. By any standards the most important book on architecture ever written. Early Roman discussion of aesthetics of building, construction methods, orders, sites, and every other aspect of architecture has inspired, instructed architecture for about 2,000 years. Stands behind Palladio, Michelangelo, Bramante, Wren, countless others. Definitive Morris H. Morgan translation. 68 illustrations. xii + 331pp. 20645-9 Paperbound $3.00

THE FOUR BOOKS OF ARCHITECTURE, Andrea Palladio. Translated into every major Western European language in the two centuries following its publication in 1570, this has been one of the most influential books in the history of architecture. Complete reprint of the 1738 Isaac Ware edition. New introduction by Adolf Placzek, Columbia Univ. 216 plates. xxii + 110pp. of text. 9½ x 12¾.
21308-0 Clothbound $12.50

STICKS AND STONES: A STUDY OF AMERICAN ARCHITECTURE AND CIVILIZATION, Lewis Mumford.One of the great classics of American cultural history. American architecture from the medieval-inspired earliest forms to the early 20th century; evolution of structure and style, and reciprocal influences on environment. 21 photographic illustrations. 238pp. 20202-X Paperbound $2.00

THE AMERICAN BUILDER'S COMPANION, Asher Benjamin. The most widely used early 19th century architectural style and source book, for colonial up into Greek Revival periods. Extensive development of geometry of carpentering, construction of sashes, frames, doors, stairs; plans and elevations of domestic and other buildings. Hundreds of thousands of houses were built according to this book, now invaluable to historians, architects, restorers, etc. 1827 edition. 59 plates. 114pp. 7⅞ x 10¾.
22236-5 Paperbound $4.00

DUTCH HOUSES IN THE HUDSON VALLEY BEFORE 1776, Helen Wilkinson Reynolds. The standard survey of the Dutch colonial house and outbuildings, with constructional features, decoration, and local history associated with individual homesteads. Introduction by Franklin D. Roosevelt. Map. 150 illustrations. 469pp. 6⅝ x 9¼. 21469-9 Paperbound $5.00

ADVENTURES OF AN AFRICAN SLAVER, Theodore Canot. Edited by Brantz Mayer. A detailed portrayal of slavery and the slave trade, 1820-1840. Canot, an established trader along the African coast, describes the slave economy of the African kingdoms, the treatment of captured negroes, the extensive journeys in the interior to gather slaves, slave revolts and their suppression, harems, bribes, and much more. Full and unabridged republication of 1854 edition. Introduction by Malcom Cowley. 16 illustrations. xvii + 448pp. 22456-2 Paperbound $3.50

MY BONDAGE AND MY FREEDOM, Frederick Douglass. Born and brought up in slavery, Douglass witnessed its horrors and experienced its cruelties, but went on to become one of the most outspoken forces in the American anti-slavery movement. Considered the best of his autobiographies, this book graphically describes the inhuman treatment of slaves, its effects on slave owners and slave families, and how Douglass's determination led him to a new life. Unaltered reprint of 1st (1855) edition. xxxii + 464pp. 22457-0 Paperbound $3.50

THE INDIANS' BOOK, recorded and edited by Natalie Curtis. Lore, music, narratives, dozens of drawings by Indians themselves from an authoritative and important survey of native culture among Plains, Southwestern, Lake and Pueblo Indians. Standard work in popular ethnomusicology. 149 songs in full notation. 23 drawings, 23 photos. xxxi + 584pp. 6⅝ x 9⅜. 21939-9 Paperbound $5.00

DICTIONARY OF AMERICAN PORTRAITS, edited by Hayward and Blanche Cirker. 4024 portraits of 4000 most important Americans, colonial days to 1905 (with a few important categories, like Presidents, to present). Pioneers, explorers, colonial figures, U. S. officials, politicians, writers, military and naval men, scientists, inventors, manufacturers, jurists, actors, historians, educators, notorious figures, Indian chiefs, etc. All authentic contemporary likenesses. The only work of its kind in existence; supplements all biographical sources for libraries. Indispensable to anyone working with American history. 8,000-item classified index, finding lists, other aids. xiv + 756pp. 9¼ x 12¾. 21823-6 Clothbound $30.00

TRITTON'S GUIDE TO BETTER WINE AND BEER MAKING FOR BEGINNERS, S. M. Tritton. All you need to know to make family-sized quantities of over 100 types of grape, fruit, herb and vegetable wines; as well as beers, mead, cider, etc. Complete recipes, advice as to equipment, procedures such as fermenting, bottling, and storing wines. Recipes given in British, U. S., and metric measures. Accompanying booklet lists sources in U. S. A. where ingredients may be bought, and additional information. 11 illustrations. 157pp. 5⅝ x 8⅛. 22090-7 **Paperbound $2.00**

GARDENING WITH HERBS FOR FLAVOR AND FRAGRANCE, Helen M. Fox. How to grow herbs in your own garden, how to use them in your cooking (over 55 recipes included), legends and myths associated with each species, uses in medicine, perfumes, etc.—these are elements of one of the few books written especially for American herb fanciers. Guides you step-by-step from soil preparation to harvesting and storage for each type of herb. 12 drawings by Louise Mansfield. xiv + 334pp. 22540-2 Paperbound $2.50

JIM WHITEWOLF: THE LIFE OF A KIOWA APACHE INDIAN, Charles S. Brant, editor. Spans transition between native life and acculturation period, 1880 on. Kiowa culture, personal life pattern, religion and the supernatural, the Ghost Dance, breakdown in the White Man's world, similar material. 1 map. xii + 144pp.
22015-X Paperbound $1.75

THE NATIVE TRIBES OF CENTRAL AUSTRALIA, Baldwin Spencer and F. J. Gillen. Basic book in anthropology, devoted to full coverage of the Arunta and Warramunga tribes; the source for knowledge about kinship systems, material and social culture, religion, etc. Still unsurpassed. 121 photographs, 89 drawings. xviii + 669pp. 21775-2 Paperbound $5.00

MALAY MAGIC, Walter W. Skeat. Classic (1900); still the definitive work on the folklore and popular religion of the Malay peninsula. Describes marriage rites, birth spirits and ceremonies, medicine, dances, games, war and weapons, etc. Extensive quotes from original sources, many magic charms translated into English. 35 illustrations. Preface by Charles Otto Blagden. xxiv + 685pp.
21760-4 Paperbound $4.00

HEAVENS ON EARTH: UTOPIAN COMMUNITIES IN AMERICA, 1680-1880, Mark Holloway. The finest nontechnical account of American utopias, from the early Woman in the Wilderness, Ephrata, Rappites to the enormous mid 19th-century efflorescence; Shakers, New Harmony, Equity Stores, Fourier's Phalanxes, Oneida, Amana, Fruitlands, etc. "Entertaining and very instructive." *Times Literary Supplement.* 15 illustrations. 246pp. 21593-8 Paperbound $2.00

LONDON LABOUR AND THE LONDON POOR, Henry Mayhew. Earliest (c. 1850) sociological study in English, describing myriad subcultures of London poor. Particularly remarkable for the thousands of pages of direct testimony taken from the lips of London prostitutes, thieves, beggars, street sellers, chimney-sweepers, street-musicians, "mudlarks," "pure-finders," rag-gatherers, "running-patterers," dock laborers, cab-men, and hundreds of others, quoted directly in this massive work. An extraordinarily vital picture of London emerges. 110 illustrations. Total of lxxvi + 1951pp. 6⅝ x 10.
21934-8, 21935-6, 21936-4, 21937-2 Four volumes, Paperbound $16.00

HISTORY OF THE LATER ROMAN EMPIRE, J. B. Bury. Eloquent, detailed reconstruction of Western and Byzantine Roman Empire by a major historian, from the death of Theodosius I (395 A.D.) to the death of Justinian (565). Extensive quotations from contemporary sources; full coverage of important Roman and foreign figures of the time. xxxiv + 965pp. 20398-0, 20399-9 Two volumes, Paperbound $7.00

AN INTELLECTUAL AND CULTURAL HISTORY OF THE WESTERN WORLD, Harry Elmer Barnes. Monumental study, tracing the development of the accomplishments that make up human culture. Every aspect of man's achievement surveyed from its origins in the Paleolithic to the present day (1964); social structures, ideas, economic systems, art, literature, technology, mathematics, the sciences, medicine, religion, jurisprudence, etc. Evaluations of the contributions of scores of great men. 1964 edition, revised and edited by scholars in the many fields represented. Total of xxix + 1381pp. 21275-0, 21276-9, 21277-7 Three volumes, Paperbound $10.50

INCIDENTS OF TRAVEL IN YUCATAN, John L. Stephens. Classic (1843) exploration of jungles of Yucatan, looking for evidences of Maya civilization. Stephens found many ruins; comments on travel adventures, Mexican and Indian culture. 127 striking illustrations by F. Catherwood. Total of 669 pp.
20926-1, 20927-X Two volumes, Paperbound $5.50

INCIDENTS OF TRAVEL IN CENTRAL AMERICA, CHIAPAS, AND YUCATAN, John L. Stephens. An exciting travel journal and an important classic of archeology. Narrative relates his almost single-handed discovery of the Mayan culture, and exploration of the ruined cities of Copan, Palenque, Utatlan and others; the monuments they dug from the earth, the temples buried in the jungle, the customs of poverty-stricken Indians living a stone's throw from the ruined palaces. 115 drawings by F. Catherwood. Portrait of Stephens. xii + 812pp.
22404-X, 22405-8 Two volumes, Paperbound $6.00

A NEW VOYAGE ROUND THE WORLD, William Dampier. Late 17-century naturalist joined the pirates of the Spanish Main to gather information; remarkably vivid account of buccaneers, pirates; detailed, accurate account of botany, zoology, ethnography of lands visited. Probably the most important early English voyage, enormous implications for British exploration, trade, colonial policy. Also most interesting reading. Argonaut edition, introduction by Sir Albert Gray. New introduction by Percy Adams. 6 plates, 7 illustrations. xlvii + 376pp. 6½ x 9¼.
21900-3 Paperbound $3.00

INTERNATIONAL AIRLINE PHRASE BOOK IN SIX LANGUAGES, Joseph W. Bátor. Important phrases and sentences in English paralleled with French, German, Portuguese, Italian, Spanish equivalents, covering all possible airport-travel situations; created for airline personnel as well as tourist by Language Chief, Pan American Airlines. xiv + 204pp. 22017-6 Paperbound $2.25

STAGE COACH AND TAVERN DAYS, Alice Morse Earle. Detailed, lively account of the early days of taverns; their uses and importance in the social, political and military life; furnishings and decorations; locations; food and drink; tavern signs, etc. Second half covers every aspect of early travel; the roads, coaches, drivers, etc. Nostalgic, charming, packed with fascinating material. 157 illustrations, mostly photographs. xiv + 449pp. 22518-6 Paperbound $4.00

NORSE DISCOVERIES AND EXPLORATIONS IN NORTH AMERICA, Hjalmar R. Holand. The perplexing Kensington Stone, found in Minnesota at the end of the 19th century. Is it a record of a Scandinavian expedition to North America in the 14th century? Or is it one of the most successful hoaxes in history. A scientific detective investigation. Formerly *Westward from Vinland.* 31 photographs, 17 figures. x + 354pp. 22014-1 Paperbound $2.75

A BOOK OF OLD MAPS, compiled and edited by Emerson D. Fite and Archibald Freeman. 74 old maps offer an unusual survey of the discovery, settlement and growth of America down to the close of the Revolutionary war: maps showing Norse settlements in Greenland, the explorations of Columbus, Verrazano, Cabot, Champlain, Joliet, Drake, Hudson, etc., campaigns of Revolutionary war battles, and much more. Each map is accompanied by a brief historical essay. xvi + 299pp. 11 x 13¾. 22084-2 Paperbound $7.00

THE PHILOSOPHY OF THE UPANISHADS, Paul Deussen. Clear, detailed statement of upanishadic system of thought, generally considered among best available. History of these works, full exposition of system emergent from them, parallel concepts in the West. Translated by A. S. Geden. xiv + 429pp.
21616-0 Paperbound $3.50

LANGUAGE, TRUTH AND LOGIC, Alfred J. Ayer. Famous, remarkably clear introduction to the Vienna and Cambridge schools of Logical Positivism; function of philosophy, elimination of metaphysical thought, nature of analysis, similar topics. "Wish I had written it myself," Bertrand Russell. 2nd, 1946 edition. 160pp.
20010-8 Paperbound $1.50

THE GUIDE FOR THE PERPLEXED, Moses Maimonides. Great classic of medieval Judaism, major attempt to reconcile revealed religion (Pentateuch, commentaries) and Aristotelian philosophy. Enormously important in all Western thought. Unabridged Friedländer translation. 50-page introduction. lix + 414pp.
(USO) 20351-4 Paperbound $4.50

OCCULT AND SUPERNATURAL PHENOMENA, D. H. Rawcliffe. Full, serious study of the most persistent delusions of mankind: crystal gazing, mediumistic trance, stigmata, lycanthropy, fire walking, dowsing, telepathy, ghosts, ESP, etc., and their relation to common forms of abnormal psychology. Formerly *Illusions and Delusions of the Supernatural and the Occult.* iii + 551pp. 20503-7 Paperbound $4.00

THE EGYPTIAN BOOK OF THE DEAD: THE PAPYRUS OF ANI, E. A. Wallis Budge. Full hieroglyphic text, interlinear transliteration of sounds, word for word translation, then smooth, connected translation; Theban recension. Basic work in Ancient Egyptian civilization; now even more significant than ever for historical importance, dilation of consciousness, etc. clvi + 377pp. 6½ x 9¼.
21866-X Paperbound $4.95

PSYCHOLOGY OF MUSIC, Carl E. Seashore. Basic, thorough survey of everything known about psychology of music up to 1940's; essential reading for psychologists, musicologists. Physical acoustics; auditory apparatus; relationship of physical sound to perceived sound; role of the mind in sorting, altering, suppressing, creating sound sensations; musical learning, testing for ability, absolute pitch, other topics. Records of Caruso, Menuhin analyzed. 88 figures. xix + 408pp.
21851-1 Paperbound $3.50

THE I CHING (THE BOOK OF CHANGES), translated by James Legge. Complete translated text plus appendices by Confucius, of perhaps the most penetrating divination book ever compiled. Indispensable to all study of early Oriental civilizations. 3 plates. xxiii + 448pp. 21062-6 Paperbound $3.50

THE UPANISHADS, translated by Max Müller. Twelve classical upanishads: Chandogya, Kena, Aitareya, Kaushitaki, Isa, Katha, Mundaka, Taittiriyaka, Brhadaranyaka, Svetasvatara, Prasna, Maitriyana. 160-page introduction, analysis by Prof. Müller. Total of 670pp. 20992-X, 20993-8 Two volumes, Paperbound $7.50

PLANETS, STARS AND GALAXIES: DESCRIPTIVE ASTRONOMY FOR BEGINNERS, A. E. Fanning. Comprehensive introductory survey of astronomy: the sun, solar system, stars, galaxies, universe, cosmology; up-to-date, including quasars, radio stars, etc. Preface by Prof. Donald Menzel. 24pp. of photographs. 189pp. $5\frac{1}{4} \times 8\frac{1}{4}$.
21680-2 Paperbound $2.50

TEACH YOURSELF CALCULUS, P. Abbott. With a good background in algebra and trig, you can teach yourself calculus with this book. Simple, straightforward introduction to functions of all kinds, integration, differentiation, series, etc. "Students who are beginning to study calculus method will derive great help from this book." Faraday House Journal. 308pp. 20683-1 Clothbound $2.50

TEACH YOURSELF TRIGONOMETRY, P. Abbott. Geometrical foundations, indices and logarithms, ratios, angles, circular measure, etc. are presented in this sound, easy-to-use text. Excellent for the beginner or as a brush up, this text carries the student through the solution of triangles. 204pp. 20682-3 Clothbound $2.00

BASIC MACHINES AND HOW THEY WORK, U. S. Bureau of Naval Personnel. Originally used in U.S. Naval training schools, this book clearly explains the operation of a progression of machines, from the simplest—lever, wheel and axle, inclined plane, wedge, screw—to the most complex—typewriter, internal combustion engine, computer mechanism. Utilizing an approach that requires only an elementary understanding of mathematics, these explanations build logically upon each other and are assisted by over 200 drawings and diagrams. Perfect as a technical school manual or as a self-teaching aid to the layman. 204 figures. Preface. Index. vii + 161pp. $6\frac{1}{2} \times 9\frac{1}{4}$. 21709-4 Paperbound $2.50

THE FRIENDLY STARS, Martha Evans Martin. Classic has taught naked-eye observation of stars, planets to hundreds of thousands, still not surpassed for charm, lucidity, adequacy. Completely updated by Professor Donald H. Menzel, Harvard Observatory. 25 illustrations. 16 x 30 chart. x + 147pp. 21099-5 Paperbound $2.00

MUSIC OF THE SPHERES: THE MATERIAL UNIVERSE FROM ATOM TO QUASAR, SIMPLY EXPLAINED, Guy Murchie. Extremely broad, brilliantly written popular account begins with the solar system and reaches to dividing line between matter and nonmatter; latest understandings presented with exceptional clarity. Volume One: Planets, stars, galaxies, cosmology, geology, celestial mechanics, latest astronomical discoveries; Volume Two: Matter, atoms, waves, radiation, relativity, chemical action, heat, nuclear energy, quantum theory, music, light, color, probability, antimatter, antigravity, and similar topics. 319 figures. 1967 (second) edition. Total of xx + 644pp. 21809-0, 21810-4 Two volumes, Paperbound $5.75

OLD-TIME SCHOOLS AND SCHOOL BOOKS, Clifton Johnson. Illustrations and rhymes from early primers, abundant quotations from early textbooks, many anecdotes of school life enliven this study of elementary schools from Puritans to middle 19th century. Introduction by Carl Withers. 234 illustrations. xxxiii + 381pp.
21031-6 Paperbound $4.00

HOW TO KNOW THE WILD FLOWERS, Mrs. William Starr Dana. This is the classical book of American wildflowers (of the Eastern and Central United States), used by hundreds of thousands. Covers over 500 species, arranged in extremely easy to use color and season groups. Full descriptions, much plant lore. This Dover edition is the fullest ever compiled, with tables of nomenclature changes. 174 full-page plates by M. Satterlee. xii + 418pp. 20332-8 Paperbound $3.00

OUR PLANT FRIENDS AND FOES, William Atherton DuPuy. History, economic importance, essential botanical information and peculiarities of 25 common forms of plant life are provided in this book in an entertaining and charming style. Covers food plants (potatoes, apples, beans, wheat, almonds, bananas, etc.), flowers (lily, tulip, etc.), trees (pine, oak, elm, etc.), weeds, poisonous mushrooms and vines, gourds, citrus fruits, cotton, the cactus family, and much more. 108 illustrations. xiv + 290pp. 22272-1 Paperbound $2.50

HOW TO KNOW THE FERNS, Frances T. Parsons. Classic survey of Eastern and Central ferns, arranged according to clear, simple identification key. Excellent introduction to greatly neglected nature area. 57 illustrations and 42 plates. xvi + 215pp. 20740-4 Paperbound $2.00

MANUAL OF THE TREES OF NORTH AMERICA, Charles S. Sargent. America's foremost dendrologist provides the definitive coverage of North American trees and tree-like shrubs. 717 species fully described and illustrated: exact distribution, down to township; full botanical description; economic importance; description of subspecies and races; habitat, growth data; similar material. Necessary to every serious student of tree-life. Nomenclature revised to present. Over 100 locating keys. 783 illustrations. lii + 934pp. 20277-1, 20278-X Two volumes, Paperbound $7.00

OUR NORTHERN SHRUBS, Harriet L. Keeler. Fine non-technical reference work identifying more than 225 important shrubs of Eastern and Central United States and Canada. Full text covering botanical description, habitat, plant lore, is paralleled with 205 full-page photographs of flowering or fruiting plants. Nomenclature revised by Edward G. Voss. One of few works concerned with shrubs. 205 plates, 35 drawings. xxviii + 521pp. 21989-5 Paperbound $3.75

THE MUSHROOM HANDBOOK, Louis C. C. Krieger. Still the best popular handbook: full descriptions of 259 species, cross references to another 200. Extremely thorough text enables you to identify, know all about any mushroom you are likely to meet in eastern and central U. S. A.: habitat, luminescence, poisonous qualities, use, folklore, etc. 32 color plates show over 50 mushrooms, also 126 other illustrations. Finding keys. vii + 560pp. 21861-9 Paperbound $4.50

HANDBOOK OF BIRDS OF EASTERN NORTH AMERICA, Frank M. Chapman. Still much the best single-volume guide to the birds of Eastern and Central United States. Very full coverage of 675 species, with descriptions, life habits, distribution, similar data. All descriptions keyed to two-page color chart. With this single volume the average birdwatcher needs no other books. 1931 revised edition. 195 illustrations. xxxvi + 581pp. 21489-3 Paperbound $5.00

AMERICAN FOOD AND GAME FISHES, David S. Jordan and Barton W. Evermann. Definitive source of information, detailed and accurate enough to enable the sportsman and nature lover to identify conclusively some 1,000 species and sub-species of North American fish, sought for food or sport. Coverage of range, physiology, habits, life history, food value. Best methods of capture, interest to the angler, advice on bait, fly-fishing, etc. 338 drawings and photographs. l + 574pp. $6\frac{5}{8}$ x $9\frac{3}{8}$.
22196-2 Paperbound $5.00

THE FROG BOOK, Mary C. Dickerson. Complete with extensive finding keys, over 300 photographs, and an introduction to the general biology of frogs and toads, this is the classic non-technical study of Northeastern and Central species. 58 species; 290 photographs and 16 color plates. xvii + 253pp.
21973-9 Paperbound $4.00

THE MOTH BOOK: A GUIDE TO THE MOTHS OF NORTH AMERICA, William J. Holland. Classical study, eagerly sought after and used for the past 60 years. Clear identification manual to more than 2,000 different moths, largest manual in existence. General information about moths, capturing, mounting, classifying, etc., followed by species by species descriptions. 263 illustrations plus 48 color plates show almost every species, full size. 1968 edition, preface, nomenclature changes by A. E. Brower. xxiv + 479pp. of text. $6\frac{1}{2}$ x $9\frac{1}{4}$.
21948-8 Paperbound $6.00

THE SEA-BEACH AT EBB-TIDE, Augusta Foote Arnold. Interested amateur can identify hundreds of marine plants and animals on coasts of North America; marine algae; seaweeds; squids; hermit crabs; horse shoe crabs; shrimps; corals; sea anemones; etc. Species descriptions cover: structure; food; reproductive cycle; size; shape; color; habitat; etc. Over 600 drawings. 85 plates. xii + 490pp.
21949-6 Paperbound $4.00

COMMON BIRD SONGS, Donald J. Borror. $33\frac{1}{3}$ 12-inch record presents songs of 60 important birds of the eastern United States. A thorough, serious record which provides several examples for each bird, showing different types of song, individual variations, etc. Inestimable identification aid for birdwatcher. 32-page booklet gives text about birds and songs, with illustration for each bird.
21829-5 Record, book, album. Monaural. $3.50

FADS AND FALLACIES IN THE NAME OF SCIENCE, Martin Gardner. Fair, witty appraisal of cranks and quacks of science: Atlantis, Lemuria, hollow earth, flat earth, Velikovsky, orgone energy, Dianetics, flying saucers, Bridey Murphy, food fads, medical fads, perpetual motion, etc. Formerly "In the Name of Science." x + 363pp.
20394-8 Paperbound $3.00

HOAXES, Curtis D. MacDougall. Exhaustive, unbelievably rich account of great hoaxes: Locke's moon hoax, Shakespearean forgeries, sea serpents, Loch Ness monster, Cardiff giant, John Wilkes Booth's mummy, Disumbrationist school of art, dozens more; also journalism, psychology of hoaxing. 54 illustrations. xi + 338pp.
20465-0 Paperbound $3.50

MATHEMATICAL PUZZLES FOR BEGINNERS AND ENTHUSIASTS, Geoffrey Mott-Smith. 189 puzzles from easy to difficult—involving arithmetic, logic, algebra, properties of digits, probability, etc.—for enjoyment and mental stimulus. Explanation of mathematical principles behind the puzzles. 135 illustrations. viii + 248pp.
20198-8 Paperbound $2.00

PAPER FOLDING FOR BEGINNERS, William D. Murray and Francis J. Rigney. Easiest book on the market, clearest instructions on making interesting, beautiful origami. Sail boats, cups, roosters, frogs that move legs, bonbon boxes, standing birds, etc. 40 projects; more than 275 diagrams and photographs. 94pp.
20713-7 Paperbound $1.00

TRICKS AND GAMES ON THE POOL TABLE, Fred Herrmann. 79 tricks and games—some solitaires, some for two or more players, some competitive games—to entertain you between formal games. Mystifying shots and throws, unusual caroms, tricks involving such props as cork, coins, a hat, etc. Formerly *Fun on the Pool Table.* 77 figures. 95pp. 21814-7 Paperbound $1.25

HAND SHADOWS TO BE THROWN UPON THE WALL: A SERIES OF NOVEL AND AMUSING FIGURES FORMED BY THE HAND, Henry Bursill. Delightful picturebook from great-grandfather's day shows how to make 18 different hand shadows: a bird that flies, duck that quacks, dog that wags his tail, camel, goose, deer, boy, turtle, etc. Only book of its sort. vi + 33pp. 6½ x 9¼. 21779-5 Paperbound $1.00

WHITTLING AND WOODCARVING, E. J. Tangerman. 18th printing of best book on market. "If you can cut a potato you can carve" toys and puzzles, chains, chessmen, caricatures, masks, frames, woodcut blocks, surface patterns, much more. Information on tools, woods, techniques. Also goes into serious wood sculpture from Middle Ages to present, East and West. 464 photos, figures. x + 293pp.
20965-2 Paperbound $2.50

HISTORY OF PHILOSOPHY, Julián Marias. Possibly the clearest, most easily followed, best planned, most useful one-volume history of philosophy on the market; neither skimpy nor overfull. Full details on system of every major philosopher and dozens of less important thinkers from pre-Socratics up to Existentialism and later. Strong on many European figures usually omitted. Has gone through dozens of editions in Europe. 1966 edition, translated by Stanley Appelbaum and Clarence Strowbridge. xviii + 505pp. 21739-6 Paperbound $3.50

YOGA: A SCIENTIFIC EVALUATION, Kovoor T. Behanan. Scientific but non-technical study of physiological results of yoga exercises; done under auspices of Yale U. Relations to Indian thought, to psychoanalysis, etc. 16 photos. xxiii + 270pp.
20505-3 Paperbound $2.50

Prices subject to change without notice.
Available at your book dealer or write for free catalogue to Dept. GI, Dover Publications, Inc., 180 Varick St., N. Y., N. Y. 10014. Dover publishes more than 150 books each year on science, elementary and advanced mathematics, biology, music, art, literary history, social sciences and other areas.